DATE			

BAKER & TAYLOR

FARM MANAGEMENT

McGraw-Hill Series in Agricultural Economics

CONSULTING EDITOR

Peter J. Barry, *University of Illinois*

Halcrow, Spitze, and Allen-Smith: Food and Agricultural Policy: Economics and Politics
Kay and Edwards: Farm Management
Looney and Uchtmann: Agricultural Law
Norton and Alwang: Introduction to Economics of Agricultural Development
Seitz, Nelson, and Halcrow: Economics of Resources, Agriculture, and Food

FARM MANAGEMENT

THIRD EDITION

Ronald D. Kay

Texas A&M University

William M. Edwards

Iowa State University

McGRAW-HILL, INC.

New York   St. Louis   San Francisco   Auckland   Bogotá
Caracas   Lisbon   London   Madrid   Mexico City   Milan
Montreal   New Delhi   San Juan   Singapore   Sydney   Tokyo   Toronto

This book was set in Times Roman by The Clarinda Company.
The editors were Anne C. Duffy and John M. Morriss;
the production supervisor was Richard A. Ausburn.
The cover was designed by Rafael Hernandez.
Project supervision was done by The Total Book.
R. R. Donnelley & Sons Company was printer and binder.

FARM MANAGEMENT

This book is printed on acid-free paper.

234567890 DOH DOH 90987654

ISBN 0-07-033868-X

Photo Credits

U.S. Department of Agriculture: pages iv, 0, 4, 26, 66, 84, 88, 106, 118, 138, 158, 170, 192, 216, 236, 260, 326, 330, 354, 384; **Iowa State University, University Relations, ISU Photo Services:** pages 22, 48, 212, 404, 426; **Agricultural Communications Department, Texas A & M University:** 0, 280, 300.

Library of Congress Cataloging-in-Publication Data

Kay, Ronald D.
    Farm management / Ronald D. Kay, William M. Edwards. —3rd ed.
      p.   cm.—(McGraw-Hill series in agricultural economics)
    Includes bibliographical references and index.
    ISBN 0-07-033868-X
    1. Farm management.   I. Edwards, William M.   II. Title.
III. Series.
S561.K36   1994
630′.68—dc20
                             93-3785

ABOUT THE AUTHORS

Ronald D. Kay is a professor in the department of agricultural economics at Texas A&M University where he has taught farm management for over 20 years. He was raised on a farm in southwest Iowa and received his B.S. and Ph.D. in agricultural economics from Iowa State University. Dr. Kay has experience as both a professional farm manager and a farm management consultant, and he maintains an active interest in a farming operation. He is a member of a number of professional organizations including the American Society of Farm Managers and Rural Appraisers where he is a certified instructor for several of the Society's educational courses. Dr. Kay has written a number of publications and was the sole author of the first two editions of *Farm Management*.

William M. Edwards is a professor of economics at Iowa State University, from where he received his Ph.D. He grew up on a family farm in south-central Iowa. Since 1974, he has taught undergraduate courses and carried out extension programs in farm management at Iowa State University. In 1992 he received the Iowa State University Foundation Award for Mid-career Achievement in Extension. Dr. Edwards has also worked with small farmer credit programs and farm management education in Latin America and Eastern Europe.

CONTENTS

PREFACE

The basic principles and concepts of farm management change slowly over time. However, the environment in which they are applied, the specific problems which need to be solved, and the data available for their solution are constantly evolving. These factors, plus the continued widespread use of the second edition, have created the need for a third edition. We hope it will be useful as we train the managers who will be managing our farms and ranches in the twenty-first century.

Users of the second edition will find most of the same topics included here. However, in response to a number of suggestions from reviewers, the order of presentation has been changed, and many chapters have undergone major revision. In addition, all chapters have been updated to provide readers with data and information that are as current as possible. William M. Edwards of Iowa State University is the new coauthor. This addition has resulted in the inclusion of some new viewpoints, examples, and points of discussion throughout the text.

This book is divided into five parts. As with previous editions, an effort was made to make each part as independent as possible, allowing them to be taught in any order. Part One discusses the decision-making process and the unique management environment in which agriculture functions. Part Two presents the basic information needed to measure management performance. It discusses how to collect and organize accounting data and how to construct and analyze an income statement and balance sheet.

The economic principles and planning tools needed to develop basic management skills are presented in Part Three. Business management decisions are guided by economic principles applied in the context of agricultural production. Different forms of budgeting are illustrated along with their application to the planning and decision-making process.

More advanced management topics are contained in Part Four, including farm business organization, managing risk and income taxes, analyzing long-term investments, and whole farm business analysis. The final portion of the book, Part Five, discusses alternatives for acquiring the key resources needed in farming, such as capital, land, labor, and machinery.

Any general theme or pattern to the changes made in this edition often comes from one or more of the following areas:

1. Increased attention to financial management and the prudent use of credit, including lessons learned during recent periods of financial stress in agriculture.
2. Use of innovative business structures for acquiring resources and managing risk, including leasing, contracting, and various types of joint ventures.
3. Concern with the long-run sustainability of agricultural resources and the environmental effects of farming practices.
4. Increased capacity for collecting and analyzing production and financial data, and the need for more efficient and consistent organization and interpretation of it.
5. The dual nature of farming and ranching both as a way of life and as a business which must compete with other businesses for resources and markets.

The recommendations, guidelines, and terminology developed by the Farm Financial Standards Task Force were followed when we developed the material in Part Two. We attempted to maintain consistency with these throughout the remainder of the text. The chapter on production and inventory records was eliminated from this edition, in recognition of the increased emphasis on record keeping found in most agricultural production courses today. Using personal computers has now become commonplace for students. This has permitted a discussion of applications and illustrations of computer use in farm management to be incorporated into many of the chapters.

The discussion of economic principles was divided into two chapters to distinguish more clearly between the input-product relationships and the principles of substitution among inputs and among products. Enterprise and partial budgeting have also been placed into separate chapters to increase the flexibility of the order in which topics can be covered. The discussions of income tax management, farm business organization, and labor management have been revised to reflect changes in the laws and regulations affecting them.

Additional forms of acquiring the use of farm resources, such as contracting, credit lines, and joint ventures with outside investors, have also been included. Finally, a glossary of key terms has been added to provide a handy reference for students.

Instructors will find that the review questions and references at the end of each chapter have been updated and expanded. In addition, an instructor's manual has been created to provide additional study questions, teaching suggestions, additional references, instructional aids, and example exercises for illustrating key concepts. Naturally, these can be adapted to any geographical area and the production environment that is familiar to students in any particular class.

The ideas and suggestions of the students and instructors who have used the first two editions have been very valuable in guiding the revision process. Suggestions for future improvements are always welcome. A special thanks goes to the following McGraw-Hill reviewers for their many ideas and comments: Peter J. Barry, University of Illinois; Robert O. Burton, Kansas State University; John R. Campbell,

Oklahoma State University; Pritam S. Dhillon, Rutgers University; James B. Kliebenstein, Iowa State University; James D. Libbin, New Mexico State University; Wesley N. Musser, Pennsylvania State University; Arnold W. Oltmans, North Carolina State University; Raymond Joe Schatzer, Oklahoma State University; Scott M. Swinton, Michigan State University; and L. Allen Torell, New Mexico State University.

Ronald D. Kay
William M. Edwards

FARM
MANAGEMENT

PART ONE

MANAGEMENT

HIGHLIGHTS

Management is an important factor in the success of any business: Farm and ranch businesses are no exception. Accepting this statement as true, is management becoming more important? To be successful, do the farm and ranch managers of today need more management skills and need to spend more time on management than their parents and grandparents? Although there may be differences of opinion, most observers would answer "Yes" to both questions.

This response comes from recognizing that production agriculture in the United States and other industrialized countries has changed and continues to change along the following lines: more and more mechanization, continued adoption of new technologies, growing capital investment per worker, large amounts of borrowed capital, increasing farm size, new marketing techniques, and increased risk. These factors create new management problems but also opportunities for managers with skills in these areas.

The importance of management can also be illustrated by asking and trying to answer some other questions. Why do some farmers make more money than others? Why do some farm businesses grow and expand while others struggle to maintain their very existence? Why do some farmers consistently get higher yields from their crops and livestock given the same soil quality and livestock facilities? Farm business records from many states show the top 25 percent of the farms to be very profitable while the bottom 25 percent show very little profit and often are operating at a loss. Why the difference? Observation and analysis often lead to the same conclusion: The difference is due to management. It is easy to agree on the last statement but more difficult to come up with specific examples without a detailed examination of individual farms.

Differences in management can show up in three areas: production, marketing/purchasing, and financing. Production differences include the choice of agricultural commodities to be produced and the details of how they are to be produced. Marketing and purchasing activities include the when, where, and how of purchasing inputs and selling commodities. Differences here are reflected in quality, quantity, and prices paid and received. Financing covers not only borrowing money, and the related questions of when, where, and how much, but also the entire area of how to acquire the resources necessary to produce agricultural commodities. There is also risk to be considered in all three areas, and how managers recognize and handle risk can have a major impact on profit and its variability. There are many who would agree that agriculture faces more risk today than in the past. This provides yet another argument for the increased importance of management in the agriculture of today.

If management is so important, we must ask even more questions. What exactly is management? What is it that managers do? What functions do managers perform? What knowledge and skills are needed to be a successful manager? The answers to the first three questions are discussed in Chapter 1. Getting answers to the last question will require reading the remainder of the book.

MANAGEMENT

CHAPTER OBJECTIVES

    1 To understand management and its importance

    2 To identify and discuss some of the more important functions of management

    3 To identify some unique characteristics of farm and ranch management problems

    4 To note how these characteristics affect farm and ranch management decision making

    5 To discuss common goals of farm and ranch management and how they affect decision making

    6 To analyze the steps in the decision-making process

    7 To show how different types of decisions affect the decision-making process

Farm and ranch managers face a never-ending task. They are continually bombarded by new information that affects how their business is organized, what commodities are produced, how they are produced, what inputs should be used, how much of each input should be used, how to finance the business, and how and when to market their production. This new information must be considered when making new decisions and will often mean old decisions must be reconsidered. Successful managers cannot simply memorize answers to problems. They must learn to continually rethink their decisions as economic and environmental conditions change.

What are the types and sources of this new information? Price information changes almost daily. The cause of these changes can be weather, government programs and policies, exports, international events, and other factors that affect the supply and demand situation for agricultural commodities. Both production and marketing decisions are affected by price changes, and these changes are a major source of the risk that exists in agricultural production. New technology provides a constant source of new information. This information comes from the development of new seed varieties, new chemicals for weed and insect control, new animal health products and feed additives, and new, improved, and larger machinery with more and more electronic controls and monitors. In addition, there are changes in the general economy, government programs and farm policies, income tax rules, and environmental regulations. All of these factors are sources of new information that the manager should and, in the case of environmental regulations for example, *must* take into account when making management decisions.

These examples of new information illustrate the continually changing environment affecting the farm manager of yesterday, today, and tomorrow. It is easy to make some broad observations of how some managers have responded to the information about these new economic and technological changes. There has been a large increase in average farm size and a corresponding decrease in the number of farms over the past several decades. The acreage of some crops has increased substantially while the acreage of others has remained constant or declined. Livestock production, particularly poultry and swine, is becoming more and more concentrated in climate-controlled buildings.

These and other changes in agricultural production have resulted from the collective effects of thousands of decisions made by farm and ranch managers as they have responded to new information about changes in technology, prices, and other economic factors. However, some managers have apparently not responded to these changes or have not made the correct response. Table 1-1 provides some evidence. Notice that farms with a net farm income in the top 20 percent of all farms had a net farm income many times higher than farms in the lowest 20 percent of the profit range and about two and one-quarter times that of the average farm. The high-profit farms had a return on investment nearly twice that of the average farm and a respectable return on equity. These and the other differences shown are substantial but cannot be fully explained by the differences in the productivity of land or other resources. There must be another explanation, and differences in the management ability of the farm operators is one possibility.

Farmers in southwest Minnesota are not unique. Similar differences in profitability can be found in data from many other states. The data in Table 1-1 only point out the importance of management and the manager's response to new information, changes, and risk. As new information becomes available, managers must be prepared to evaluate it, incorporate it into their decision making, and make a proper response. Obviously, some managers are better at doing this than others. Without a timely and correct response, a manager cannot expect to survive in a rapidly changing agricultural economy. Present and future managers will need to improve and continually update their management skills if they are to be effective and profitable managers in the years to come.

TABLE 1-1   COMPARISON OF LOW- AND HIGH-PROFIT FARMS IN SOUTHWESTERN MINNESOTA, 1991

Item	All farms (Average)	Lowest 20% (Average)	Highest 20% (Average)
Net accrual farm income	$55,824	$6,916	$126,751
Labor and management earnings	$11,281	$(32,889)	$52,242
Return on investment	5%	−4%	9%
Return on equity	3%	−20%	11%
Change in net worth	$4,699	$(26,742)	$39,203
Corn yield (bu)	124.59	117.39	128.95
Soybean yield (bu)	38.23	33.65	41.30

Source: 1991 Annual Report Southwestern Minnesota Farm Business Management Association, Economic Report ER92-3, Department of Agricultural and Applied Economics, April, 1992.

WHAT IS MANAGEMENT?

The discussion so far has pointed out the importance of management without defining the term. What is management? What do managers do? How do they spend their time? What are they trying to accomplish? How does management differ from labor? Or does it? Management is a widely used term but one that is subject to many individual definitions. A general discussion will serve to provide a broad understanding of management.

Some of the more common definitions of management use phrases such as "making decisions to increase profits," "making the best use of available resources," "is concerned with meeting goals," and "using, managing, or allocating resources." The references to using resources are similar to a definition of economics, which is often defined as "the study of the allocation of scarce resources."

This small sampling of phrases used in definitions of management does provide some insight into management and the things managers do. First, they imply the existence of a *goal* or goals. Managers must either establish these goals or be sure they clearly understand the business owner's goals if the manager is an employee. Second, there are *resources* to use or allocate. This means the manager must identify the finite amount available and then properly allocate or use this amount to meet the goals. Third, the need to allocate or use resources implies more than one possible use for them. The manager must identify all possible *alternatives,* analyze them, and then select those representing the best use of the resources. All of these steps indicate the need for the manager to be making decisions.

FUNCTIONS OF MANAGEMENT

Another approach to discussing management is to list the functions of management. Common functions are planning, organizing, coordinating, controlling, staffing, directing, supervising, and implementing. These functions do not lead to a definition of management but provide another set of ideas on "what managers do." They also illustrate the broad scope of management and its complexity.

Three functions are often identified as being basic or fundamental to management. They are planning, implementation, and control.

Planning   Planning may be the most fundamental and important of the functions. It means deciding on a course of action, policy, or procedure. Not much will happen without a plan. The organizing function might be considered part of planning.

Implementing   Once a plan is developed, it must be implemented. This includes the acquisition of the resources and materials necessary to put the plan into effect as well as overseeing the entire process. Coordinating, staffing, directing, and supervising could fit under this function.

Control   The control function can be thought of as monitoring results, recording information, and taking corrective action. It monitors the results of the plan to see if it is being followed and producing the desired results. If not, it should provide an early warning so adjustments can be made. Outcomes and other related data should be recorded as this becomes a source of new and often improved information to use when making adjustments and for improving future plans.

Figure 1-1 illustrates the flow of action from planning through implementation to control. It also shows that information obtained from the control function can be used for revising, modifying, and making future plans. This circular flow implies a continuous process of planning, implementation, and control followed by improved planning based on new information. The process of continual improvement and refinement of the decision can continue through many cycles.

FARM AND RANCH MANAGEMENT

Is farm and ranch management greatly different from the management of other types of businesses? Should it be a separate, distinct subject or discipline? Or is management the same for all types of businesses with no differences in the methods and principles used? Some argue that the basic management functions, principles, and techniques needed are the same but only applied to different types of businesses. Others say a typical farm or ranch business has some unique characteristics that affect the management principles and techniques used.

Some of the more obvious characteristics are differences in size, type of business or business organization, products produced, and their location outside urban areas. Farms

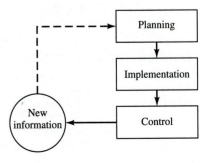

FIGURE 1-1   Management flowchart based on three functions of management.

and ranches are often considered to be subject to more risk in terms of output and price variability and the effect of weather on the entire production process. Also, except on some large farms, there is little opportunity for specialization in labor and management tasks. However, these and other differences are not readily apparent when definitions of business management and farm and ranch management are compared.

There are many different textbook definitions of farm and ranch management but several common points run through most of them. One of the more concise definitions is: "Farm management is concerned with the decisions which affect the profitability of the farm business."[1] This broad definition contains several important points. First, it identifies profitability as a major goal of the business but not necessarily to the exclusion of other goals. Second, this definition specifically identifies decisions and decision making as part of the management activities.

Other definitions contain many of the same concepts or ideas. There is usually some reference to *decisions* or *decision making* as being important in management. Also, some mention is usually made of *goals* or *objectives*. This may be done in general terms or a more specific goal such as profit maximization may be identified. Finally, some mention of the *organization* and *operation* of the farm or ranch business is included in many definitions.

Farm and ranch management can be thought of, then, as being a decision-making process. It is a continual process because of the continual changes taking place in our economy and in the individual business. The decisions are concerned with allocating the limited resources of land, labor, and capital among alternative and competing uses. This allocation process forces the manager to identify goals to guide and direct the decision making.

GOALS AND THEIR IMPORTANCE

A manager's first job is to establish goals for the business. In the case of a hired manager, the business owner may establish and communicate them to the manager. Whether the manager establishes goals or receives them from the business owner, it is important that they exist and be fully understood. Without goals there is no way to make management decisions nor measure their results. Goals are the gauges used to determine if one alternative use of resources is better than another.

When establishing goals, it is important to keep the following points in mind:

1 Goals should be written. This allows everyone involved to see and agree on them and provides a record for review at later dates.

2 Goals should be specific. "To increase profit by $20,000 per year" is better than "to increase profit." A manager can determine when a specific goal has been reached. This provides a sense of accomplishment and a time to think about defining new goals.

3 Goals should be measurable. The $20,000 per year goal is measurable, and each year the manager can measure progress toward the goal as well as what remains before reaching it.

[1]Emery N. Castle, M. H. Becker, and A. Gene Nelson, *Farm Business Management,* 3d ed., Macmillan Publishing Co., New York, 1987, p. 3.

4 Goals should have a timetable. "To increase profit by $20,000 per year in *2 years*" is better than a goal with an open-ended or unspecified completion date. The deadline helps keep the manager focused on the goal.

Because of their close and direct involvement with the farm business, a farm operated by a family unit can have more than one set of goals. There can be personal goals as well as business goals, and each individual within the family unit may have different goals within each set. In these situations, it is important to use a family conference or similar method to discuss and hopefully agree on at least the business goals. Without an agreement, everyone may go in different directions with no business goals being reached.

Since individuals and the businesses they manage are different, many potential goals exist. Surveys of farm operators have identified the following goals:

- Survival, stay in business, do not go broke, avoid foreclosure
- Maximize profit, get the best return on investment
- Increase or maintain standard of living, attain a desirable standard of living
- Increase net worth, steady increase in net worth
- Reduce debt, become free of debt
- Attain at least a minimum profit each year, avoid years of low profit, maintain a stable income
- Pass the entire farm on to the next generation
- Increase leisure, free time
- Increase farm size, expand, add acreage
- Maintain the quality of soil and water resources

These surveys have also found that rarely does a single goal exist; farm operators often have several or multiple goals. A two-step process is indicated: (1) Identify the important goals and (2) establish their priority or ranking. Some goals may be in conflict, which makes the ranking process even more important. Another job for the manager is to evaluate the trade-offs among conflicting goals.

Each of the goals listed may rank first for some individual depending on time and circumstances. Goals can and do change with changes in age, financial condition, family status, and other factors. Also, long-run goals may be different than short-run goals. Profit maximization is often assumed to be the goal of all business owners, particularly in economics. However, farm operators often rank survival or staying in business as their number one goal, and others may be ranked above profit maximization. However, it is important to note that profit plays a direct or at least an indirect role in meeting some of the other possible goals.

A business cannot survive for long without making a profit. Profit is needed to increase family living expenses, increase net worth, decrease borrowing, and to expand. However, several of the possible goals listed imply some sort of risk minimization or risk avoidance that puts these goals in conflict with profit maximization. Many of the most profitable production plans and strategies *over time* are also the most variable in profit, that is, have the most risk. A highly variable profit may greatly reduce the chances for survival and conflicts with the desire for a stable income and avoiding years of low or negative profit. For these and other reasons, profit maximization is not always the most important goal for all farm operators.

Nonetheless, it is the goal assumed throughout the remainder of this text unless stated otherwise. It has the advantage of being easily measured, quantified, and compared across different businesses. As mentioned, it also contributes to or is necessary to meet several of the other potential goals. However, the reader should always remember that goals are very personal, very individual, and will vary from individual to individual. It is a good idea to know and understand someone's goals before being too critical of his or her decisions, methods, and actions.

PROBLEM TYPES AND CHARACTERISTICS

The discussion so far suggests that management is a problem-solving and decision-making activity. What types of economic or management problems are found on farms and ranches? What are the basic characteristics of these problems? As an example, consider just the production problems a manager must solve. These problems fall into one of three types, each of which can be put into the form of a question:

How much to produce? Production is determined primarily by the number of inputs used and input levels. A manager is faced with the problems of how much fertilizer and irrigation water to use, seeding rates, feeding levels, labor and machinery use, and determining rates and levels for other inputs. The level of production and profit will be determined by the input levels selected.

How to produce? Many agricultural products can be produced in a number of ways. Beef can be produced with a high-grain or a high-roughage ration. Hogs can be produced with a large capital investment in buildings and little labor or with less investment and more labor. Crops can be produced with large machinery and little labor or smaller machinery and more labor. A manager must select the appropriate combination of inputs that will minimize the cost of producing a given quantity of some output.

What to produce? This problem involves selecting the combination of crops and livestock to be produced. Should the business produce only crops, only livestock, or some combination? Which crops or crop rotation? Which livestock? The manager must select from among the many alternatives that combination that will maximize profit or best meet some other goal.

Every production problem a farm manager faces relates to one or a combination of these three questions. These production problems are also economic problems and, as such, have the three basic characteristics of an economic problem:

1 Goals to be attained

2 A limited amount of resources available

3 A number of alternative ways to use the limited resources in attempting to attain the goals

Goals

Goals and their importance were discussed earlier in this chapter but should be emphasized again. Not only are they a characteristic of an economic problem they also provide a focus and direction for the entire management process. Goal attainment is the engine that drives management. Without goals, the business has no direction and goes nowhere.

Limited Resources

A manager must consider the resources available for attaining the goals that have been set. Limits are placed on goal attainment because most managers have a finite amount of resources available. In a farm or ranch business, goal attainment is confined within some limits set by the amount of land, labor, and capital available. These resources may change over time, but they are never available in infinite amounts. The level of management skills available or the expertise of the manager may be another limiting resource. Identifying current resource limits and acquiring additional resources, including management skills, are part of the continuing responsibilities of the farm manager.

Alternative Uses

If the limited resources could only be used one way to produce one agricultural product, the manager's job would be much easier. The usual situation allows the limited resources to be used several different ways to produce each of a number of different products. In other words, the manager is faced with a number of alternative uses for the limited resources and must make decisions on how to allocate these resources in order to maximize profit. The emphasis should be on maximizing profit for the entire business and not for just one of the alternatives.

In the more arid regions of the western United States, the land resource is such that the only alternative may be to use it as pasture for livestock production. But even in this situation, the manager must still decide whether to use the pasture for cow/calf production, for grazing stocker steers during the summer, or in some areas, for sheep and goat production. Other areas of the country have land suitable for both crop and livestock production, and a larger number of alternatives exist. As the number of alternative uses for the limited resources increases, so does the complexity of the manager's problem.

THE DECISION-MAKING PROCESS

The allocation of limited resources among a number of alternative uses requires a manager to make decisions. This is one of the reasons many definitions of management and farm management contain some mention of decision making. Without decisions nothing will happen. Even allowing things to drift along as they are implies a decision, probably not a good decision but a decision nevertheless.

The process of making a decision can be formalized into a logical and orderly series of steps. If not already done, a goal or goals must be established before beginning the process. Assuming this has been done, the decision-making process then consists of the following steps:

1 Identify and define the problem.
2 Collect data and information.
3 Identify and analyze alternative solutions.
4 Make the decision—select the best alternative.
5 Implement the decision.
6 Monitor and evaluate the results.
7 Accept the responsibility for the decision.

Following these steps will not make every decision a perfect decision. It will, however, help any manager make a decision in a logical and organized manner that will result in better decisions.

Identifying and Defining the Problem

Many problems confront a farm or ranch manager. Earlier in this chapter some problems were identified in terms of deciding how much to produce, how to produce, and what to produce. These are basic problems faced by all managers. Problems also result from identifying something that is not as it should be. This may be a goal that is not being attained or a deficiency in the organization or operation of the business identified by finding a difference between *what is* and *what should be.* For example, a farmer may have a cotton yield 100 pounds per acre lower than the average for other farmers in the same county on the same soil type. This difference between what is, the farm yield, and what it should be, at least the county average yield, identifies a problem that needs attention.

A manager must constantly be on the alert to identify problems and to identify them as quickly as possible. Most problems will not go away by themselves and represent an opportunity to increase the profitability of the business through wise decision making. Once identified, the problem should be concisely defined. Good problem definition will minimize the time required to complete the remainder of the decision-making steps.

Collecting Data and Information

Once a problem has been identified and properly defined, the next step should be to gather data, information, and facts and to make observations that pertain to the specific problem. A concise definition of the problem will help identify the type of data needed and prevent time being wasted gathering information that is not useful to the particular problem. Data may be obtained from a number of sources, including the local county extension office, bulletins and pamphlets from state experiment stations and agricultural colleges, dealers, salespersons of agricultural inputs, radio, TV, farm magazines, and neighbors. An important source of data and information is an accurate and complete set of past records kept for an individual business. When available, this is often the best source. Whatever the source, the relative accuracy and reliability of the information obtained should be considered.

Decisions typically require information about future events since producing crops and livestock takes time. Information about future prices and yields is often needed and may require the decision maker to formulate some "estimates" or "expectations." Past prices and yields provide a starting point but will often need to be adjusted based on the best current information and the judgment of the decision maker.

It is important to make a distinction between data and information. Data can be thought of as an unorganized collection of facts and numbers obtained from various sources. In this form the data may have very little use. To be useful, these data and facts need to be organized, sorted, and analyzed. Information can be thought of as the final product obtained from analyzing data in such a way that useful conclusions and results are obtained. Not all data need to be organized and summarized into useful information, but most are not useful until some type of analysis is made.

Gathering data and facts and transforming them into information can be a never-ending task. A manager may never be satisfied with the accuracy and reliability of the data and the resulting information. However, this step must be terminated at some point to make it possible to move to the third step. It is important to remember that gathering data has a cost in terms of both time and money. Too much time spent gathering and analyzing data may result in a cost that cannot be justified by the extra income received from continual refinement of data collection and information processing.

Identifying and Analyzing Alternatives

Once the relevant information is available, the manager can begin listing alternatives that are potential solutions to the problem. Several may become apparent during the process of collecting data and transforming data into information. Others may take considerable time and thought, and all possible alternatives should be considered. This is the time to brainstorm and list any idea that comes to mind. Custom, tradition, or habit should not restrict the number or types of alternatives considered.

Each alternative should be analyzed in a logical and organized manner to ensure accuracy and to prevent something from being overlooked. The principles and procedures discussed in Part Three provide the basis for sound analytical methods. In some cases, additional information may be needed to complete the analysis. Good judgment and practical experience may have to substitute for information that is unavailable or available only at an additional cost that is greater than the additional return from its use.

Making the Decision

Choosing the best solution to a problem is not always easy nor is the best solution always obvious. Sometimes the best solution is to do nothing or to go back, redefine the problem, and go through the decision-making steps again. These are legitimate decisions, but they should not be used to avoid making a decision when a promising alternative is available.

After carefully analyzing each alternative, one may not appear to be definitely better than any other. The alternative that best meets the established goal(s) would normally be selected. With profit maximization as the primary goal, the alternative resulting in the largest profit or increase in profit would be chosen. However, the selection is often complicated by uncertainty about the future, particularly future prices. If several alternatives have nearly the same potential effect on profit, the manager must then assess the chances or probability that each will have the expected or identified outcome and the risk associated with it.

Making decisions is never easy, but it is what people must do when they become managers. Most decisions will be made without all the desired information. An alternative must be selected from a list of alternatives, all of which have some disadvantages and some risk. Just because a decision is difficult is no reason to postpone making it. Many opportunities have been lost by delay and hesitation.

Implementing the Decision

Nothing will happen and no goals will be met by simply making a decision. That decision must be correctly and promptly implemented, which means taking some action. Resources may need to be acquired, detailed plans made, a timetable constructed, and everything communicated to employees. At this point the manager needs organizational skills. Remember that not implementing a decision has the same result as not making a decision.

Monitoring and Evaluating the Results

Not every decision will be perfect. Therefore, managers must be prepared for and expect less than perfect results. Even when a perfect decision is made based on the information known at the time, changes will occur over time. The longer it takes for the results of the decision to become known, the more likely the results will be something unexpected. In other words, a good decision can have bad results. Good managers may not always make perfect decisions that have exactly the planned outcome, but they will react quickly when the results of their decisions are not as expected. To do this requires monitoring the results of the decision with an eye toward modifying or changing it.

Managers must set up a system to monitor and measure the results so any deviation from the expected can be quickly identified. This system should also include collecting any data that become available. Careful observation and good records will provide new data to be analyzed. The results of this analysis provide new information to use in modifying or correcting the decision and making future decisions. Evaluating decisions is a way to "learn from your past mistakes."

Bearing Responsibility

Responsibility for the outcome of a decision rests with the decision maker. A reluctance to bear responsibility may explain why some individuals find it so difficult to make a decision. However, since it is difficult for managers to avoid decision making, it follows that they must bear the responsibility. It is part of the job.

CLASSIFYING DECISIONS

The decisions made by farm and ranch managers can be classified in a number of ways. One classification system would be to consider decisions as either *organizational* or *operational* in nature. Organizational decisions are those in the general areas of developing plans for the business, acquiring the necessary resources, and implementing the overall plan. Examples of such decisions are how much land to purchase or lease, how much capital to borrow, and what types of crops and livestock to raise. Organizational decisions tend to be long-run decisions that are not modified or reevaluated more than once a year.

Operational decisions are made more frequently and relate to the many details necessary to implement the farm business plan. They need to be made on a daily, weekly, or monthly basis and are repeated more often than organizational decisions. They follow the routines and cycles of agricultural production. Examples of operational decisions are selecting fertilizer and seeding rates, making changes in livestock feed rations, selecting planting and harvesting dates, marketing grain and livestock, and developing daily work schedules.

Decisions can also be classified as either *strategic* or *tactical.* Strategic decisions are broad in nature, long run, and thus made or changed infrequently. They are those made while developing or modifying the long-run "master plan." Examples would be purchasing additional land, installing an irrigation system, or changing the long-run crop rotation. Tactical decisions are those involved with the details necessary to implement the master plan. They are more short run, limited in scope, and made more often. Examples would be deciding on seeding rates and dates, feed rations, fertilizer levels, and the timing of field operations. The distinction between strategic and tactical decisions is very similar to that between organizational and operational decisions.

Decisions can have a number of characteristics that provide another classification system. One such list of decision characteristics is[2]

1 Importance
2 Frequency
3 Imminence
4 Revocability
5 Available alternatives

Each of these characteristics may affect how the decision is made and how the manager applies the steps in the decision-making process to a specific problem.

Importance Given the many decisions made by farm and ranch managers, some will be more important than others. This means some should be given a higher priority

---

[2]Emery N. Castle, M. H. Becker, and A. Gene Nelson, *Farm Business Management,* 3d ed., Macmillan Publishing Co., New York, 1987, pp. 7–8.

than others. Importance can be measured in several ways. The number of dollars involved in the decision or the size of the potential gain or loss would be one way. Decisions risking only a few dollars might be made rather routinely with little time spent gathering data and proceeding through the steps in the decision-making process.

Decisions involving a large amount of capital and potential profit or loss need to be analyzed more carefully. They can easily justify more time spent on gathering data and analyzing possible alternatives. Examples would be the purchase of additional land, establishing an irrigation system, and constructing a new total-confinement hog building.

Frequency   Some decisions may be made only once in a lifetime, such as the decision to choose farming or ranching as a vocation. Other decisions must be made almost daily, such as livestock feeding times, milking times, and the amount of feed to be fed each day. Such frequently made decisions should be based on some rule of thumb or other predetermined method. If much time was allocated to making these routine decisions, the manager would accomplish very little else. Even though some sort of standard routine, rule of thumb, or output from a computerized decision aid is used, the manager must be aware of the cumulative effects of a small error in these decisions. Because these decisions do occur frequently, small errors in the decision-making process can accumulate into a substantial amount over a period of time.

Imminence   A manager is often faced with making some decisions before a certain deadline or very quickly to avoid a potential loss. Other decisions may have no deadline, and there may be little or no penalty for delaying the decision until more information is obtained and more time spent analyzing the alternatives. When prompt action is required, the manager may have to proceed through the decision-making steps quickly and without complete information. The approach to any decision will depend on the amount of time available.

Revocability   Some decisions can be easily reversed or changed if observation indicates the first decision was not the best. An example would be livestock feeding or milking times, which can be changed rather quickly and easily. Managers may spend very little time making the initial decision in these situations, as future observation may allow corrections to be made quickly and at very little cost.

Other decisions may not be reversible or can be changed only at a very high cost. Examples would be the decision to drill a new irrigation well or to construct a new building. Once the decision is made to go ahead with these projects and they are completed, the choice is either use them or abandon them. It may be very difficult or impossible to recover the money invested. These nonreversible decisions justify much more of the manager's time being used to move carefully through the steps in the decision-making process.

Number of Available Alternatives   Some decisions have only two possible alternatives. They are of the yes or no and buy or don't buy type. The manager may find these decisions easier and less time-consuming than others that have a large number of alternative solutions or courses of action. Where a large number of alternatives exist, the manager may be forced to spend considerable time identifying the alternatives and analyzing each one.

THE DECISION-MAKING ENVIRONMENT

The manager of any business faces the problem of making decisions, but the manager of a farm or ranch makes decisions in a somewhat unique environment. One factor is the limitation placed on a manager's decisions by the biological and physical laws of nature. Managers soon find there are some things that cannot be changed by their decisions. Nothing can be done to shorten the gestation period in livestock production, there is a limit on how much feed a pig can consume in a day, and crops require some minimum time to reach maturity. The manager must be aware of the limits placed on decision making by these biological and physical factors.

In a large corporation the stockholders own it, and the board of directors sets policies and goals and hires managers to achieve them. It is generally easy to identify three distinct groups: owners, management, and labor. These distinct groups do not exist on the typical farm or ranch where one individual or a small family group owns the business and provides the management as well as most or all of the labor. This makes it difficult to separate the management activity from labor because the same individual(s) are involved. It also sets up the possibility that a constant need for labor "to get a job done" places management in a secondary role with decisions constantly delayed or ignored.

Production agriculture is often used as an example of a perfectly competitive industry. This means that each individual farm or ranch is only one of many and represents a very small part of the total industry. Therefore, the individual manager cannot affect either the prices paid for resources or the prices received for products sold. Prices are determined by national and international supply and demand factors over which an individual manager has very little control except possibly through some type of collective action.

The discussion so far illustrates some of the risks involved in producing agricultural commodities. These and other risks come from a number of sources and can be classified in a number of ways. A general classification of the sources of risk might be

- Production risk
- Market risk
- Financial risk

Production risk is the variability in crop yields, percent calf crop, weaning weights, rate of gain, and so forth due to weather, diseases, insects, and other factors beyond the direct control of the manager. These factors are often unpredictable as to their frequency, timing, and severity.

Market risk comes from the unpredictable and variable nature of prices. This is primarily the prices farmers receive for their commodities but, to a somewhat lesser degree, includes prices paid for production inputs. The latter tend to be less variable and generally known at the time the purchase decision is made, but changes over time affect costs and therefore profit.

Financial risk includes fluctuating interest rates, availability of new loans, and variable cash flows caused by production and market risk. The result is uncertainty about the ability to obtain new loans and to repay debt on time. It would also include changes in asset values due to changes in the price of land and other asset prices. This affects borrowing ability, and large decreases in asset values can make a business insolvent.

Decision making is not impossible in the presence of risk, only more difficult. Risk management becomes another task for the farm manager. The variability of the factors discussed makes it difficult to determine what price and yield values should be used when analyzing alternatives. Given the time lag between making a decision and having the commodity ready for sale, there may have been a large change in prices and other factors. This is only one of the reasons why a manager should not expect every decision to be perfect. However, it also illustrates why risk is a factor to consider in nearly every decision a manager makes.

SUMMARY

Possible definitions and the importance and functions of management were discussed from several viewpoints in this chapter. Decision making is an important element in management, and the steps in the decision-making process were discussed in some detail. The establishment of goals takes first priority when taking over management of a business. Goals provide the direction and focus for the entire decision-making process.

Farm and ranch management is unique relative to the management of other businesses. Farm and ranch managers operate in a different environment than other managers. They may use the same management principles and techniques, but they face different constraints and apply them under a different set of circumstances.

QUESTIONS FOR REVIEW AND FURTHER THOUGHT

1 What is your own definition of management? Of farm management?
2 Do farm and ranch managers need different management skills than managers of other businesses? If so, which skills?
3 Would a successful farm manager be successful managing some other business and vice versa?
4 Do above-average managers have some readily identifiable characteristics not found in below-average managers? Name some. Is it education? Training? Experience? Personality? Other factors?
5 Why are goals important? List some examples of both short-term and long-term goals for a farm or ranch business.
6 Which of the common goals of farm and ranch managers might be in conflict with each other?
7 What are your personal goals for the next week? Next year? Next 5 years?
8 Why is it so important to evaluate the results of a decision?
9 List and discuss the steps in the decision-making process. Which steps would be part of the planning function of management? Implementation? Control?

10 Would you classify the following decisions as organizational or operational?
   a Deciding if a field was too wet to till
   b Deciding to sell wheat today or wait
   c Deciding whether to take a partner into the business
   d Deciding whether to hire a full-time livestock manager for the business

REFERENCES

Boehlje, Michael D., and Vernon R. Eidman: *Farm Management,* John Wiley & Sons, New York, 1984, chap. 1.

Castle, Emery N., Manning H. Becker, and A. Gene Nelson: *Farm Business Management,* 3d ed., Macmillan Publishing Co., New York, 1987, chap. 1.

Gessaman, Paul H., and Kathy Prochaska-Cue: *Goals for Family and Business Financial Management,* Nebraska Coop. Extension Service Publications CC 312, CC 313, and CC 314. July, 1985.

Giles, Tony, and Malcolm Stansfield: *The Farmer as Manager,* 2d ed., CAB International, Oxon, United Kingdom, 1990.

Kadlec, John E.: *Farm Management,* Prentice-Hall, Inc., Englewood Cliffs, NJ, 1985, chaps. 1, 2.

Mu'min, Ridgely A., and Ralph E. Hepp: *Evaluating Managerial Effectiveness,* Agricultural Economics Report No. 507, Michigan State University, East Lansing, 1988.

PART TWO

TWO

MEASURING MANAGEMENT PERFORMANCE

HIGHLIGHTS

Two of several points made in Chapter 1 were: (1) the importance of setting goals and (2) control is one of the functions of management. Part Two will continue with the assumption of a profit maximization goal. It will explore how to measure profit and other financial characteristics of a farm or ranch business so the manager can determine how well and to what degree the goal is being attained.

This discussion is also related to the control function of management. Control is a monitoring system to see if the business is following its plan and how far along it is toward meeting its goal. Many of the same records needed to measure profit and the financial status of the business are also needed to perform the control function of management. They provide not only a method of measuring how well the business is doing but also how well the manager or management is doing.

This discussion introduces the need for a record-keeping or accounting system for the farm or ranch business. There are many choices and options in the design and implementation of a farm record system, and they often range from the very simple to the very complex. The right system for any business will depend on many factors including the size of the business, the form of business organization, the amount of borrowing done, lender requirements, and what specific financial reports are needed and in what detail.

Chapter 2 discusses some of these choices and options in a farm record system. It also covers the purposes and parts of a farm record system plus some general concepts such as depreciation and depreciation methods and valuation of assets. Chapters 3 and 4 cover the two most common financial reports, the balance sheet and the income statement. A balance sheet is not intended to measure profit but to record the financial condition of the business at a point in time. It provides the data needed to evaluate the overall financial health of the business.

Chapter 4 introduces the income statement, which provides an estimate of profit or net farm income for an accounting period. The accuracy of the reported profit depends on many factors including the type of record system employed and the effort put forth to keep good records. Many of the choices and options discussed in Chapter 2 will be shown along with their effect on the estimated profit. The emphasis will be on understanding what it takes to accurately measure profit or net farm income. Without an accurate measurement, the effects of past management decisions may be distorted, and inaccurate information will be used to make decisions in the future.

2

ACQUIRING AND ORGANIZING MANAGEMENT INFORMATION

CHAPTER OBJECTIVES

1 To appreciate the importance and value of establishing a good farm or ranch accounting system

2 To review the purposes of an accounting system

3 To show the difference between single- and double-entry accounting

4 To discuss cash accounting, accrual accounting, and the differences between these two methods

5 To outline some of the financial records that can be obtained from a good accounting system

6 To illustrate the different methods that can be used to value farm and ranch property for accounting purposes

7 To define depreciation and show how to use several methods to compute it

A business with poor or no records can be likened to a ship in the middle of the ocean that has lost the use of its rudder and navigational aids: It does not know where it has been, where it is going, or how long it will take to get there. Records tell the manager where the business has been and where it now is on the path to making profits and creating financial stability. They are in one respect the manager's "report card" as they show the results of management decisions over past time periods. Records may not directly show where a business is going, but they can provide considerable information to correct or amend past decisions and to improve future decision making. In that way they at least influence the future direction of the business.

For a number of reasons, farm and ranch records have traditionally been rather poorly kept. Even the better record systems have not been totally consistent with the standards the accounting profession follows for other types of businesses. The financial problems on farms and ranches during the 1980s again focused attention on the poor records kept by many farmers and ranchers, the many different styles and formats of financial reports being used, differences in terminology, and inconsistent treatment of some accounting transactions unique to agriculture.

One result was the formation of the Farm Financial Standards Task Force to address these problems. Its report included recommendations to bring farm records into closer compliance with basic accounting principles. The task force also recommended more standardized and consistent financial reports, uniform financial terminology, and suggested approved methods to handle some of the more difficult and unusual accounting transactions unique to farms and ranches.[1]

PURPOSE AND USE OF RECORDS

Several uses for records have already been mentioned. A more detailed and expanded list of the purpose and use of farm records is possible. One such list might be:

1 Measure profit and assess financial condition.
2 Provide data for business analysis.
3 Assist in obtaining loans.
4 Measure the profitability of individual enterprises.
5 Assist in the analysis of new investments.
6 Prepare income tax returns.

This is not a complete list of all possible reasons for keeping and using farm records. Other possible uses are for establishing insurance needs, estate planning and valuation, monitoring inventories, dividing landlord/tenant expenses, and developing marketing plans. However, the six uses in the list are the more common and will be discussed in more detail.

Measure Profit and Assess Financial Condition   These two reasons for keeping and using farm records are among the more important. Profit is estimated by developing an income statement, which will be the topic for Chapter 4. The financial condition of the business as shown on a balance sheet will be covered in detail in Chapter 3.

Provide Data for Business Analysis   After the income statement and balance sheet are prepared, the next logical step is to use this information to do an in-depth business analysis. There is a difference between just making "a profit" and having "a profitable" business. Is the business profitable? How profitable? Just *how* sound is the financial condition of the business? The answers to these and related questions require

---

[1]The list of references for this chapter contains the citation for the report from the Farm Financial Standards Task Force.

more than just preparing an income statement and balance sheet. A financial analysis of the business can provide information not only on the results of past decisions but information that can be very useful when making current and future decisions.

Assist in Obtaining Loans   Lenders need and require financial information about the farm business to assist them in their lending decisions. After the financial difficulties of the 1980s, many agricultural lenders are requiring more and better farm records. Good records can greatly increase the odds of getting a loan approved and for the amount requested.

Measure the Profitability of Individual Enterprises   A farm or ranch showing a profit may have several different enterprises. It is possible that one or two of the enterprises are producing all or most of the profit, and one or more of the other enterprises are actually losing money. A record system can be designed that will show revenue and expense not only for the entire business but for each individual enterprise. With this information, the unprofitable or least profitable enterprises can be eliminated and resources redirected to use in the more profitable ones.

Assist in the Analysis of New Investments   A decision to commit a large amount of capital to a new investment can be difficult and may require a large amount of information in the analysis. The records from the past operation of the business can be an excellent source of information to assist in analyzing the potential investment. For example, records on the same or similar investments can provide data on expected profitability, expected life, and typical repairs over its life.

Prepare Income Tax Returns   IRS regulations *require* keeping records that permit the proper reporting of taxable income and expenses. This can often be done with a minimal set of records and a set that is not adequate for management purposes. A more detailed management accounting system can also have income tax benefits. It may identify additional deductions and exemptions, for example, and allow better management of taxable income from year to year. This may reduce income taxes paid over time. In case of an IRS audit, good records are invaluable for proving and documenting all income and expenses.

FARM BUSINESS ACTIVITIES

In the design of a farm accounting system, it is useful to keep in mind the three types of business activities that must be incorporated into the system. Figure 2-1 indicates that an accounting system must be able to handle transactions relating not only to the *production* activities of the business but also the *investment* and *financing* activities.

Production Activities   Accounting transactions for production activities are those related to the production of crops and livestock. Revenue from their sale and other

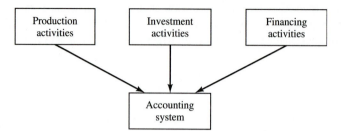

FIGURE 2-1   Farm business activities to be included in an accounting system.

farm revenue such as government payments and custom work performed would be included here. Expenses incurred in producing that revenue—such as feed, fertilizer,
chemicals, fuel, interest, and depreciation—also are part of the production activities
that need to be recorded in the accounting system.

Investment Activities   These activities are those related to the purchase, depreciation, and sale of long-lived assets. Examples would be land, buildings, machinery, and
breeding livestock. Records kept on these assets should include purchase date, purchase price, annual depreciation amounts, book value, current market value, sale date,
sale price, and gain or loss when sold.

Financing Activities   Often a new investment will require borrowing money to finance the purchase. Financing activities are all transactions connected with borrowing
money and the payment of interest and principal on debts of all kinds. They would
also include operating money borrowed to finance the production activities for the
year as well as accounts payable at farm supply stores.

Dividing the farm business activities into these three types illustrates the broad
range of transactions that should be recorded in any accounting system. It also shows
some of the interrelationships between these activities. Interest expense comes from financing activities but is really a production or operating expense. Depreciation results
from the investment in a depreciable asset, but it is also a production or operating expense. Therefore, a good accounting system must be able to not only record all of the
various types of transactions but also integrate them into each appropriate part of the
system.

OPTIONS IN CHOOSING AN ACCOUNTING SYSTEM

Before anyone can begin entering transactions into an accounting system, a number of
decisions must be made about the type of system to be used. A number of options are
available and generally fall into the following areas:

   1 What accounting period should be used?
   2 Should it be a cash or accrual system?
   3 Should it be a single- or double-entry system?
   4 Should it be a manual or computerized system?

It is often difficult to make certain types of changes in an accounting system once one is established, in use, and users are familiar with it. Therefore, considerable thought should be given and advice obtained when making the initial selection of an accounting system.

Accounting Period

An accounting period can be either a *calendar year* or a *fiscal year*. With a calendar year accounting period, all transactions occurring between January 1 and December 31 of each year are organized and summarized into financial reports. Reporting can be done for shorter periods such as quarters, but everything would still be consolidated into annual reports. A fiscal year is any 12-month period that begins on some date other than January 1. Accounting can be done on a fiscal year basis both for management purposes and for income tax purposes.

It is generally recommended that a firm's accounting period follow the production cycle of the major enterprises and end at a time when business activities are slow. For most crop production activities and some livestock, a December 31 ending date fits this recommendation, and most farmers and ranchers use a calendar year accounting period. However, winter wheat, citrus, and winter vegetables are examples of crops where intensive production or harvesting activities may be under way on December 31. These producers may want to consider a fiscal year accounting period that ends after harvesting is completed. Large dairies and commercial feedlots with continuous feeding activities would have a difficult time finding a month when business transactions are much slower than any other month. They could just as well use a calendar year or a fiscal year accounting period.

Cash versus Accrual Accounting

This topic will be covered again when discussing income taxes in Chapter 14. However, the discussion here will be restricted to accounting for management purposes and not for income taxes. While the concepts are the same in either case, the advantages and disadvantages of each accounting method may be different depending on the use for management or for income tax purposes.

Cash Accounting The cash accounting method recognizes and records revenue only when cash is actually received. Expenses are recorded only at the time they are paid. As the name implies, cash accounting is basically a "cash in, cash out" system. Accounting transactions are recorded only when cash is received or paid out. Depreciation expense and any products or services received or delivered in place of cash revenue or expenses would be the only exceptions.

Cash accounting is simple, easy to use, and widely used by farmers and ranchers. However, it has some serious disadvantages as a provider of information for management decision making. The cash accounting method can result in very misleading estimates of profit or net farm income. As one example, consider a first-year crop farmer who uses and pays for all seed, fertilizer, fuel, and chemicals during the crop season but decides to store this first crop. The result is large cash expenses, *no* cash revenue, and a loss for the year. Cash accounting includes no procedure for recognizing the value of the crop in storage until it is sold and cash received.

This inability to match revenue and the expenses incurred to produce that revenue all within the same accounting period prevents cash accounting from providing an accurate estimate of net farm income. Supplies can be purchased and used in one accounting period but not paid for until another or they can be paid for in one and not used until another. Crops produced in one accounting period may not be sold until another. Livestock raised and fattened in one accounting period may show a large increase in value, but no revenue will be recognized until the accounting period in which they are sold and cash received. These are just several examples of the distortions that can occur with cash accounting.

Accrual Accounting Accrual accounting recognizes revenue when it is "earned" and expenses when they are "incurred." The differences between the "received" and "paid" of cash accounting and the "earned" and "incurred" of accrual accounting can be substantial. A major difference is the ability of accrual accounting to recognize all revenue earned in an accounting period whether or not it was converted to cash. All expenses incurred in producing that revenue are then recognized whether or not a cash payment was made during that accounting period. The purpose is to match expenses with revenue within the same accounting period. This process results in several non-cash revenue and expense items being included on the accrual income statement.

For the beginning crop farmer example discussed, accrual accounting would recognize the value of the stored crop as revenue. Any expenses incurred in producing that crop, even though they were not paid within the accounting period, would be recorded as expenses. Accrual accounting includes a number of other factors in the computation of net farm income. However, this example illustrates its ability to present a much more accurate picture of the true revenue, expenses, and net farm income for the accounting period.

Combination System Some farmers and ranchers attempt to combine the easy-to-use features of cash accounting with the improved accuracy of accrual accounting. The procedure is to use cash accounting but include changes in inventory values in the system. These changes are usually recognized only at the end of the accounting period or whenever an income statement is prepared and not tracked throughout the period. Changes in inventory value come from different quantities and/or prices for grain in storage and market livestock on hand at the beginning of the year versus the end of the year. These changes often represent the single largest difference in the estimate of net farm income obtained from the two accounting methods. Including inventory changes in the computations does not eliminate all the differences but is an important first step.

While many farmers and ranchers only adjust for changes in inventory values, several other adjustments should be made. Additional adjustments for changes in accounts receivable, accounts payable, prepaid expenses, and accrued expenses complete the process. When all factors are properly and accurately used in the adjustment process, the resulting net farm income can be as accurate as that obtained from using a complete accrual system. Obviously, any interim estimate of net farm income made during the accounting period without the adjustments would not be accurate. This system does eliminate the need to record many accrual-type transactions during the accounting period. However, there is still a need for accurate beginning and ending values for inventory, accounts payable, accounts receivable, prepaid expenses, and accrued expenses. This adjustment process will be explained in Chapter 4.

Single versus Double Entry

With a single-entry cash system only one entry is made in the books to record a receipt or expenditure. A sale of wheat would simply have the dollar amount recorded under the Grain Sales column in the ledger and a check written to pay for feed would have the amount entered under the Feed Expense column. The other side of the transaction is always assumed to be cash and affect the balance in the checking account. In practice, the checkbook register might be thought of as the other entry, but one that is not included in the ledger.

A double-entry system records changes in the values of assets and liabilities as well as revenue and expenses. There must be equal and offsetting entries for each transaction. This system will result in more transactions being recorded during an accounting period, but it has two important advantages:

1 Improved accuracy as the accounts can be kept in balance more easily.

2 The ability to produce complete financial statements, including a balance sheet, at any time, directly from data already recorded in the system.

The improved accuracy of double-entry accounting comes from the two offsetting entries, which means *debits* must equal *credits* for each transaction recorded. It also means the basic accounting equation of

$$\text{Assets} = \text{liabilities} + \text{owner's equity}$$

will be maintained. The double-entry system maintains the current values of assets and liabilities within the accounting system. This is what allows the generation of financial statements directly from the accounting system without any need for outside information.

Table 2-1 shows some examples of how different business events would be entered using a double-entry accrual system versus a single-entry cash system. Several things are apparent from a study of this table. It is clear that a double-entry accrual system requires more entries in addition to the two entries made for each transaction. However, the payoff for that extra effort comes from maintaining current balances for each of the asset and liability accounts. This can be seen by the entries affecting cash, inventory, account payable, and account receivable. The accrual system entries also record revenue and expenses in the year "earned" or "incurred" and not necessarily in the year cash was received for a sale or paid out for a purchase.

TABLE 2-1   EXAMPLES OF ENTRIES USING ACCRUAL OR CASH ACCOUNTING

Business event	Entry using double-entry accrual accounting	Entry using single-entry cash accounting
Purchased feed and charged to account	Increase feed expense and increase account payable	None
Paid feed bill	Decrease cash and decrease account payable	Enter feed expense
Placed harvested grain in storage	Increase grain income and increase inventory value	None
Sold grain from storage	Increase cash and decrease inventory value	Enter grain income
Purchased and paid for next year's seed corn	Decrease cash and increase prepaid expense	Enter seed expense
Planted seed corn in following year	Increase seed expense and decrease prepaid expense	None
Paid insurance premium for next 12 months	Decrease cash and increase prepaid insurance	Enter full amount as insurance expense
Adjust insurance expense at end of accounting period	Increase insurance expense and decrease prepaid insurance	None
To recognize interest expense accrued since last payment	Increase interest expense and increase accrued interest	None
Did custom work for neighbor	Increase both custom work income and account receivable	None
Received payment from neighbor	Increase cash and decrease account receivable	Enter custom work income

Manual versus Computerized System

A complete, accurate, and useful set of records can be kept with a manual accounting system. There are a number of record books available from office supply stores, agricultural lenders, and state extension services, for example. Most manual systems used on farms and ranches are single-entry cash systems, but it is possible to use a manual, double-entry accrual system. The latter would require a good understanding of accounting principles both when recording transactions and when organizing financial statements at the end of an accounting period.

Computerized farm accounting programs range from inexpensive, single-entry cash-based systems to very detailed double-entry accrual systems. The latter include very complete records on employee payroll, inventory, accounts payable, and accounts receivable, for example. Some allow both cash and accrual accounting so the first can be used for income tax purposes and the latter for management purposes.

A lack of familiarity with both computers and accounting principles has kept many farmers and ranchers from moving to computerized accounting systems. However, computers are now a very inexpensive resource and becoming easier to use. Many double-entry accrual accounting programs designed for farm and ranch use do a good job of minimizing the need to have a good accounting background. It would not take much time or study for someone with a basic knowledge of farm records and computers to be able to use some of the better computerized farm accounting systems.

Computerized accounting is sometimes promoted as saving time. It is unlikely that switching to a computerized system will save time when entering transactions. This is particularly true if the move is made from cash to accrual accounting at the same time. Any time saved with a computerized system comes not when entering transactions but from less time spent looking for errors and generating financial reports. Both interim and end-of-year reports will typically require no more than pressing several keys on the keyboard and waiting for the printer to print the results. Even if they save no time, computerized farm accounting systems can easily be justified on the basis of the improved quality and quantity of financial information they provide. The next step is to be sure this information is properly used.

OUTPUT FROM AN ACCOUNTING SYSTEM

Any accounting system should have the capability of producing some basic financial reports. Computerized accrual systems can generate many different reports. Figure 2-2 expands on Figure 2-1 to show the possible products from an accounting system. The balance sheet and income statement are shown first for two reasons. First, they are the two most common reports to come out of an accounting system and second, they are the subjects of the next two chapters. Some of the other possible reports are often necessary as well as useful but may not be available from all systems nor widely used.[2]

Balance Sheet   The balance sheet is the report that shows the financial condition of the business at a point in time—its assets, liabilities, and equity. A detailed discussion of this report and its use will be covered in Chapter 3.

Income Statement   An income statement is a report of revenue and expenses ending with an estimate of net farm income. This report will be discussed in detail in Chapter 4.

Transaction Journal   This is a record of all financial transactions including check and deposit numbers, dates, payees and payors, amounts, and descriptions. A check register is a form of transaction journal but may not contain all of this information. This journal is used to make entries into the general ledger and to provide an audit trail.

General Ledger   The general ledger contains the different financial accounts for the business and the balances in these accounts. Balances in the revenue and expense accounts are used to prepare an income statement; balances in the asset, liability, and net income accounts are used to prepare a balance sheet.

[2]The 12 possible reports discussed here come from J. F. Guenthner and R. L. Wittman: *Selection and Implementation of a Farm Record System,* Western Regional Extension Publication WREP 99, Cooperative Extension Service, University of Idaho, 1986.

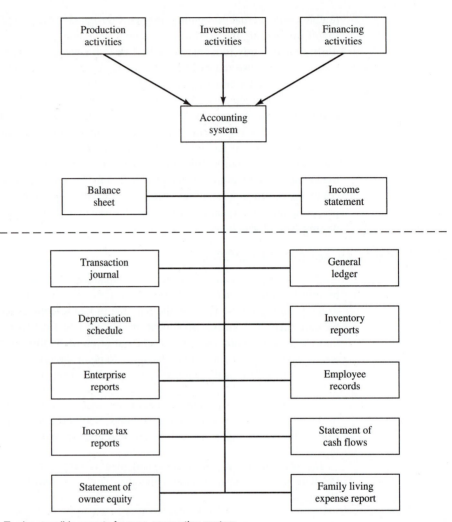

FIGURE 2-2   Twelve possible reports from an accounting system.

Depreciation Schedule   A depreciation schedule is a necessary part of any accounting system. Annual depreciation on all depreciable assets must be computed and recorded as an expense before an income statement can be produced. This is true whether a cash or accrual system is used. This schedule should contain a description of the asset, purchase date, cost, depreciation method used, accumulated or total depreciation to date, and current book value (cost minus accumulated depreciation). In addition to annual depreciation, this schedule provides information necessary to compute any gain or loss when the asset is sold. An example of a depreciation schedule is shown in Table 2-2.

A depreciation schedule can be done manually, and many farm record books will contain pages for completing it. The results of manual computations can then be en-

TABLE 2-2 DEPRECIATION SCHEDULE

Item	Date purchased	Cost or basis	Salvage value	Life	Depreciation method	19___ Depreciation	19___ Book value	19___ Depreciation	19___ Book value	19___ Depreciation	19___ Book value	19___ Depreciation	19___ Book value

tered into a computerized system at the end of the accounting period. Many of the more advanced computer accounting programs have the ability to prepare a schedule and automatically enter the annual depreciation directly into the general ledger. There are also some "stand alone" computer depreciation programs. Given that computing depreciation on a large number of assets can be tedious, time-consuming, and subject to error, a computer depreciation program of some kind can be a good investment.

Inventory Report   This is a useful report, particularly for large crop farms and feedlots. It tracks the quantity and value of crops and livestock on hand through the recording of purchases, sales, births, deaths, amounts harvested, and amounts fed. This report is useful for monitoring feed availability and usage, for developing a marketing program, and for monitoring any inventory pledged as collateral for a loan.

Enterprise Reports   These are basically income statements for each individual enterprise. They are very useful for determining which enterprises are contributing the most profit to the business and therefore candidates for expansion. Enterprises identified as unprofitable become candidates for elimination.

Employee Records   Any business with employees must keep considerable data related to each employee. This includes not only information such as hours worked but financial data on gross pay, deductions for income taxes and social security, and so forth. A number of payroll reports must be filed in a timely manner with both state and federal agencies. All of this payroll related work can be done by hand, but there are a number of computer programs designed specially to compute and record gross pay, deductions, and net pay. If the payroll program is part of a general accounting program, all of this information can be automatically entered into the general ledger.

Income Tax Reports   The information from any farm accounting system must be sufficient to prepare the farm tax return. In some systems it will be necessary to take values from the accounting reports and enter them on Schedule F, Form 1040. It is helpful if the general ledger accounts are named and organized in the same way as on this schedule. Some computer accounting systems can compile and print the tax information in the same format making it easy to copy the data. Other programs can, with the right type of printer, actually duplicate Schedule F and print a completed return.

Statement of Cash Flows   This statement summarizes all sources and uses of cash during the accounting period and is useful when analyzing the business activities during that period. When prepared on a monthly basis, it allows comparison of actual cash flows with budgeted cash flows. It is also important as a source of data when completing a cash flow budget for the next accounting period.

Statement of Owner Equity   Financial transactions during the accounting period will affect the net worth or equity of the business. This statement identifies and summarizes the sources of these changes.

Family Living Expense Report   Although not really a part of the business financial activities, it is desirable to keep detailed records of family living expenses. This is particularly true for any expenditures that may be deductible on income tax returns. Again, this can be done manually or included as part of the farm accounting system provided care is taken to be sure business and personal records are not mixed. There are also rather inexpensive computer programs designed solely for recording, summarizing, and analyzing personal expenses and investments.

VALUATION OF ASSETS

While beginning, maintaining, or improving an accounting system, it is often necessary to determine the value of business assets. This dollar value is the only unit of measurement common to the wide range and types of property found on a typical farm or ranch. Once property is identified, acquired, or produced, some value must be placed on it before an entry can be made in the accounting system.

There are several valuation methods that can be used, and the choice will depend on the type of asset and the purpose of the valuation. Whichever method is used, the accounting concepts of conservatism and consistency should be kept in mind. *Conservatism* cautions against placing too high a value on any asset while *consistency* stresses using the same valuation method or methods over time. Use of these concepts makes financial statements directly comparable from year to year and prevents an overly optimistic portrayal of the firm's financial condition. The following are commonly used valuation methods.

Market Value   This method values an asset using its current market price. It may sometimes be called the fair market value or net market price method. Any normal marketing charges such as transportation, selling commissions, and fees are subtracted to find the *net* market value. This method can be used for many types of property but works particularly well for items that could or will be sold in a relatively short period of time as a normal part of the business activities and for which current market prices are available. Examples would be hay, grain, feeder livestock, stocks, and bonds.

Cost   Items that have been purchased can be valued at their original cost. This method works well for items that have been purchased recently and for which cost records are still available. Feed, fertilizer, supplies, and purchased feeder livestock can be valued this way as can land. Items such as buildings and machinery, which normally lose value or depreciate over time, should not be valued with this method. Livestock and crops that have been raised cannot be valued at cost as there is no purchase price to use.

Lower of Cost or Market   This valuation method requires valuing an item at both its cost and its market value and then using whichever value is lower. This is a conservative method, as it minimizes the chance of placing too high a value on any item. Using this method, property that is increasing in value because of inflation will have a value equal to its original cost. Valuing such property at cost eliminates any increase in value over time caused solely by inflation or a general increase in prices. When prices have decreased since the item was purchased, this method results in valuation at market value.

Farm Production Cost   Items produced on the farm can be valued at their farm production cost. This cost is equal to the accumulated costs of producing the item but should not include profit nor any opportunity costs associated with the production. Grain, hay, silage, and raised livestock can be valued by this method if good cost of production or enterprise records are available. Established but immature crops growing in a field are generally valued this way with the value set equal to the actual, direct production expenses incurred to date. It is not appropriate to value a growing crop using expected yield and price because poor weather, a hailstorm, or lower prices could drastically change the value before harvest and final sale. This procedure is another example of conservatism in valuation. However, a detailed set of enterprise accounts would be needed to use this method.

Cost Less Depreciation   Property that provides services to a business over a period of years but loses value over time because of age, use, or obsolescence should be valued at the original cost less all previous depreciation. Examples would be machinery, buildings, fences, and purchased breeding livestock. Each year the item's value is reduced by the amount of depreciation for that year. Therefore, the value in the current time period is equal to the original cost less the accumulated depreciation from purchase date to the current time. The resulting value is referred to as the item's *book value*.

DEPRECIATION AND DEPRECIATION METHODS

An understanding of depreciation and the methods that can be used in its computation is essential for maintaining, using, and analyzing farm records. The depreciation schedule has been mentioned as one of the necessary products from an accounting system, and cost less depreciation is a common valuation method.

Depreciation is also a business expense and can be viewed as such from two different but related viewpoints. First, it represents a loss in value because the item is used in the business to produce income. Second, it is an accounting procedure to spread the original cost over the item's useful life. It is not appropriate or correct to deduct the full purchase price as an expense in the year of purchase, as the item will be used to generate income for several years. Instead, the purchase price less salvage value is allocated or "spread" over time through the business expense called depreciation.

Before several methods for estimating annual depreciation are discussed, several additional terms need to be introduced. *Useful life* is the expected number of years the item will be used in the business. It may be the age at which the item will be completely worn out if the manager expects to own it that long, or it may be a shorter period if it will be sold before then. The key to determining a useful life is the number of years the manager *expects* to own the item and therefore an estimate and subject to error.

Another estimate is *salvage value,* or terminal value, which is the item's value at the end of its assigned useful life. Salvage value may be zero if the item will be owned until completely worn out and will have no junk or scrap value at that time. A positive salvage value should be assigned to an item if it will have some value as scrap or will be sold before it is completely worn out. In the latter case the salvage value should be its estimated market value at the end of its assigned useful life. In general, the shorter the useful life the higher the salvage value.

Straight Line   The straight-line method of calculating depreciation is widely used. This easy-to-use method gives the same annual depreciation for each full year of an item's life.

Annual depreciation can be computed from the equation

$$\text{Annual depreciation} = \frac{\text{cost} - \text{salvage value}}{\text{useful life}}$$

Straight-line depreciation can also be computed by an alternative method using the equation

$$\text{Annual depreciation} = (\text{cost} - \text{salvage value}) \times R$$

where R is the annual straight-line percentage rate found by dividing 100 percent by the useful life (100 percent ÷ useful life).

For example, assume the purchase of a machine for $10,000 that is assigned a $2,000 salvage value and a 10-year useful life. The annual depreciation using the first equation would be

$$\frac{\$10,000 - \$2,000}{10 \text{ years}} = \$800$$

Using the second equation, the percentage rate would be 100 percent ÷ 10, or 10 percent, and the annual depreciation is

$$(\$10,000 - \$2,000) \times 10\% = \$800$$

The result is the same for either procedure, and the total depreciation over 10 years would be $800 × 10 years = $8,000, reducing the machine's book value to its salvage value of $2,000.

Declining Balance   There are a number of variations or types of declining balance depreciation. The basic equation for all types is

$$\text{Annual depreciation} = (\text{book value at beginning of year}) \times R$$

where R is a constant percentage value or rate. The same R value is used for each year of the item's life and is multiplied by the book value, which declines each year by an amount equal to the previous year's depreciation. Therefore, annual depreciation declines each year with this method. Notice that the percentage rate is multiplied by each year's book value and *not* cost minus salvage value as was done with the straight-line method.

The various types of declining balance come from the determination of the R value. For all types the first step is to compute the straight-line percentage rate, which was 10 percent in our previous example. The declining balance method then uses a multiple of the straight-line rate such as 200 (or double), 175, 150, or 125 percent as the R value. If double declining balance is chosen, R would be 200 percent or two times the straight-line rate. The R value would be determined in a similar manner for the other variations of declining balance depreciation.

Using the previous example, the double declining balance rate would be 2 times 10 percent, or 20 percent, and the annual depreciation would be computed in the following manner:

Year 1: $\$10,000 \times 20\% = \$2,000$
Year 2: $\$\ 8,000 \times 20\% = \$1,600$
Year 3: $\$\ 6,400 \times 20\% = \$1,280$

.   .   .   .
.   .   .   .
.   .   .   .

Year 7: $\$\ 2,622 \times 20\% = \$524$
Year 8: $\$\ 2,098 \times 20\% = \$420$ (but this amount would reduce the book value below the $2,000 salvage value; so only $98 of depreciation can be taken)
Year 9 and 10:   No remaining depreciation

This example is not unusual, as double declining balance will often result in the total allowable depreciation being taken before the end of the useful life and depreciation must stop when the book value equals salvage value. Notice also that the declining balance method will never reduce the book value to zero. With a zero salvage value it is necessary to switch to straight-line depreciation at some point to get all the allowable depreciation or to take all remaining depreciation in the last year.

If a 150 percent declining balance is used, R is one and one-half times the straight-line rate, or 15 percent for this example. The annual depreciation for the first 3 years would be

Year 1: $\$10,000 \times 15\% = \$1,500$
Year 2: $\$8,500 \times 15\% = \$1,275$
Year 3: $\$7,225 \times 15\% = \$1,084$

Notice that each year's depreciation is smaller than if the double declining balance is used, and therefore the book value declines at a slower rate.

Sum-of-the-Year's Digits   The annual depreciation using the sum-of-the-year's digits method is computed from the equation

$$\text{Annual depreciation} = (\text{cost} - \text{salvage value}) \times \frac{RL}{SOYD}$$

where RL = remaining years of useful life as of the beginning of the year for which
    depreciation is being computed

SOYD = sum of all the numbers from 1 through the estimated useful life. For ex-
    ample, for a 5-year useful life SOYD would be $1 + 2 + 3 + 4 + 5 = 15$ and
    would be 55 for a 10-year useful life[3]

Continuing with the same example used in the previous sections, the SOYD would
be 55 (the sum of the numbers 1 through 10). The annual depreciation would be computed in the following manner:

$$\text{Year} \quad 1: \quad (\$10,000 - \$2,000) \times \frac{10}{55} = \$1,454.55$$

$$\text{Year} \quad 2: \quad (\$10,000 - \$2,000) \times \frac{9}{55} = \$1,309.09$$

$$\text{Year} \quad 3: \quad (\$10,000 - \$2,000) \times \frac{8}{55} = \$1,163.64$$

$$\cdot \qquad \cdot \qquad \cdot \qquad \cdot \qquad \cdot$$
$$\cdot \qquad \cdot \qquad \cdot \qquad \cdot \qquad \cdot$$
$$\cdot \qquad \cdot \qquad \cdot \qquad \cdot \qquad \cdot$$

$$\text{Year} \quad 10: \quad (\$10,000 - \$2,000) \times \frac{1}{55} = \$145.45$$

Notice the annual depreciation is highest in the first year and declines by a constant
amount each year thereafter.

Comparing Depreciation Methods   Figure 2-3 graphs the annual depreciation
for each depreciation method based on a $10,000 machine with a $2,000 salvage value
and a 10-year life. The annual depreciation over time is considerably different for each
method. Double declining balance and sum-of-the-year's digits have a higher annual
depreciation in the early years than straight line, with the reverse being true of the later
years. For this reason double declining balance and sum-of-the-year's digits are referred to as "fast" or "accelerated" depreciation methods, with double declining balance being the "fastest" method.

It is also important to note that the choice of depreciation method does not
change the total depreciation taken or allowable over the useful life. In the example
used, there is only $8,000 of depreciation regardless of the method selected. The
depreciation methods affect only the pattern or distribution of depreciation over
time. A final choice of the most appropriate method will depend on the type of
property and the use to be made of the resulting book value. For example, the actual market value of automobiles, tractors, and other motorized machinery tends to

---

[3]A quick way to find the sum-of-the-year's digits (SOYD) is from the equation $[(n)(n+1)]/2$, where n is
the useful life.

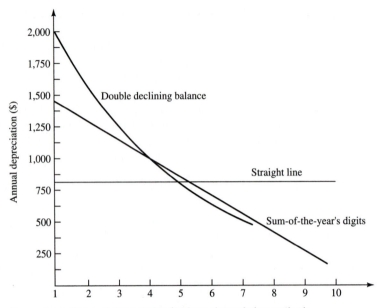

FIGURE 2-3   Comparison of annual depreciation for three depreciation methods.

decline most rapidly during the first few years of life and more slowly in the later years. If it is important for depreciation on these items to approximate their actual decline in value, one of the "fast" depreciation methods should be used. For property such as fences and buildings that have little or no market value and provide a rather uniform flow of services to the business over time, straight-line depreciation would be appropriate.

Items purchased during the year should have the first year's depreciation prorated according to the length of time the item was actually owned. For example, a tractor purchased on April 1 would be eligible for $\frac{9}{12}$ of a full year's depreciation the first year and a pickup purchased on October 1 would get $\frac{3}{12}$ of a year's depreciation the first year. Any time a partial year's depreciation is taken the first year, there will be only a partial year's depreciation remaining in the last year of the item's useful life.

Economic versus Tax Depreciation   Economic depreciation can be thought of as the decline in an item's value due to a decline in its ability to produce income now and in the future. This happens due to a reduction in its annual productive capacity and fewer years of life remaining to produce income. The standard depreciation methods discussed are used in accounting to estimate this decline in value and the amount of depreciation to include in expenses for the accounting period.

Like most business expenses, depreciation is also a deductible expense for income tax purposes. However, the tax regulations contain different methods for computing depreciation for income taxes. The Modified Accelerated Cost Recovery System

(MACRS) uses a combination of declining balance and straight-line methods, zero salvage value for all items, and rather short useful lives. Annual depreciation for tax purposes may, therefore, be somewhat "fast" compared to the true economic depreciation. (See Chapter 14 for a discussion of MACRS.)

Since depreciation must be computed for tax purposes anyway, it is easy and convenient to use the tax values throughout the accounting system. However, if the tax depreciation is higher than the true economic depreciation, there are several results. First, there is a higher depreciation expense during the first few years an asset is owned and therefore a lower net farm income or profit. Second, any item valued using cost less tax depreciation will have a lower book value, which will cause a lower net worth or equity for the business.

The effects and implications of these results will be more clearly understood after reading Chapters 3 and 4. However, it is clear that this misstatement of values can lead to poor conclusions about the financial operations and condition of the business and therefore poor management decisions. To prevent this, it may be necessary to complete two depreciation schedules. One would estimate economic depreciation and be part of the records used for business analysis and management decision making. The other schedule would contain tax depreciation and be used only for preparing the income tax return.

SUMMARY

This chapter discussed the importance, purpose, and use of records as a management tool. Records provide the information needed to measure how well the business is doing in terms of meeting goals. They also provide feedback so the results of past decisions can be evaluated as well as the decision-making ability of the manager. Finally, individual farm records are perhaps the single best source for information needed to make current and future decisions.

Any accounting or record system used must be able to handle transactions from the production, investment, and financing activities of the farm business. The choices of accounting period, cash or accrual accounting, single or double entry, and manual or computerized system are important. They affect the quantity, quality, and accuracy of the information provided by the accounting system as well as the time required to maintain the records. The output required or desired from an accounting system must also be considered when making these choices.

Any accounting system will from time to time require values for business assets. Several possible valuation methods were discussed along with the usual methods for computing economic depreciation for accounting purposes. Depreciation for tax purposes uses different methods, and these values may not be the best choice to use in an accounting system whose results are used to make management decisions.

QUESTIONS FOR REVIEW AND FURTHER THOUGHT

1 Can farmers and ranchers use a fiscal year accounting period? If so, should they? Under what conditions?
2 How would one produce a balance sheet if the accounting was done using a single-entry cash system?
3 Is it possible to use double-entry with a cash accounting system? If so, what are the advantages and disadvantages?
4 Is it possible to use single-entry with an accrual system? Why or why not?
5 Check advertising material for several computerized farm accounting programs. Are they cash or accrual systems? Single or double entry? How many of the 12 outputs from an accounting system discussed in this chapter are available from each program? Are any additional outputs available?
6 Place an "X" under the column(s) to indicate whether each business event is an operating, investment, or financing activity.

Event	Operating	Investment	Financing
Pay cash for repairs	_____	_____	_____
Borrow $10,000	_____	_____	_____
Pay interest on loan	_____	_____	_____
Charge $1,000 of feed	_____	_____	_____
Equipment depreciates	_____	_____	_____
Sold corn for $6,000	_____	_____	_____
Purchase pickup with a $15,000 loan	_____	_____	_____
Paid principal and interest on mortgage	_____	_____	_____

7 Assume a new tractor is purchased on January 1 for $62,000 and given a salvage value of $10,000 and a useful life of 8 years. What would the annual depreciation be for the first 2 years under each depreciation method?

	Year 1	Year 2
Straight line	_____	_____
Double declining balance	_____	_____
Sum-of-year's digits	_____	_____

8 For Problem 7 what would the tractor's book value be at the end of year 2 under each depreciation method?
9 What makes a depreciation method "fast" or "slow"? Give an example of each.
10 What are the advantages and disadvantages of each valuation method? Give examples of property that might be valued by each method. What types of property could or could not be valued by each method?

REFERENCES

Armbruster, David B.: *Introduction to Agricultural Accounting,* Red Wing Business Systems, Red Wing, MN, 1983.

Castle, Emery N., Manning H. Becker, and A. Gene Nelson: *Farm Business Management,* 3d ed., Macmillan Publishing Co., New York, 1987, chap. 3.

Guenthner, J. F., and R. L. Wittman: *Selection and Implementation of a Farm Record System,* Western Regional Extension Publication WREP 99, Cooperative Extension Service, University of Idaho, 1986.

Hobson, Barry: *Accounting Farmer Style,* Doane Information Services, St. Louis, 1992.

James, Sydney C., and Everett Stoneberg: *Farm Accounting and Business Analysis,* 3d ed., Iowa State University Press, Ames, IA, 1986, chaps. 1, 2, 6.

Libbin, James D., and Lowell B. Catlett: *Farm and Ranch Financial Records,* Macmillan Publishing Co., New York, 1987, chaps. 1–4.

Nolting, Greg, and George T. Allton: *Computerize Your Farm Accounting,* Doane Information Services, St. Louis, 1989.

Recommendations of the Farm Financial Standards Task Force, American Bankers Association, Washington, D.C., 1991.

3

THE BALANCE SHEET
AND ITS ANALYSIS

CHAPTER OBJECTIVES

1 To discuss the purpose of a balance sheet

2 To illustrate the format and structure of a balance sheet

3 To outline the problems of valuing assets and the recommended procedures for different types of assets

4 To show the differences between a cost basis and a market basis balance sheet

5 To define owner's equity or net worth and show its importance

6 To analyze a firm's solvency and liquidity through the use of a number of financial ratios derived from the balance sheet

Chapter 2 introduced the *balance sheet* and the *income statement* as two of the products or outputs from an accounting system. They are both part of a complete set of financial statements but are meant to serve different purposes. A balance sheet summarizes the financial condition of the business *at a point in time* while an income statement summarizes those financial transactions that affected revenue and expenses *over a period of time*. The purpose of an income statement is to provide an estimate of net farm income or profit while the balance sheet concentrates on estimating net worth or owner's equity by valuing and organizing assets and liabilities.

Most transactions affect both the balance sheet and the income statement. Since transactions can occur daily, the balance sheet can change daily. That is why the point in time concept is emphasized when discussing the balance sheet. While intended for different purposes, there is a key relationship, or connection, between these two financial statements. However, a discussion of this relationship will have to be postponed until the end of Chapter 4.

PURPOSE AND USE OF A BALANCE SHEET[1]

A balance sheet is a systematic organization of everything "owned" and "owed" by a business or individual at a given point in time. Anything of value owned by a business or individual is called an *asset,* and any debt or other financial obligation owed to someone else is referred to as a *liability.* Therefore, a balance sheet is a listing of assets and liabilities concluding with an estimate of *net worth* or *owner's equity.* These two terms mean the same thing, which is the difference between total assets and total liabilities. The "balance" in balance sheet comes from the requirement that the ledger be in balance through the basic accounting equation of

$$\text{Assets} = \text{liabilities} + \text{owner's equity}$$

A rearrangement of this equation allows finding owner's equity once assets and liabilities are known:

$$\text{Owner's equity} = \text{assets} - \text{liabilities}$$

A balance sheet can be completed any time during an accounting period. One of the advantages of a computerized accounting system is the ease with which an interim balance sheet can be prepared. However, most balance sheets are prepared at the end of the accounting period, which will be December 31 on most farms and ranches. This procedure allows a single balance sheet to be both an end of the year for one accounting period and a beginning of the year statement for the next accounting period. For purposes of comparison and analysis, it is necessary to have a balance sheet available for both the beginning and end of each year.

Measuring the financial position of a business at a point in time is done primarily through the use of two concepts.

1 *Solvency,* which measures the liabilities of the business relative to the amount of owner's equity invested in the business. It also provides an indication of the ability to pay off all financial obligations or liabilities if all assets were sold, that is, assets are greater than liabilities. If assets are not greater than liabilities, the business is insolvent or bankrupt.

2 *Liquidity,* which measures the ability of the business to meet financial obligations as they come due without disrupting the normal operations of the business. Liquidity measures the ability to generate cash in the amounts needed and at the time it is needed. These cash requirements and possible sources of cash are generally measured only over the next accounting period, making liquidity a short-run concept.

---

[1]This chapter makes every effort to adopt the terminology, definitions, and balance sheet format recommended by the Farm Financial Standards Task Force (hereafter FFSTF). (See references at the end of this chapter.)

BALANCE SHEET FORMAT

A condensed and general format for a balance sheet is shown in Table 3-1. Assets are shown on the left side or top part of a balance sheet and liabilities are placed on the right of or below assets.

Assets

An asset has value for one or both of two reasons. First, it can be sold to generate cash or, second, it can be used to produce other goods that in turn can be sold for cash at some future time. Goods that have already been produced, such as grain and feeder livestock, can be sold quickly and easily without disrupting future production activities and are called *liquid assets.* Assets such as machinery, breeding livestock, and land are owned primarily to produce another commodity that can then be sold to produce cash income. Selling them to generate cash to meet immediate cash needs would affect the firm's ability to produce future income so they are less liquid or illiquid. These assets are also more difficult to sell quickly and easily at their full market value.

Current Assets   Accounting principles require current assets to be separated from other assets on a balance sheet. Current assets are the more liquid assets and those that will either be used up or sold within the next year as part of normal business activities. Cash on hand and checking and savings account balances are current assets and are the most liquid of all assets. Other current assets include readily marketable stocks and bonds, accounts and notes receivable (which represent money owed to the business because of loans granted or services rendered), and inventories of feed, grain, supplies, and feeder livestock. The latter are livestock held primarily for sale and not for breeding purposes.

Noncurrent Assets   Any asset that is not classified as a current asset is, by default, a noncurrent asset. On a farm or ranch this would include primarily machinery and equipment, breeding livestock, buildings, and land.

TABLE 3–1   GENERAL FORMAT OF A BALANCE SHEET

Assets		Liabilities	
Current assets	$XXX	Current liabilities	$XXX
Noncurrent assets	XXX	Noncurrent liabilities	XXX
		Total liabilities	$XXX
		Owner's equity	XXX
Total assets	$XXX	Total liabilities and owner's equity	$XXX

Liabilities

A liability is an obligation or debt owed to someone else. It represents an outsider's claim against one or more of the business assets.

Current Liabilities Current liabilities must be separated from all other liabilities for the balance sheet to conform to accounting principles. These are financial obligations that will become due and payable within 1 year from the date of the balance sheet and will therefore require cash be available in these amounts within the next year. Examples would be accounts payable at farm supply stores for goods and services received but not yet paid for and the full amount of principal and accumulated interest on any short-term loans or notes payable. Short-term loans are those requiring complete payment of the principal in 1 year or less. These would typically be loans used to purchase crop production inputs and market livestock.

Loans obtained for the purchase of machinery, breeding livestock, and land are typically for a period longer than one year. Principal payments may extend for 3 to 5 years for machinery and 20 years or more for land. However, a principal payment is typically due annually or semiannually, and these payments will require cash within the next year. Therefore, all principal payments due within the next year whether they are for short-term loans or for other loans are included as current liabilities.

The point in time concept requires identifying all liabilities existing as of the date of the balance sheet. In other words, what obligations would have to be met if the business sold out and ceased to exist on this date? Some expenses tend to accrue daily but are only paid once or twice a year. Interest and property taxes are examples. To properly account for these, "accrued expenses" are included as current liabilities. Included in this category would be interest on all loans that has accumulated from the last interest payment to the date of the balance sheet. Accrued property taxes would be handled in a similar manner, and there may be other accrued expenses such as wages and employee tax withholdings that have been incurred but not yet paid. Income taxes on farm income are typically paid several months after the close of an accounting period. Therefore, a balance sheet for the end of an accounting period should also show accrued income taxes or income taxes payable as a current liability.

Noncurrent Liabilities These include all obligations that do not have to be paid in full within the next year. As discussed, any principal due within the next year would be shown as a current liability, and the remaining balance on the debt would be listed as a noncurrent liability. Care must be taken to be sure the current portion of these liabilities has been deducted and only the amount remaining to be paid *after* the next year's principal payment is recorded as a noncurrent liability.

Owner's Equity

Owner's equity represents the amount of money left for the owner of the business should the assets be sold and all liabilities paid as of the date of the balance sheet. It is also called *net worth*. Equity can be found by subtracting total liabilities from total assets and is therefore the "balancing" amount, which causes total assets to exactly equal total liabilities plus owner's equity. Owner's equity is the owner's current investment or equity in the business. It is properly listed as a liability as it is money due the owner upon liquidation of the business.

Owner's equity can and does change for a number of reasons. One common and periodic change comes from using assets to produce crops and livestock, and the profit from these production activities then used to purchase additional assets or to reduce liabilities. This production process takes time, and one of the reasons for comparing a balance sheet for the beginning of the year with one for the end of the year is to study the effects of the year's production on owner's equity and the composition of assets and liabilities. Owner's equity will also change if there is a change in an asset's value, a gift or inheritance is received, or an asset is sold for more or less than its balance sheet value.

However, changes in the composition of assets and liabilities may not cause a change in owner's equity. For example, if $10,000 cash is used to purchase a new machine, owner's equity does not change. There is now $10,000 less current assets (cash) but an additional $10,000 of noncurrent assets (the machine). Total assets remain the same and therefore so will owner's equity. If $10,000 is borrowed to purchase this machine, both assets and liabilities will increase by the same amount leaving the difference or owner's equity the same as before. This purchase will not affect owner's equity until its depreciation and resulting loss in value is recognized. Using the $10,000 to make a principal payment on a loan will also have no effect on equity. Assets have been reduced by $10,000 but so have liabilities. Equity will remain the same.

These examples illustrate an important point. The owner's equity in a business changes only when the owner puts additional personal capital into the business, withdraws capital from the business, or when the business shows a profit or loss. Changes in asset values due to inflation or disinflation also affect equity if assets are valued at market value. However, many business transactions only change the mix or composition of assets and liabilities and do not affect owner's equity.

Alternate Format

Farm and ranch balance sheets have traditionally included three categories of both assets and liabilities while the accounting profession uses two, as shown in Table 3-1. The FFSTF restates the accounting principle of separating current from noncurrent items on a balance sheet. However, it does not restrict all noncurrent items to a single category. A three-category balance sheet is permitted if the preparer believes such a format would add information and usefulness to the balance sheet.

The usual categories for the traditional farm or ranch balance sheet are shown in Table 3-2. Both current assets and current liabilities are separated from other assets and liabilities as required by accounting principles. These categories would each contain exactly the same items as mentioned in the discussion of Table 3-1. The difference in the two formats comes in the division of noncurrent assets and liabilities into two categories, intermediate and fixed.

Intermediate assets are generally defined as those less liquid than current assets and with a life greater than 1 year but less than about 10 years. Machinery, equipment, and breeding livestock would be the usual intermediate assets found on farm and ranch

TABLE 3–2 FORMAT OF A THREE-CATEGORY BALANCE SHEET

Assets		Liabilities	
Current assets	$XXX	Current liabilities	$XXX
Intermediate assets	XXX	Intermediate liabilities	XXX
Fixed assets	XXX	Long-term liabilities	XXX
		Total liabilities	$XXX
		Owner's equity	$XXX
		Total liabilities and	
Total assets	$XXX	owner's equity	$XXX

balance sheets. Fixed assets are the least liquid and have a life greater than 10 years. Land and buildings are the usual fixed assets.

Intermediate liabilities are debt obligations where repayment of principal is over a period of more than 1 year and up to as long as 10 years. Some principal and interest would typically be due each year, and the current year's principal payment would be shown as a current liability as discussed earlier. Most intermediate liabilities would be loans where the money was used to purchase machinery, breeding livestock, and other intermediate assets. Long-term liabilities are debt obligations where the repayment period is for a length of time exceeding 10 years. Farm mortgages and contracts for the purchase of land are the usual long-term liabilities and the repayment period may be 20 years or longer.

Intermediate assets and liabilities are often a substantial and important part of both total assets and total liabilities on farms and ranches. This may explain why the three-category balance sheet came into use in agriculture. Given its longtime use and user's familiarity with the three-category format, it may be some time before the two-category balance sheet is widely adopted. However, the FFSTF encourages the use of the two-category format and predicts a movement away from the use of a three-category balance sheet.

ASSET VALUATION AND RELATED PROBLEMS

The general approach to asset valuation and the proper method to use for specific assets have long been debated. Much of the discussion is whether or not agriculture should use a cost basis or a market basis balance sheet. A cost basis balance sheet is required when following basic accounting principles. It values all assets using the cost, cost less depreciation, or farm production cost methods discussed in Chapter 2. The one general exception would be inventories of grain and market livestock. Stored grain can be valued at market value less selling costs provided it meets several conditions, which it normally would.[2] While market livestock are not specifically mentioned in accounting principles, the FFSTF uses the argument that there is little difference between grain and market livestock particularly when the livestock are nearly ready for

[2]These conditions are: (a) a reliable, determinable, and realizable market price, (b) relatively small and known selling expenses and, (c) ready for immediate delivery.

market. They recommend market valuation for market livestock even on a cost basis balance sheet.

A market basis balance sheet would have all assets valued at market value less selling costs. Long-run inflation could cause land owned for a number of years to have a market value higher than its cost. Inflation and fast depreciation methods would also result in machinery and breeding livestock market values being higher than their book values. Therefore, a market basis balance sheet would typically show a higher total asset value and consequently a higher equity than one using cost valuation. An exception would be during periods of falling asset values.

Arguments for using *cost basis* balance sheets include conformity to generally accepted accounting principles, conservatism, and their direct comparability with balance sheets from other types of businesses using cost basis. Any changes in equity reflect net income that has been retained in the business. *Market basis* balance sheets have the advantage of being a more accurate reflection of the current financial condition of the business and the value of collateral available to secure loans. Since lenders have been the primary users of farm and ranch balance sheets, market basis valuation has been in general use.

The FFSTF considered both types of balance sheets and concluded that information on both cost and market values are needed to properly analyze the financial condition of a farm or ranch business. They state that acceptable formats would be: (1) a market basis balance sheet with cost information included as footnotes or shown in supporting schedules or (2) a double-column balance sheet with one column containing cost values and the other market values.

Table 3-3 contains the FFSTF recommendations for valuing certain assets on both types of balance sheets. Cost valuation as used in this table represents one of three valuation methods: cost, cost less depreciation, or farm production cost. Cost less depreciation (or book value) would be used for all depreciable assets such as machinery, purchased breeding livestock, and buildings. Farm production cost or the accumulated direct expenses incurred to date would be the value used for the investment in growing

TABLE 3–3 VALUATION METHODS FOR COST BASIS AND MARKET BASIS BALANCE SHEETS

Asset	Cost Basis	Market Basis
Marketable securities	Cost	Market
Inventories of grain and market livestock	Market	Market
Accounts receivable	Cost	Cost
Prepaid expenses	Cost	Cost
Investment in growing crops	Cost	Cost
Purchased breeding livestock	Cost	Market
Raised breeding livestock	Cost or a base value	Market
Machinery and equipment	Cost	Market
Buildings and improvements	Cost	Market
Land	Cost	Market

crops. Inventories of current assets are valued at market on a cost basis balance sheet for the reasons discussed earlier.

Raised breeding livestock are a difficult valuation problem. Accounting principles require and the FFSTF recommends a farm production cost method. However, this requires segregating and accumulating all costs associated with raising each animal from birth to productive age. When the animal becomes productive, it is added to the depreciation schedule, and the total accumulated expense is depreciated just as if it were a purchase price. The FFSTF recognized that, while recommended, this is a procedure most farmers and ranchers would not use because of the difficulty of separating these costs from other livestock costs. As an alternative, they recommend using a fixed base value for each age and type of raised breeding livestock. This value should approximate the cost of raising the animal but would remain fixed over time. Changes in the total value of raised breeding livestock would then occur only when there was a change in the number of animals. A change in market value would have no effect.

Note that not every asset is valued at market even on a market basis balance sheet. Accounts receivable and prepaid expenses are to be valued at cost, which is simply their actual dollar amount. Investment in growing crops is valued as on a cost basis balance sheet at an amount equal to the direct expenses incurred for that crop to date. The crop is not yet harvested, not ready for market, and still subject to production risks. Valuing it at "expected" market value would be a very optimistic approach and not in compliance with conservative accounting principles.

BALANCE SHEET EXAMPLE

Table 3-4 is an example of a balance sheet with a format and headings that follow the FFSTF recommendations. It uses the double-column format to present both cost and market values on a single balance sheet. For simplicity, only farm business assets and liabilities are included. Many farm and ranch balance sheets will include personal as well as business assets and liabilities. It is often difficult to separate them, and lenders often prefer they be combined for their analysis. This presentation of a complete balance sheet with both cost and market values includes some new headings and concepts that need to be discussed along with a review of the asset valuation process.

Asset Section

With inventories valued at market in both cases, there is often little or no difference between the total value of current assets under the two valuation methods. Marketable securities, such as stocks and bonds that can be easily sold, can have market values that differ substantially from their original cost. In this example the $1,200 difference between the two valuations of current assets is due entirely to the marketable securities.

Most of the difference in asset values between cost- and market-based balance sheets will show up in the noncurrent asset section. A combination of inflation and rapid depreciation can result in the book values for machinery, equipment, purchased breeding livestock, and buildings being much less than their market values. The market value of raised breeding livestock can be, and hopefully is, higher than the cost of

TABLE 3-4   BALANCE SHEET FOR I. M. FARMER, DECEMBER 31, 19XX

Assets		
Current assets:	*Cost*	*Market*
Cash/checking account	$5,000	$5,000
Marketable securities	1,000	2,200
Inventories		
Crops	40,000	40,000
Livestock	52,000	52,000
Supplies	4,000	4,000
Accounts receivable	1,200	1,200
Prepaid expenses	500	500
Investment in growing crops	7,600	7,600
Other current assets	0	0
Total current assets	$111,300	$112,500
Noncurrent assets:		
Machinery and equipment	67,500	95,000
Breeding livestock (purchased)	48,000	60,000
Breeding livestock (raised)	12,000	24,000
Buildings & improvements	27,000	50,000
Land	288,000	400,000
Other noncurrent assets	0	0
Total noncurrent assets	$442,500	$629,000
Total assets	$553,800	$741,500
Current liabilities:		
Accounts payable	6,000	6,000
Notes payable within 1 year	15,000	15,000
Current portion of term debt	28,000	28,000
Accrued interest	15,700	15,700
Income taxes payable	8,000	8,000
Current portion—deferred taxes	15,020	15,260
Other accrued expenses	900	900
Total current liabilities	$88,620	$88,860
Noncurrent liabilities:		
Notes payable		
Machinery	20,000	20,000
Breeding livestock	40,000	40,000
Real estate debt	175,000	175,000
Noncurrent portion—deferred taxes	—	45,000
Total noncurrent liabilities	$235,000	$280,000
Total liabilities	$323,620	$368,860
Owner equity:		
Contributed capital	50,000	50,000
Retained earnings	180,180	180,180
Valuation adjustment	—	142,460
Total equity	$230,180	$372,640
Total liabilities and owner equity	$553,800	$741,500

raising them. Land that has been owned for a number of years during which there was only moderate inflation can still have a market value considerably higher than its original cost. All of these factors can combine to make the noncurrent asset value with market valuation much higher than with cost valuation. Of course, it is also possible for market values to be less than cost during periods when asset values are declining.

Liability Section

There is little difference in the valuation of the usual liabilities on a cost- or market-based balance sheet. However, there are several entries related to income taxes in the liability section of this balance sheet that have not been discussed before. *Income taxes payable* under current liabilities represent taxes due on any taxable net farm income for the past year. Taxes on farm income are generally paid several months after the year ends, but they must be paid. Since they are caused by the past year's activity but are not yet paid, they are like an account payable, which would still have to be paid even if the business ceased to operate as of December 31. This is true regardless of the valuation method being used.

The current portion of *deferred income taxes* represents taxes that would be paid on the sale of current assets, less any current liabilities which would be tax deductible expenses. They are called deferred taxes as the assets have not been sold nor the expenses paid. Therefore, no taxes are payable at this time. If cash accounting is used for tax purposes, taxes are deferred into a future accounting period when the assets are converted into cash and the expenses actually paid. However, since the assets exist and the expenses have been incurred, there is no question there will be taxable events in the future.

The noncurrent portion of deferred taxes arises from the difference between cost and market values for noncurrent assets. Market values are generally higher than cost and result in a balance sheet presenting a stronger financial position than one done on a cost basis. If the assets were sold for this market value, the business would have to pay income taxes on the difference between market value and the asset's cost or tax basis. Ignoring these taxes on a market balance sheet results in an owner's equity higher than that which would actually result from a complete liquidation of the business. Therefore, noncurrent deferred taxes are included on a market-based balance sheet and are an estimate of the income taxes that would result from a liquidation of the assets at their market value. They are called deferred taxes as they have not been incurred at this time and are deferred until the time the assets are sold.

Owner's Equity Section

Owner's equity has three basic sources: (1) capital contributed to the business by its owner(s), (2) earnings or business profit that has been left in the business rather than withdrawn and, (3) any change caused by fluctuating market values when market valuation rather than cost is used. The FFSTF recommends showing all three sources of equity rather than simply combining them into one value. This provides additional information for anyone analyzing the balance sheet.

The example in Table 3-4 shows $50,000 of contributed capital. This is the value of any personal cash or property the owner used to start the business and any that might have been contributed since that time. In the absence of any further contributions or withdrawals charged against contributed capital, this value will remain the same on all future balance sheets.

Any before-tax net farm income that is not used for family living expenses, income taxes, or withdrawn for other purposes remains in the business. These retained earnings end up in the form of additional assets (not necessarily cash), decreased liabilities, or some combination of the two. Retained earnings, and therefore owner's equity, will increase any year net farm income is greater than the combined total of income taxes paid and withdrawals for family living expenses. If these two items are greater than net farm income, retained earnings and owner's equity will decrease. In the example a total of $180,180 has been retained since the business began. Retained earnings will be discussed in further detail in Chapter 4.

On a cost basis balance sheet there is no valuation adjustment. Owner's equity will change only when there are changes in contributed capital or retained earnings. Whenever a market-based balance sheet contains an asset valued at more than its cost, it creates equity that is neither contributed capital nor retained earnings. For example, during periods of high inflation the market value of land may increase substantially. A market-based balance sheet would include an increase in the land value each year, which would cause an equal increase in equity. However, this increase is due only to the ownership of the land and not from its direct use in producing agricultural products. Any differences between cost and market values, which can be either positive or negative, should be shown as a valuation adjustment in the equity section of the balance sheet. This makes it easy for an analyst to determine what part of total equity on a market-based balance sheet is a result of valuation differences.

BALANCE SHEET ANALYSIS

As mentioned earlier in this chapter, a balance sheet is used to measure the financial condition of a business and, more specifically, its liquidity and solvency. Analysts often want to compare the relative financial condition of different businesses and of the same business over time. Differences in business size cause potentially large differences in the dollar values on a balance sheet and therefore problems comparing their relative financial condition. A large business can have serious liquidity and solvency problems as can a small business, but the difficulty is measuring the size of the problem *relative* to the size of the business.

To get around this problem, ratios are often used in balance sheet analysis. They provide a standard procedure for analysis and permit comparison over time and between businesses of different sizes. A large business and a small one would have substantial differences in the dollar values on their balance sheets, but the same ratio value would indicate the same *relative* degree of financial strength or weakness. Ratio values can be used as goals and easily compared against the same values for other businesses. In addition, many lending institutions use ratio analysis from balance sheet information to make lending decisions and to monitor the financial progress of their customers.

Most farm and ranch balance sheets are used for loan purposes where market values are of the most interest to the lender. Therefore, the analysis of the balance sheet in Table 3-4 will use market values.

Analyzing Liquidity

An analysis of liquidity concentrates on current assets and current liabilities. The latter represent the need for cash over the next 12 months and the former the sources of cash. Liquidity is a relative rather than an absolute concept as it is difficult to state that a business is or is not liquid. Based on an analysis, however, it is possible to say that one business is more or less liquid than another.

Current Ratio This is one of the more common measures of liquidity and is computed from the equation

$$\text{Current ratio} = \frac{\text{current assets}}{\text{current liabilities}}$$

The current ratio for the balance sheet example in Table 3-4 would be

$$\text{Market basis:} \quad \frac{\$112,500}{\$\ 88,860} = 1.27$$

This ratio measures the amount of current assets relative to current liabilities. A value of 1 means current liabilities are just equal to current assets and, while there are sufficient current assets to cover current liabilities, there is no safety margin. Asset values can and do change with changes in market prices so values larger than 1 are preferred to provide an allowance for price changes and other factors. The larger the ratio value the more liquid the business and vice versa.

Working Capital Working capital is the difference between current assets and current liabilities:

$$\text{Working capital} = \text{current assets} - \text{current liabilities}$$

This equation computes the dollars that would remain after selling all current assets and paying all current liabilities. It therefore is an indication of the margin of safety for liquidity. I. M. Farmer's balance sheet in Table 3-4 shows working capital of $23,640.

Working capital is a dollar value and not a ratio. This makes it difficult to use working capital to compare the liquidity of businesses of different sizes. Larger businesses would be expected to have larger current assets and liabilities and would need a larger working capital to have the same *relative* liquidity as a smaller business. Therefore, it is important to relate the amount of working capital to the size of the business.

Debt Structure This measure computes current liabilities as a percentage of total liabilities. The equation is

$$\text{Debt structure} = \frac{\text{current liabilities}}{\text{total liabilities}} \times 100$$

A high value indicates a large proportion of the total liabilities are due within the next year, which could indicate a liquidity problem. However, the debt structure percent must be evaluated only after checking the amount of current liabilities and current assets. It is possible for the debt structure percent to be very high and still be safe if most of the assets are also current or total liabilities are very small. The I. M. Farmer example indicates a debt structure percent of 24.1 percent for market valuation.

Analyzing Solvency

Solvency measures the relative relationships between assets, liabilities, and equity. It is a way to analyze the business debt and to see if all liabilities could be paid off by the sale of all assets. The latter requires assets to be greater than liabilities, indicating a solvent business. However, solvency is generally discussed in relative terms by measuring the degree by which assets exceed liabilities.

Three ratios are commonly used to measure solvency and are recommended by the FFSTF. Each of them uses two of the three items mentioned earlier, making them all related to one another. Any one of these ratios will, when properly computed and analyzed, provide full information about solvency. However, all three have been in common use and some individuals prefer one over another.

Debt/Asset Ratio   The debt/asset ratio is computed from the equation

$$\text{Debt/asset ratio} = \frac{\text{total liabilities}}{\text{total assets}}$$

and measures what part of total assets is owed to lenders. This ratio should have a value less than 1 and smaller values are preferred. Notice that a debt/asset ratio of 1 means debt or liabilities equal assets, and therefore equity is zero. Ratios greater than 1 would be obtained for an insolvent business. I. M. Farmer has a debt/asset ratio of 0.50.

Equity/Asset Ratio   This ratio is computed from the equation

$$\text{Equity/asset ratio} = \frac{\text{total equity}}{\text{total assets}}$$

and measures what part of total assets is financed by the owner's equity capital. Higher values are preferred but the equity/asset ratio cannot exceed 1. A value of 1 is obtained when equity equals assets, which means liabilities are zero. An insolvent business would have a negative equity/asset ratio as equity would be negative. For the example in Table 3-4 the equity/asset ratio is 0.50 using the market values.

Debt/Equity Ratio   The debt/equity ratio is also called the leverage ratio by some analysts and is computed from the equation

$$\text{Debt/equity ratio} = \frac{\text{total liabilities}}{\text{total equity}}$$

TABLE 3-5   SUMMARY OF I. M. FARMER'S FINANCIAL CONDITION

Measure	Market basis
Liquidity	
Current ratio	1.27
Working capital	$23,640
Debt structure	24.1%
Solvency	
Debt/asset ratio	0.50
Equity/asset ratio	0.50
Debt/equity ratio	0.99

This ratio compares the proportion of financing provided by lenders with that provided by the business owner. When the debt/equity ratio is equal to 1, lenders and the owner are providing an equal portion of the financing. Smaller values are preferred and the debt/equity ratio will approach zero as liabilities approach zero. Very large values result from very small equity, which means an increasing chance of insolvency. I. M. Farmer has a debt/equity ratio of 0.99.

Summary of Analysis

Table 3-5 summarizes the values obtained in the analysis of the I. M. Farmer balance sheet. These values show current assets to be only 27 percent higher than current liabilities, and about 24 percent of all liabilities are due within the next 12 months. Using market values, the business financing is about equally divided between borrowed and equity capital.

Performing an analysis of this balance sheet using the cost values would show two general results. First, there would be little difference in the liquidity measures as there is little difference in the values of current assets and current liabilities under the two valuation methods. Second, cost values, being lower, would show a weaker solvency position for each measure used. These results would be typical. Market valuation will have little effect on liquidity measures, but it has the potential for a large impact on any solvency measure. This illustrates the importance of knowing how assets were valued on any balance sheet being analyzed and why the FFSTF recommends including information on both cost and market values. Providing complete information allows a more thorough analysis and eliminates any possible confusion about which valuation method was used.

SUMMARY

A balance sheet shows the financial position of a business at a point in time. It does this by presenting an organized listing of all assets and liabilities belonging to the business. The difference between assets and liabilities is owner's equity, which represents the investment the owner has in the business.

An important consideration when constructing and analyzing balance sheets is the method used to value assets. They can be valued using cost methods, which would be the more conservative method, or by using market valuations. The latter would generally result in higher asset values and therefore a higher owner's equity. There are advantages to each method and the FFSTF recommends providing both cost and market values on a farm or ranch balance sheet. This provides full information to the user of the balance sheet.

Two factors, liquidity and solvency, are used to analyze the financial position of a business. Liquidity considers the ability to generate the cash needed to meet cash requirements over the next year and to do so without disrupting the production activities of the business. Solvency measures the debt structure of the business and whether all liabilities could be paid by selling all assets, that is, assets greater than liabilities. Several ratios are used to measure liquidity and solvency and to analyze the relative strength of the business in these areas.

QUESTIONS FOR REVIEW AND FURTHER THOUGHT

1 True or false. If the debt/equity ratio increases, the debt/asset ratio will also increase.
2 True or false. A business with a higher working capital will also have a higher current ratio.
3 Use your knowledge of balance sheets and ratio analysis to complete the following abbreviated balance sheet. The current ratio = 2.01 and the debt/equity ratio = 1.0

Assets		Liabilities	
Current assets	$80,000	Current liabilities	—————
Noncurrent assets	—————	Noncurrent liabilities	—————
		Total liabilities	—————
		Owner's equity	$100,000
		Total liabilities	
Total assets	—————	plus equity	—————

4 Can a business be solvent but not liquid? Liquid but not solvent? How?
5 Does a balance sheet show the annual net farm income for a farm business? Why or why not?
6 True or false. Assets + liabilities = equity.
7 Assume you are an agricultural loan officer for a bank and a customer requests a loan based on the following balance sheet. Conduct a ratio analysis and give your reasons for granting or denying an additional loan. What is the weakest part of this customer's financial position?

Assets		Liabilities	
Current assets	$40,000	Current liabilities	$60,000
Noncurrent assets	$240,000	Noncurrent liabilities	$50,000
		Total liabilities	$110,000
		Owner's equity	$170,000
Total assets	$280,000	Total liabilities plus equity	$280,000

8 Why are there no noncurrent deferred income taxes on a cost basis balance sheet?

9 Assume a mistake was made and the value of market livestock on a balance sheet is $10,000 higher than it should be. How does this error affect the measures of liquidity and solvency? Would the same results occur if land had been overvalued by $10,000?

10 Would the following entries to a farm balance sheet be classified as an asset or liability? As current or noncurrent?

   a Machine shed

   b Feed bill at local feed store

   c A 20-year farm mortgage contract

   d A 36-month certificate of deposit

   e Newborn calves

REFERENCES

Hein, Norlin A., and Herman E. Workman: *Balance Sheet,* Extension Division, University of Missouri, Publication EC 946, 1982.

Libbin, James D., and Lowell B. Catlett: *Farm and Ranch Financial Records,* Macmillan Publishing Company, New York, 1987.

McGrann, James M., John Parker, and Shannon Neibergs: *User's Manual: FINYEAR Farm/Ranch Financial Statement Preparation Package,* Department of Agricultural Economics, Texas Agricultural Experiment Station, August, 1991.

Morehart, Mitchell J., Elizabeth G. Nielsen, and James D. Johnson: *Development and Use of Financial Ratios for the Evaluation of Farm Businesses,* USDA, Economic Research Service, Technical Bulletin Number 1753, October, 1988.

Oltmans, Arnold W., Danny A. Klinefelter, and Thomas L. Frey: *Agricultural Financial Reporting and Analysis,* Century Communications Inc., Niles, IL, 1992.

Recommendations of the Farm Financial Standards Task Force, American Bankers Association, Washington, D.C., 1991.

4

THE INCOME STATEMENT
AND ITS ANALYSIS

CHAPTER OBJECTIVES

1 To discuss the purpose and use of an income statement

2 To illustrate the structure and format of an income statement

3 To define the sources and types of revenue and expenses that should be included on an income statement

4 To show how profit or net farm income is computed from an income statement and what it means and measures

5 To analyze farm profitability by computing returns to assets and equity and by studying other measures of profitability

The *income statement* is a summary of revenue and expenses for a given accounting period. It is sometimes called an *operating statement,* or a *profit and loss statement,* and the Farm Financial Standards Task Force (FFSTF) also refers to it as the *statement of earnings and comprehensive income.* Its purpose is to measure the difference between revenue and expenses. A positive difference indicates a profit, or a positive net farm income, and a negative value indicates there was a loss, or a negative net farm income for the accounting period. Therefore, an income statement as it will be called in this chapter answers the question: Did the farm or ranch business have a profit or loss last year and how large was it?

The balance sheet and income statement are two different yet related financial statements for a given business. A balance sheet shows financial position at a *point* in time while an income statement is a summary of the revenue and expenses as recorded over a *period* of time. This distinction is shown in Figure 4-1. Using a calendar year ac-

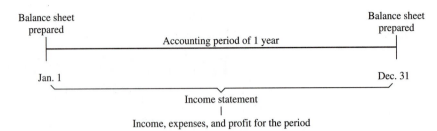

FIGURE 4-1   Relationship between balance sheet and income statement.

counting period, a balance sheet can be prepared at the end of each year, and it can also serve as the balance sheet for the beginning of the following year. The result is a record of financial position at the beginning and end of each accounting period. However, comparing these balance sheets does not permit a direct calculation of net farm income for the year. This is the purpose of the income statement.

Even though the balance sheet and income statement contain different information and have a different purpose and use, they are both financial statements for the same business. It seems only logical that revenue, expenses, and profit affect the financial position of the business and indeed they do. However, an explanation of this relationship will be delayed until the end of this chapter and after a complete discussion of the structure and components of an income statement.

IDENTIFYING REVENUE AND EXPENSES

To construct an income statement showing the difference between revenue and expenses, a necessary first step is to identify all revenue and expenses that should be included in the computation of net farm income. The discussion of cash versus accrual accounting in Chapter 2 introduced some of the difficulties that might be encountered in this seemingly simple task. This section will expand on this earlier discussion before introducing the format and construction of an income statement.

Revenue

An income statement should include all revenue earned during the accounting period but no other revenue. The problem is one of determining when revenue should be *recognized,* that is, in what accounting period was it earned. This problem is further compounded by the fact that revenue can be either *cash* or *noncash.*

When revenue is received in the form of cash for a commodity produced and sold within the same accounting period, recognition is easy and straightforward. However, revenue should also be recognized whenever an agricultural commodity is ready for sale. Inventories of grain and market livestock fit this classification, and changes in these inventories (ending value minus beginning value) are included on an accrual income statement. Accounts receivable represent earned revenue, which should be recognized, but it is revenue for which payment has not yet been received. Any change in their value from beginning to the end of the year must be included as revenue. Note that these two items represent sources of revenue where the revenue may be recognized in one accounting period but the receipt of cash will not occur until a later accounting period.

When an inventory or account receivable is recognized as revenue, it is noncash revenue at that time but something for which a cash payment will typically be received at a later date. However, payment may sometimes be received in the form of goods or services instead of cash. This noncash payment should be treated in the same manner as a cash payment. The value of grain or livestock received in payment for custom work done for a neighbor should be included in revenue as a commodity was simply received in lieu of cash. Also, products that originate on the farm and are consumed by the farm family rather than sold can be included as noncash income.

Gain or Loss on Sale of Capital Assets

The gain or loss on the sale of a capital asset is an entry that often shows up in the revenue section of an income statement. It is the difference between the sale price of a capital asset such as land and its cost. For depreciable assets such as machinery, buildings, and purchased breeding livestock, it is the difference between the selling price and the asset's book value. Gain or loss is recognized only when an asset is actually sold. Before then the market value or selling price is subject to some uncertainty.

Any gain or loss on the sale of a nondepreciable asset such as land is the direct result of an increase or decrease in the market price since its purchase. For depreciable assets, changes in market value affect gains and losses as well how accurately depreciation was estimated. If an asset is sold for exactly its book value, there is no gain or loss. The depreciation over its life has perfectly matched its decline in market value. A sale price higher than book value implies too much total depreciation has been taken (perhaps due to an unanticipated strength in market values) and that past annual depreciation expense has been too high. Recognizing the gain as a revenue adjusts for the fact that depreciation expense was not correct in past years. Selling a depreciable asset for less than its book value means it lost market value faster than what was accounted for by the annual depreciation. There should have been a higher depreciation expense in the past and an adjustment is made in the form of a "loss on sale" entry on the income statement.

Given that (1) useful lives and salvage values, or market values at the end of the useful life, are only estimates made at the time of purchase and that (2) the choice of depreciation method will affect the amount of annual depreciation, there will generally be a gain or loss to be recognized when a depreciable asset is sold.

Expenses

Once all revenue for an accounting period has been identified, the next step is to identify all expenses incurred in producing that revenue. They may be either cash or noncash expenses. For example, cash expenses would include purchases of and payment for feed, fertilizer, seed, market livestock, and fuel. Noncash expenses would include depreciation, accounts payable, accrued interest, and other accrued expenses. There is also an adjustment for prepaid expenses.

Depreciation is a noncash expense that reflects decreases in the value of assets used to produce the revenue. Accounts payable, accrued interest, and other accrued expenses such as property taxes are expenses incurred during the past accounting period but not yet paid. To properly match expenses with the revenue they helped produce, these expenses must be included on this year's income statement. They must also be subtracted from next year's income statement when the cash will actually be expended for their payment. Prepaid expenses are goods or services paid for in one year but that will not be used to produce revenue until the next year. Examples would be seed, fertilizer, chemicals, and feed purchased and paid for in December to take advantage of price discounts or income tax deductions. However, since they will not be used until next year, this expense should be deferred until then. Prepaid expenses should be *subtracted* from this year's expenses but then included as part of the next year's. This assures a consistent set of records.

Income taxes are sometimes included on an income statement making the net farm income an *after-tax* net farm income. However, the taxes due on farm income can be difficult to estimate particularly when there is also off-farm income to be included in taxable income. The examples used in this text will omit income taxes as a business expense and concentrate on estimating a *before-tax* net farm income.

INCOME STATEMENT FORMAT

The outline of an income statement shown in Table 4-1 follows one of the formats recommended by the FFSTF. In very condensed form the structure is

> Total revenue
> *Less* total expenses
> *Equals* net farm income from operations
> *Plus or minus* gain/loss on sale of capital assets
> *Equals* net farm income

Most of the entries on this income statement have already been discussed, but several items need additional discussion.

Any gain/loss on the sale of culled breeding livestock is shown in the revenue section, but other gains/losses are included in a separate section at the end of the income statement. The argument for including the former in revenue is that the sale of culled breeding livestock is a normal, expected, and usual part of the ongoing production activities of the business. While the FFSTF recommends including all gains/losses in net farm income, it recommends those on machinery, land, and buildings be shown separately. Any gains/losses from the sale of these assets are less frequent, result from investment rather than production activities, and may be very large. Therefore, "net farm income from operations" is the profit from normal, ongoing production activities. Net farm income includes this amount plus any gains/losses on sale of land and machinery. Showing these items separately toward the end of the income statement allows the user to easily determine if they had an unusually large effect on net farm income.

TABLE 4-1   INCOME STATEMENT FORMAT

Revenue:
  Cash crop sales
  Cash livestock sales
  Inventory changes:
    Crops
    Market livestock
  Livestock product sales
  Government program payments
  Change in value of raised breeding stock
  Gain/loss from sale of culled breeding stock
  Change in accounts receivable
  Other farm income   ————————
    Total revenue

Expenses:
  Purchased feed and grain
  Purchased market livestock
  Other cash operating expenses:
    Crop expenses
    Livestock expenses
    Fuel, oil
    Labor
    Repairs, maintenance
    Property taxes
    Insurance
    Other: ————————
           ————————
           ————————

  Adjustments
    Accounts payable
    Prepaid expenses
  Depreciation   ————————
    Total operating expenses
  Cash interest paid
  Change in interest payable   ————————
    Total interest expense   ————————
      Total expenses
  Net farm income from operations
  Gain/loss on sale of capital assets:
    Machinery
    Land
    Other
Net farm income

Cash interest paid, accrued interest, and total interest are shown separately from other expenses. While often categorized as an operating expense, the position of the FFSTF is that interest results from financing activities rather than production activities. Therefore, it should not be included with the direct operating expenses associated with the production of crops and livestock. This separation also makes it easy for the user to find and note the amount of interest expense when conducting an analysis of the income statement.

The other income statement format recommended by the FFSTF computes "value of farm production" as an intermediate step. In condensed form this format has the following structure

> Total revenue
> *Less* cost of purchased feed and grain
> *Less* cost of purchased market livestock
> *Equals* value of farm production
> *Less* all other expenses
> *Equals* net farm income from operations
> *Plus or minus* gain/loss on sale of capital assets
> *Equals* net farm income

Value of farm production measures the dollar value of all goods and services produced by the farm business. Any feed, grain, and market livestock purchased were produced by some other farm or ranch and should not be credited to this business. Subtracting the cost of these items from total revenue results in a "net" production value. Note that any increase in value obtained by feeding purchased feed and grain to purchased livestock will end up credited to the business being analyzed through sales revenue or an inventory increase.

Value of farm production is a useful measure by itself when analyzing a farm business and is used to compute other measures as well. Although the FFSTF does not require use of the format showing value of farm production, they do recognize its usefulness. Therefore, for the format shown in Table 4-1, they recommend the cost of purchased feed and grain and purchased market livestock be shown separate from other operating expenses. This makes it easy for the user to find the values needed to compute value of farm production.

ACCRUAL ADJUSTMENTS TO A CASH BASIS INCOME STATEMENT

It is possible to use cash basis accounting throughout the accounting period and, with some additional effort and records, compute an accrual basis net farm income. Because many farmers and ranchers use cash accounting for tax reasons, this procedure has been used for some time to convert a cash basis net farm income into what it would have been using accrual accounting.

Since a cash basis income statement would include only cash revenue, cash expenses, and depreciation, there are a number of adjustments to be made. Beginning with cash basis net farm income the adjustment process is as follows:

> Net farm income (cash basis)
> *Plus or minus* change in inventory value of crops and market livestock
> *Plus or minus* change in value of raised breeding livestock
> *Plus or minus* change in accounts receivable
> *Plus or minus* change in accounts payable
> *Plus or minus* change in prepaid expenses
> *Plus or minus* change in accrued interest

Plus or minus change in other accrued expenses
Equals net farm income (accrual basis)

Notice that the first three adjustments come from the revenue section of the income statement and the last four from the expense section.

This adjustment process is rather straightforward but does require some effort and attention to detail. First, to get the *change* in each of the adjustment items, it is necessary to know their values at the beginning of the year and at the end of the year. Since these values are not part of the cash accounting system, they must be measured and recorded in some other way. This adjustment process will result in the correct accrual net farm income only if all measurements recorded have the same values as they would have had if an accrual accounting system had been used. The most accurate method, and one that will keep all financial records consistent and correlated, is to use values from the beginning and end of the year balance sheets.

Second, care must be taken to make sure the adjustments have the correct effect on revenue or expenses and in turn on net farm income. Table 4-2 summarizes how increases or decreases in each adjustment factor will affect revenue or expense and its final effect on net farm income. The revenue adjustments all follow the same expected pattern as do the expense adjustments with one exception. Prepaid expenses are those expenses paid for in one year but that will not produce revenue until the following

TABLE 4-2   EFFECTS OF ACCRUAL ADJUSTMENTS ON CASH BASIS NET FARM INCOME

Adjustment	Revenue	Expenses	Accrual net farm income
Inventory			
Increase	Positive	None	Increase
Decrease	Negative	None	Decrease
Raised breeding livestock			
Increase	Positive	None	Increase
Decrease	Negative	None	Decrease
Accounts receivable			
Increase	Positive	None	Increase
Decrease	Negative	None	Decrease
Accounts payable			
Increase	None	Positive	Decrease
Decrease	None	Negative	Increase
Prepaid expenses			
Increase	None	Negative	Increase
Decrease	None	Positive	Decrease
Accrued Interest			
Increase	None	Positive	Decrease
Decrease	None	Negative	Increase
Other accrued expenses			
Increase	None	Positive	Decrease
Decrease	None	Negative	Increase

year. Since they relate to next year and not the current year, they affect expenses and net farm income differently.

As an example, assume $2,000 of fertilizer was purchased and paid for in the fall of 19X1 to be used for a 19X2 crop. With cash basis accounting this cash expenditure would be entered as a fertilizer expense in 19X1. There is now a $2,000 expense in 19X1 that will not produce revenue until 19X2 so it represents an *increase* in prepaid expenses of $2,000. This increase is subtracted from 19X1 expenses, which increases net farm income for that year. If there are no prepaid expenses in 19X2, there will be a *decrease* of $2,000 in prepaid expenses. A decrease is added to 19X2 expenses, which will in turn reduce 19X2 net farm income. The result of these adjustments is to get the $2,000 prepaid expense out of 19X1 expenses (since it did not produce any revenue that year) and into 19X2 expenses. This is the year where it produced revenue. The accounting principle of matching expenses with the revenue it produces is met by this adjustment procedure.

ANALYSIS OF NET FARM INCOME

Table 4-3 contains a complete accrual basis income statement for I. M. Farmer. It shows a net farm income from operations of $46,800 and a net farm income of $47,900. The business shows a profit for the year, but is it a "profitable" business? Profitability is concerned with the size of the profit *relative* to the size of the business, or the value of the resources used to produce the profit. A business can have a profit but have a poor profitability rating if this profit is small relative to the size of the business. For example, two farms with the same net farm income are not equally profitable if one used twice as much capital as the other to produce that profit.

Profitability is a measure of how efficient the business is in using its resources to produce profit or net farm income. The FFSTF recommends four measures of profitability: (1) net farm income, (2) rate of return on assets, (3) rate of return on equity, and (4) operating profit margin ratio. In addition to the rates of return on assets and equity, it is also possible to compute a return to operator labor and management. This is another common measure of profitability.

Net Farm Income

As shown on the income statement, net farm income is the amount by which revenue exceeds expenses plus any gain or loss on the sale of capital assets. It can also be thought of as the amount available to provide a return to the farmer for the unpaid labor, management, and equity capital used to produce that net farm income. As discussed earlier, it is an absolute, dollar amount making it difficult to use net farm income by itself as a measure of profitability. It should be considered more as a starting point for analyzing profitability than as a good measure of profitability itself.

Rate of Return on Assets

Return on assets (ROA) is the term used by the FFSTF, but the same concept has been called *return to capital,* or *return on investment* (ROI). Return on assets measures

TABLE 4-3 INCOME STATEMENT FOR I. M. FARMER FOR YEAR ENDING DECEMBER 31, 19XX

Revenue:		
Cash crop sales	$132,500	
Cash livestock sales	63,400	
Inventory changes:		
Crops	8,500	
Market livestock	(9,200)	
Livestock product sales	0	
Government program payments	3,400	
Change in value of raised breeding stock	1,200	
Gain/loss from sale of culled breeding stock	600	
Change in accounts receivable	0	
Other farm income	0	
Total revenue		$200,400
Expenses:		
Purchased feed and grain	12,000	
Purchased market livestock	28,000	
Other cash operating expenses:		
Crop expenses	47,500	
Livestock expenses	6,500	
Fuel, oil	3,200	
Labor	7,500	
Repairs, maintenance	3,600	
Property taxes	2,800	
Insurance	2,000	
Other: *Utilities*	2,400	
Adjustments		
Accounts payable	(400)	
Prepaid expenses	0	
Accrued expenses	800	
Depreciation	8,200	
Total operating expenses		124,100
Cash interest paid	28,500	
Change in interest payable	1,000	
Total interest expense		29,500
Total expenses		153,600
Net farm income from operations		46,800
Gain/loss on sale of capital assets:		
Machinery	1,100	
Land	0	
Other	0	1,100
Net farm income		47,900

profitability with a ratio obtained by dividing the dollar return to capital by the average farm asset value for the year. The latter is found by averaging the beginning and ending total asset values from the farm's balance sheets. Expressing ROA as a percentage allows an easy comparison with the same values from other farms, over time for the same farm, and with returns from other investments. The equation is

$$\text{Rate of return on assets } (\%) = \frac{\text{return to assets}}{\text{average farm asset value}} \times 100$$

The return to assets is the dollar return to both debt and equity capital so net farm income must be adjusted. Any interest on debt capital was deducted as an expense when calculating net farm income from operations. This interest must be added back to net farm income before the return to assets is computed. This makes the result independent of the type and amount of financing and what net farm income would have been if there had been no debt capital and therefore no interest expense. For I. M. Farmer, the computations are

Net farm income from operations	$46,800
Plus interest expense	29,500
Equals adjusted net farm income	$76,300

This is what Farmer's net farm income from operations would have been if there had been no borrowed capital, or 100% equity.

It is also necessary to adjust net farm income for the unpaid labor and management provided by the farm operator and the farm family. Without this adjustment that part of net farm income credited to assets would also include what labor and management contributed toward earning the net farm income. Therefore, a return to unpaid labor and management must be subtracted from adjusted net farm income to find the actual return to assets in dollars. Since there is no way of knowing exactly what part of net farm income was earned by labor and management, opportunity costs are used as estimates. Assuming an opportunity cost of $20,000 for the unpaid labor provided by I. M. Farmer and other family members and $5,000 for management, the calculations are

Adjusted net farm income	$76,300
Less opportunity cost of unpaid labor	−20,000
Less opportunity cost of management	− 5,000
Equals return to assets	$51,300

The final step is to convert this dollar return to assets into a percentage of total assets, which is the dollar value of all capital invested in the business. An immediate problem is which total asset value, cost or market basis? Although either can be used, the market basis value is generally used as it represents the current investment in the business and allows a better comparison of ROA's between farms. It also makes the resulting ROA comparable to the return these assets could earn in other investments should the assets be converted into cash at current market values and invested elsewhere. However, cost basis provides a better indicator of the return on the actual funds invested and is the better indicator for trend analysis.

I. M. Farmer's balance sheet in Chapter 3 shows a market basis total asset value of $741,500 on December 31, 19XX. If the total asset value on January 1, 19XX was $710,000, the average asset value for 19XX was $725,750, which is found by averaging the two values. Therefore, the rate of return on Farmer's assets for 19XX was

$$\text{ROA} = \frac{\$51,300}{\$725,750} = 0.0707 \times 100 = 7.07\%$$

This ROA of 7.07 percent represents the return on capital invested in the business after adjusting net farm income for the opportunity cost of labor and management. The "profitability" of the farm can now be judged by comparing this ROA to those from similar farms, the returns from other possible investments, the opportunity cost of the farm's capital, and past ROAs for the same farm.

Several things should be noted whenever ROA is computed and analyzed. First, the FFSTF recommends using net farm income from operations rather than net farm income. Any gains or losses on sale of capital assets included in the latter value are sporadic in nature, can be very large, and are not income generated by the use of assets in the normal production activities of the business. Therefore, they should not be included in any calculation of how well assets are used in generating profit. Second, the opportunity costs of labor and management are estimates and changing them will affect ROA. Third, any comparisons of ROAs should be done only after making sure they were all computed the same way and that the same method was used to value assets. Market valuation is recommended for comparison purposes, cost valuation for checking trends on the same farm. Fourth, any personal or nonfarm assets that might be included on the balance sheet should be subtracted, just as any nonfarm earnings such as interest on personal savings accounts should not be included in net farm income. The ROA computed should be for farm earnings generated by farm assets. Finally, ROA is an *average* return and not a marginal return. It should not be used when making decisions about investing in additional assets where the marginal return is the important value.

Rate of Return on Equity

The return on assets is the return on all assets or capital invested in the business. On most farms and ranches, there is a mixture of both debt and equity capital. Another important measure of profitability is the return on equity (ROE), which is the return on the owner's share of the capital invested. Should the business be liquidated and the liabilities paid off, only the equity capital would be available for alternative investments.

The calculation of return on equity begins directly with net farm income from operations. No adjustment is needed for any interest expense. Interest is the payment for the use of borrowed capital, and it must be deducted as an expense before the return on equity is computed. This has already been done when computing net farm income from operations. However, the opportunity cost of operator and family labor and management must again be subtracted so the ROE does not include their contribution to earning net farm income. Continuing with the I. M. Farmer example, the dollar return on equity would be

Net farm income from operations	$46,800
Less opportunity cost of unpaid labor	−20,000
Less opportunity cost of management	− 5,000
Equals return on equity	$21,800

The rate of return on equity is computed from the equation

$$\text{Rate of return on equity } (\%) = \frac{\text{return on equity}}{\text{average equity}} \times 100$$

An average equity for the year is used in the divisor and is typically the average of beginning and ending market or cost basis equity for the year. Return on equity is also typically expressed as a percent to allow an easy comparison across farms and with other investments. Assuming an average market basis equity of $360,000, I. M. Farmer's return on equity would be

$$\text{ROE} = \frac{\$21,800}{\$360,000} = 0.0606 \times 100 = 6.06\%$$

The ROE can be either greater or less than the return on assets depending on the ROA's relationship to the average interest rate on borrowed capital. If the ROA is greater than the interest rate paid on borrowed capital, this extra margin or return above interest accrues to equity capital, making its return greater than the average return on total assets. Conversely, if the return on assets is less than the interest rate on borrowed capital, the return on equity will be less than the return on assets. Some of the equity capital's earnings had to be used to make up the difference when the interest was paid, thereby lowering the ROE. With i being the interest rate on debt, these relationships can be summarized as follows:

If ROA $> i,$ then ROE $>$ ROA
If ROA $< i,$ then ROE $<$ ROA

Operating Profit Margin Ratio

This ratio computes operating profit as a percent of total revenue. The higher the result the more profit the business is making per dollar of revenue. The first step in computing the operating profit margin ratio is to find the operating profit. The process is

Net farm income from operations
Plus interest expense
Less opportunity cost of unpaid labor
Less opportunity cost of management
Equals operating profit

Interest is added back to net farm income to eliminate the effect of financing on operating profit. This allows the operating profit margin ratio to focus strictly on the profit made from producing agricultural commodities without regard to the amount of financing, which can vary substantially from farm to farm. Eliminating this variable permits a valid comparison of this ratio across different farms. To recognize that unpaid labor and management contributed to earning the profit, their opportunity costs are subtracted. This makes the results comparable to those from businesses where all labor and management is hired, and these expenses have already been deducted in the computation of net farm income.

The equation for the operating profit margin ratio is

$$\text{Operating profit margin ratio} = \frac{\text{operating profit}}{\text{total revenue}} \times 100$$

I. M. Farmer's operating profit margin ratio would be

$$\frac{\$46{,}800 + 29{,}500 - 20{,}000 - 5{,}000}{\$200{,}400} = 0.256 \times 100 = 25.6\%$$

This means that, on the average, for every dollar of revenue 25.6 cents remained as profit after paying the operating expense necessary to generate that dollar. Farms with a low operating profit margin ratio may wish to concentrate on improving this ratio before expanding production. It does little good to increase total revenue if there is little or no profit per dollar of revenue. Total profit or net farm income is related to both the operating profit margin ratio, a measure of profitability, *and* total revenue, which measures volume of business.

Return to Labor and Management

Net farm income was described as the amount available to provide a return to unpaid labor, management, and equity capital. A rate of return on equity was computed in an earlier section and, in a like manner, it is possible to compute a return to labor and management. This dollar amount is that part of net farm income remaining to pay for operator labor and management after all capital (total asset value) is paid a return equal to its opportunity cost.

The procedure is similar to that used to compute returns on assets and equity except the result is expressed in dollars and not as a ratio or percentage. Return to labor and management is computed as follows:

> *Adjusted* net farm income
> *Less* opportunity cost of all capital
> *Equals* return to labor and management

If the opportunity cost of I. M. Farmer's capital is assumed to be 8 percent, the opportunity cost on all capital (average asset value) is $725,750 × 8 percent, or $58,060. Therefore, after all capital is assigned a return equal to its opportunity cost, I. M. Farmer's labor and management earned $76,300 − $58,060 = $18,240.

Assuming the opportunity costs of I. M. Farmer's labor and management are $20,000 and $5,000, respectively, the $18,240 indicates either capital, labor, management, or some combination did not receive a return equal to their opportunity cost. Unfortunately, there is no way to determine which did or did not have a return equal to or higher than its opportunity cost.

Return to Labor   The return to labor and management can be used to compute a return to labor alone. With the opportunity cost of capital already deducted, the only step remaining to find return to labor is to subtract the opportunity cost of management.

Return to labor and management	$18,240
Less opportunity cost of management	− 5,000
Equals return to labor	$13,240

The result is further confirmation that, in this example, net farm income was not suffi-cient to provide labor, management, and capital a return at least equal to their opportu-nity costs.

Return to Management   Management is often considered the residual claimant to net farm income for several reasons. It is difficult to estimate and, in some ways, it measures how well the manager did combining the other resources to generate a profit. Return to management for average or typical farms is often reported in various publi-cations and is computed by subtracting the opportunity cost of labor from the return to labor and management.

Return to labor and management	$18,240
Less opportunity cost of labor	−20,000
Equals return to management	$−1,760

A negative return to management is a rather common occurrence on many farms and ranches assuming, as this process does, that capital and labor earned their opportunity costs. However, the negative return means only that net farm income was not suffi-cient to provide a return to capital, labor, *and* management equal to or higher than their opportunity costs.

CHANGE IN OWNER'S EQUITY

After computing and analyzing net farm income, several questions may come to mind. What was this profit used for? Where is this money now? Did this profit affect the bal-ance sheet? If so, how? The answers to these questions are related and illustrate the re-lationship between the income statement and the beginning and ending balance sheets.

The year's profit or net farm income must end up in one of four uses: (1) family liv-ing expenses and other withdrawals, (2) payment of income and social security taxes, (3) increases in cash or other farm assets, or (4) a reduction in liabilities through prin-cipal payments on loans or payment of other liabilities. Figure 4-2 shows what hap-pens to net farm income and how it affects the balance sheet. Withdrawals from the business to pay for family living expenses, income and social security taxes, and other purposes reduce the amount of net farm income available for use in the farm business. What remains is called *retained farm earnings.*

As the name implies, retained farm earnings is that part of the farm earnings, after personal withdrawals and taxes, retained for use in the farm business. Asset and liabil-ity values will have changed but, if retained farm earnings is positive, equity must have increased. Retained farm earnings represents increased assets or decreased liabil-ities, or some other combination of changes, which will increase the *difference* be-tween assets and liabilities and therefore increase equity. It is important to note that

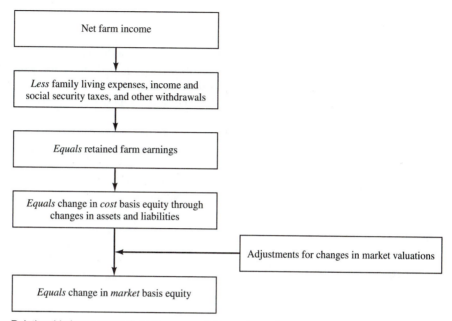

FIGURE 4-2   Relationship between net farm income and change in equity.

this change in equity may not be positive. If living expenses, taxes, and other withdrawals are greater than net farm income, retained farm earnings will be negative. In this case, assets had to be withdrawn from the farm business or additional borrowing was needed to meet living expenses and taxes.

The direct relationship between retained farm earnings and change in equity shown in Figure 4-2 applies only to a *cost basis* balance sheet. When depreciable assets are valued at book value, their decrease in value matches the depreciation expense on the income statement. Land value is unchanged and, if there are no new farm assets from gifts, inheritance, or nonfarm income, retained farm earnings will equal the change in owner's equity for the year. Under these conditions, equity on a cost basis balance sheet will increase only if there is positive retained farm earnings, that is, net farm income is greater than the sum of living expenses, taxes, and other withdrawals. This further emphasizes the necessity of making a substantial profit if the farm operator/manager wants to live well *and* increase business equity at the same time.

On a market basis balance sheet, changes in equity will come not only from retained farm earnings but also from any changes in the value of assets due to changes in their market value. As shown in Figure 4-2, the change in cost basis equity must be adjusted for any changes in market valuation of assets and related deferred income taxes. Farms or ranches with a large proportion of their assets in land can experience rapid gains or losses in equity due to fluctuations in land values. Without this adjustment, it is impossible to correlate and account for all changes in equity.

SUMMARY

An income statement organizes and summarizes revenue and expenses for an accounting period and computes net farm income for that period. First, all revenue earned during the period should be identified and recorded. Next, the same thing is done for all expenses incurred in producing that revenue. Net farm income is the amount by which revenue exceeds expenses. If records are kept using cash basis accounting, accrual adjustments should be made at the end of each accounting period to derive the accrual net farm income. A cash basis net farm income can be very misleading and management decisions should be made only on the information obtained from an accrual net farm income.

Net farm income or profit is an actual dollar amount whereas profitability refers to the size of that profit *relative* to the resources used to produce it. An analysis of profitability should be done each year to provide a means to compare results against previous years and against other farm businesses. Rate of return on assets, rate of return on equity, and operating profit margin ratio are recommended measures of profitability.

Part of net farm income goes to pay for family living expenses, income and social security taxes, and for other personal withdrawals. The remaining net farm income is called retained farm earnings because it has been retained for use in the farm business. A positive retained farm earnings ends up as an increase in assets, a decrease in liabilities, or some other combination of changes in assets and liabilities that causes equity to increase. Conversely, a negative retained farm earnings means some farm equity has been used for living expenses, taxes, and other personal withdrawals. Equity based on market values for assets will also change by the amount these market values change during the year.

QUESTIONS FOR REVIEW AND FURTHER THOUGHT

1 Why is inventory change included on an accrual income statement? What effect does an increase in inventory value have on net farm income? A decrease?

2 What are the differences between net farm income computed on a cash basis versus an accrual basis? Which is the better measure of net farm income? Why?

3 What factors determine the change in *cost* basis equity? In *market* basis equity?

4 Why are changes in land values not included with the inventory changes shown on an income statement?

5 Why are there no entries for the purchase price of new machinery on an income statement? How does the purchase of a new machine affect the income statement?

6 Use the following information to compute values for each of the items listed below.

Net farm income	$36,000	Opportunity cost of labor	$16,000
Average net worth	$220,000	Opportunity cost of management	$8,000
Average asset value	$360,000	Family living expenses	$20,000
Interest expense	$11,000	Income and social security taxes	$4,000
Total revenue	$109,500	Opportunity cost of capital	10%

   a Rate of return on assets ———%
   b Rate of return on equity ———%
   c Operating profit margin ratio ———%
   d Change in equity $———
   e Return to management $———

7 Use the following information to answer the questions:

Cash expenses	$110,000	Beginning inventory value	$42,000
Cash revenue	$167,000	Ending inventory value	$28,000
Depreciation	$8,500	Cost of new tractor	$48,000
Beginning accounts receivable	$2,200	Ending accounts receivable	$0
Beginning accounts payable	$7,700	Ending accounts payable	$1,500

   a Cash basis net farm income $———
   b Total revenue on accrual basis $———
   c Accrual basis net farm income $———

REFERENCES

Libbin, James D., and Lowell B. Catlett: *Farm and Ranch Financial Records,* Macmillan Publishing Company, New York, 1987.

McGrann, James M., John Parker, and Shannon Neibergs: *User's Manual: FINYEAR Farm/Ranch Financial Statement Preparation Package,* Department of Agricultural Economics, Texas Agricultural Experiment Station, August, 1991.

Oltmans, Arnold W., Danny A. Klinefelter, and Thomas L. Frey: *Agricultural Financial Reporting and Analysis,* Century Communications Inc., Niles, IL, 1992.

Recommendations of the Farm Financial Standards Task Force, American Bankers Association, Washington, D.C., 1991.

Reff, Tom, Arlen Leholm, and Glenn Pederson: *Your Income Statement,* North Dakota Cooperative Extension Service Publication EC819, 1983.

PART THREE

DEVELOPING BASIC MANAGEMENT SKILLS

HIGHLIGHTS

Chapter 1 identified planning as a function of management. Planning is primarily making choices and decisions, selecting the most profitable alternative from among all possible alternatives. The alternative(s) selected form the plan for the next year or longer. Planning might also be thought of as organizing, as a plan is a predetermined way to combine or organize resources to produce some combination and quantity of agricultural commodities. Land, labor, and capital do not automatically produce corn, wheat, vegetables, beef cattle, or any other product. These resources must be organized into the proper combination, the proper amounts, and at the proper time for the desired production to occur. Plans can be developed for an individual enterprise such as wheat or beef cattle or for the entire business. Both types of plans are needed.

The alternatives that best meet the manager's goals can be identified only after a thorough analysis of all possible alternatives. "Analyze alternatives" was included as one of the steps in the decision-making process. Performing this important step along the way to developing management plans requires the manager to use some analytical tools and methods. Part Three is designed to introduce the basic concepts, tools, and methods used to analyze alternatives.

Chapters 5, 6, and 7 cover economic principles and cost concepts and their application to decision making in agriculture. Economic principles are important because they provide a systematic and organized procedure to identify the best alternative when profit maximization is the goal. The decision rules derived from these principles require an understanding of some marginal concepts and different types of costs. However, once they are learned, they can be applied to many types of management problems.

Four types of budgeting are explored in Chapters 8, 9, 10, and 11. Budgeting is a formal procedure for comparing alternatives on paper before committing resources to a particular plan or course of action. Knowledge of some marginal and cost concepts is needed to prepare budgets. All budgets are forward planning tools used to analyze and develop plans for the future. They can be applied to a single enterprise, a small part of the farm business, the whole farm business, or cash flows. Economic principles combined with enterprise, partial, whole farm, and cash flow budgeting provide the farm manager with a powerful set of tools for analyzing and then selecting alternatives to be included in management plans.

5

ECONOMIC PRINCIPLES— CHOOSING PRODUCTION LEVELS

CHAPTER OBJECTIVES

1 To explain the concept of marginalism

2 To show the relationship between a variable input and an output by use of a production function

3 To describe the calculation of average physical product and marginal physical product

4 To illustrate the law of diminishing returns and its importance

5 To show how to find the profit-maximizing amount of a variable input using the concepts of marginal value product and marginal input cost

6 To show how to find the profit-maximizing amount of output to produce using the concepts of marginal revenue and marginal cost

7 To explain the use of the equal marginal principle for allocating limited resources

A knowledge of economics provides the manager with a set of principles, procedures, and rules for decision making. This knowledge is useful when making plans for organizing and operating a farm or ranch business. It also provides some help and direction when moving through the steps in the decision-making process. Once a problem has been identified and defined, the next step is to acquire data and information. The information needed to apply economic principles provides a focus and direction to this search and prevents time being wasted searching for unnecessary data. Once the data are acquired, economic principles provide some guidelines for processing data into useful information and for analyzing the potential alternatives.

Economic principles consist of a set of rules that ensures that the choice or decision made will result in maximum profit. Their application tends to follow the same three steps. They are (1) acquire physical and biological data and process it into useful information, (2) acquire price data and process it into useful information, and (3) apply the appropriate economic decision-making rule to maximize profit. The reader should look for each of these steps as each new economic principle is discussed.

MARGINALISM

Much of economics is related to the concept of marginalism or marginality. The economist and the manager are often interested in what changes will result from a change in one or more factors under their control. For example, they may be interested in how cotton yield changes from using an additional 50 pounds of fertilizer, or how 2 additional pounds of grain in the daily feed ration affects milk production from a dairy cow, or the change in profit from raising an additional 20 acres of corn and reducing soybean production by 20 acres.

The term *marginal* will be used extensively throughout this chapter. It refers to incremental changes, increases or decreases, that occur at the edge or the margin. Until the reader thoroughly understands the use of the term, it may be useful to substitute "extra" or "additional" mentally whenever the word marginal is used (remembering that the "extra" can be a negative amount or zero). It is also important to remember that any marginal change being measured or calculated is a result of or caused by a marginal change in some other factor.

To calculate a marginal change of some kind, it is necessary to find the difference between an original value and the new value, which resulted from the change in the controlling factor. In other words, the change in some value *caused by* the marginal change in another factor is needed. Throughout this text, a small triangle (actually the Greek letter delta) will be used as shorthand for "the change in." For example, Δ *corn yield* would be read as "the change in corn yield" and would be the difference in corn yield before and after some change in an input affecting yield such as seed, fertilizer, or irrigation water. Although other inputs may also be necessary, there is an assumed fixed or constant amount being used. This does not mean they are unimportant, but this assumption serves to simplify the analysis.

THE PRODUCTION FUNCTION

A basic concept in economics is the *production function.* It is a systematic way of showing the relationship between different amounts of a resource or input that can be used to produce a product and the corresponding output or yield of that product. In other agricultural disciplines the same relationship may be called response curve, yield curve, or input/output relationship. By whatever name, a production function shows the amount of output that would be produced by using different amounts of a variable input. It can be presented in the form of a table, graph, or mathematical equation.

The first two columns of Table 5-1 are a tabular presentation of a production function. Different levels of the input that can be used to produce the product are shown in the first column, assuming all other inputs are held fixed. The amount of production

TABLE 5-1 PRODUCTION FUNCTION IN TABULAR FORM

Input level	Total physical product, TPP	Average physical product, APP	Marginal physical product, MPP
0	0	0	
1	12	12.0	12.0
2	30	15.0	18.0
3	44	14.7	14.0
4	54	13.5	10.0
5	62	12.4	8.0
6	68	11.3	6.0
7	72	10.3	4.0
8	74	9.3	2.0
9	72	8.0	−2.0
10	68	6.8	−4.0

expected from using each input level is shown in the second column labeled total physical product. In economics, output or yield is generally called total physical product, which will be abbreviated TPP.

Average and Marginal Physical Products

The production function provides the basic data that can be used to derive additional information about the relationship between the input and TPP. It is possible to calculate the average amount of output or TPP produced by each unit of input at each input level. This value is called average physical product (APP) and is shown in the third column of Table 5-1; APP is calculated by the formula

$$APP = \frac{\text{total physical product}}{\text{input level}}$$

In this example at 4 units of input, the APP is $54 \div 4 = 13.5$. The production function in Table 5-1 has an APP that increases over a short range and then decreases as more than 2 units of input are used. This is only one example of how APP may change with input level. Other types of production functions have an APP that declines continuously after the first unit of input, which is a common occurrence.

The first marginal concept to be introduced is marginal physical product (MPP), shown in the fourth column of Table 5-1. Remembering that marginal means additional or extra, MPP is *the additional or extra TPP produced by using an additional unit of input.* It requires measuring changes in both output and input.

Marginal physical product is calculated as

$$MPP = \frac{\Delta \text{ total physical product}}{\Delta \text{ input level}}$$

The numerator is the *change* in TPP caused by a *change* in the variable input, and the denominator is the actual amount of change in the input. For example, Table 5-1 indicates that using 4 units of input instead of 3 causes TPP to *increase* or change +10 units. Since this change was caused by a 1 unit increase or change in the input level, MPP would be found by dividing +10 by +1 to arrive at an MPP of +10. Observing the results of using 9 units of input instead of 8 shows that the change in TPP is *negative* or −2 units. Dividing this result by the 1 unit increase (+1) in input causing the change in TPP, the MPP is −2. Marginal physical product can be positive or negative. It can also be zero if changing the input level causes no change in TPP. A negative MPP indicates too much variable input is being used *relative to* the fixed input(s) and this combination causes TPP to decline.

The example in Table 5-1 has the input increasing by increments of 1, which simplifies the calculation of MPP. Other examples and problems may show the input increasing by increments of 2 or more. In this case the denominator in the formula for MPP must be the actual total change in the input. For example, if the input is changing by increments of 4, the denominator in the equation would be 4. The result is not an exact determination of MPP for the *last* unit of input but an average MPP for the 4 units in the change. Many times this will provide sufficient information for decision making unless the change in input between two possible input levels is fairly large. In this case either more information should be obtained about expected output levels for intermediate levels of the variable input or more sophisticated mathematical techniques should be used.[1]

A Graphical Analysis

A production function and its corresponding APP and MPP can also be shown in graphical form. Figure 5-1 is a production function with the same general characteristics as the data in Table 5-1. Notice that TPP or output increases at an increasing rate as the input level is increased from zero. As the input level is increased further, TPP continues to increase, but now at a decreasing rate, and eventually begins to decline absolutely as too much variable input is used relative to the amount of fixed input(s) available.

Both Table 5-1 and Figure 5-1 illustrate several important relationships between TPP, APP, and MPP. Notice that as long as TPP is increasing at an increasing rate, MPP is increasing along with APP. At the point where TPP changes from increasing at an increasing rate to increasing at a decreasing rate, MPP reaches a maximum and then declines continuously, having a value of zero where TPP reaches its maximum. Where TPP is at its maximum, a very small change in the input level neither increases nor decreases output, and therefore MPP is zero.

Average physical product increases over a slightly longer range than MPP before beginning to decline, which serves to illustrate an interesting relationship between

---

[1]The reader who has had a course in calculus will recognize that for a production function that is known and expressed as a continuous mathematical function, the exact MPP can be found by taking the first derivative of TPP with respect to input.

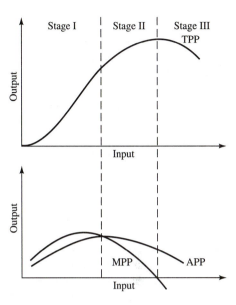

FIGURE 5-1   Graphical illustration of a production function.

APP and MPP. Notice that whenever MPP is above APP, APP will be increasing and vice versa. This can be explained by remembering that to raise (lower) any average value, the additional or marginal value used in calculating the new average must be above (below) the old average. The only way baseball players can raise their batting averages is to have a daily batting average (the marginal or additional value), which is above their current season average.

The relationships between TPP, APP, and MPP are often used to divide this particular type of production function into three stages, as shown in Figure 5-1. Stage I begins at the zero input level and continues to the point where APP is maximum and equal to MPP. Stage II begins where APP is maximum and ends where MPP is zero (or TPP is maximum). Stage III is the range of input levels where MPP is negative and TPP is declining absolutely. The importance of these stages in determining the proper amount of input to use will be discussed later.

Law of Diminishing Marginal Returns

Table 5-1 and Figure 5-1 can be used to illustrate an important law from both a theoretical and practical standpoint in economics and other agricultural disciplines. Diminishing returns can be discussed in terms of either physical production or economic returns, but the discussion here will be limited to physical production.

Diminishing marginal returns is used to describe what happens to MPP as additional input is used. The law of diminishing marginal returns states that, *as additional units of a variable input are used in combination with one or more fixed inputs, marginal physical product will eventually begin to decline.* Notice that this law is expressed in terms of marginal physical product and not average or total production.

Three properties of this law and its definition need to be emphasized. First, for diminishing returns to exist, one or more fixed inputs must be used in the production process in addition to the variable input. One acre of land or one head of livestock is often the fixed input used to define the production function and therefore to illustrate diminishing marginal returns. Second, the definition does not preclude diminishing marginal returns from beginning with the first unit of the variable input. This is often the case in agricultural applications of this law. Third, this law is based on the biological processes found in agricultural production. It results from the inability of plants and animals to provide the same response indefinitely to successive increases in nutrients or some other input.

Numerous examples of diminishing returns exist in agricultural production. As additional units of seed, fertilizer, or irrigation water are applied to a fixed acreage of some crop, the additional output or MPP will eventually begin to decline. The MPP gets smaller and smaller as the crop nears its biological capacity to utilize the input. A similar result is obtained with additional feed fed to dairy cows or to fat steers. The importance and practical significance of these examples will become apparent in the next two sections.

HOW MUCH INPUT TO USE

An important use of the information derived from a production function is in determining how much of the variable input to use. Given a goal of maximizing profit, the manager must select from all possible input levels the one that will result in the greatest profit.

Some help in this selection process can be found by referring back to the three stages in Figure 5-1. Any input level in stage III can be eliminated from consideration as additional input causes TPP to decrease and MPP to be negative. For any input level in stage III, the same output can be obtained with less input in one of the other stages. Choosing an input level in stage III is clearly irrational, and a manager should not use these input levels even if the input is available at no cost.

Stage I covers the area where adding additional units of input causes the average physical product to increase. In this stage adding another unit of input increases the productivity of all previous inputs as measured by average productivity or APP. If any input is to be used, it seems reasonable that a manager would want to use at least that input level that gives the greatest average physical product per unit of input. This point is at the boundary of stage I and stage II and represents the greatest efficiency in the use of the variable input. However, as will be shown shortly, profit can often be increased further by using more input even though APP is declining.

This discussion effectively eliminates stage I and stage III from consideration in determining the profit-maximizing input level. This leaves only stage II, which is the logical stage of production. Using only the physical information available at this point, it is not possible to determine which input level in stage II will actually maximize profit. More information, specifically price information, is needed.

TABLE 5-2   MARGINAL VALUE PRODUCT, MARGINAL INPUT COST, AND THE
OPTIMUM INPUT LEVEL
(Input Price = $12 per Unit; Output Price = $2 per Unit)

Input level	Total physical product, TPP	Marginal physical product, MPP	Total value product, TVP ($)	Marginal value product, MVP ($)	Marginal input cost, MIC ($)
0	0		0		
1	12	12	24	24	12
2	30	18	60	36	12
3	44	14	88	28	12
4	54	10	108	20	12
5	62	8	124	16	12
6	68	6	136	12	12
7	72	4	144	8	12
8	74	2	148	4	12
9	72	−2	144	−4	12
10	68	−4	136	−8	12

Marginal Value Product

Table 5-2 contains the same production function data as Table 5-1 with APP eliminated. As will be shown, APP is of little or no value in determining the profit-maximizing input level once stage I is eliminated from consideration. Two additional columns of information needed to determine the optimum input level have been added to the table using an input price of $12 per unit and an output price of $2 per unit. *Marginal value product (MVP) is the additional or marginal income received from using an additional unit of input.* It is calculated by using the equation

$$MVP = \frac{\Delta \text{ total value product}}{\Delta \text{ input level}}$$

Total value product (TVP) for each input level is found by simply multiplying the quantity of output (TPP) times its selling price.[2] For example, the MVP for moving from 2 to 3 units of inputs is found by subtracting the total value product at 2 units of input ($60) from the total value product at 3 units ($88) and dividing by the change in input (1) to get MVP of $28. The other MVP values are obtained in a similar manner.

Marginal Input Cost

Marginal input cost (MIC) is defined as *the change in total input cost or the addition to total input cost caused by using an additional unit of input.* It is calculated from the equation

---

[2] Total value product is the same as total income or gross income. However, TVP is the term used in economics when discussing input levels and input use.

$$MIC = \frac{\Delta \text{ total input cost}}{\Delta \text{ input level}}$$

Total input cost is equal to the quantity of input used times its price. For example, the MIC for moving from 2 to 3 units of input is found by subtracting the total input cost at 2 units ($24) from the cost at 3 units ($36) and dividing by the change in input (1) to find the MIC of $12.

Table 5-2 indicates that MIC is constant for all input levels. This should not be surprising if the definition of MIC is reviewed. The additional cost of acquiring and using an additional unit of input is equal to the price of the input, which is equal to MIC. This conclusion will always hold *provided* the input price does not change as more or less input is purchased.

The Decision Rule

The MVP and MIC values can now be compared and used to determine the optimum input level. The first few lines of Table 5-2 show MVP to be greater than MIC. In other words the additional income received from using one more unit of input exceeds the additional cost of that input. Therefore, additional profit is being made. These relationships exist until the input level reaches 6 units. At this point the additional income from and additional cost of another unit of input are just equal. Using more than 6 units of input causes the additional income (MVP) to be less than additional cost (MIC), which causes profit to decline as more input is used.

This discussion leads to the decision rule for determining the profit-maximizing input level. This input level is the point where

Marginal value product = marginal input cost
or   MVP = MIC

If MVP is greater than MIC, additional profit can be made by using more input. When MVP is less than MIC, more profit can be made by using less input. Note that the profit-maximizing point is *not* at the input level which maximizes TVP or total income. Profit is maximized at a lower level of input and output.

Obviously the profit-maximizing input level can also be found by calculating the total value product (total income) and total input cost at each input level and then subtracting cost from income to find the level where profit is greatest. This procedure has two disadvantages. First, it conceals the marginal effects of changes in the input level. Notice how diminishing marginal returns affect MVP as well as MPP, causing MVP to decline until it finally equals MIC. Second, it takes more time to find the new optimum input level if the price of either the input or output changes. Using the MVP and MIC procedure requires only recalculating MVPs if output price changes before the MVP = MIC rule is applied. Similarly, a change in input price requires only substituting the new MIC values for the old. The reader should experiment with different input and output prices to determine their effect on MVP, MIC, and the optimum input level.

HOW MUCH OUTPUT TO PRODUCE

The discussion in the previous section concentrated on finding the input level that maximized profit. There is also a related question. How much output should be produced to maximize profit? To answer this question directly requires the introduction of two new marginal concepts.

Marginal Revenue

Table 5-3 is the same as the previous table except for the last two columns. *Marginal revenue (MR)* is defined as *the change in income or the additional income received from selling one more unit of output.* It is calculated from the equation

$$MR = \frac{\Delta \text{ total revenue}}{\Delta \text{ total physical product}}$$

where total revenue is the same as total value product. Total revenue (TR) is used in place of total value product when discussing output levels. It is apparent from Table 5-3 that MR is constant and equal to the price of the output. This should not be surprising given the definition of MR. *Provided* the output price does not change as more or less output is sold, which is typical for an individual farmer or rancher, MR will always equal the price of the output. The additional income received from selling one more unit of output will equal the price received for that output. However, if the selling price varies with changes in the quantity of output sold, MR must be calculated using the preceding equation.

TABLE 5-3   MARGINAL REVENUE, MARGINAL COST, AND THE OPTIMUM
OUTPUT LEVEL
(Input Price = $12 per Unit; Output Price = $2 per Unit)

Input level	Total physical product, TPP	Marginal physical product, MPP	Total revenue, TR ($)	Marginal revenue, MR ($)	Marginal cost, MC ($)
0	0		0		
1	12	12	24.00	2.00	1.00
2	30	18	60.00	2.00	0.67
3	44	14	88.00	2.00	0.86
4	54	10	108.00	2.00	1.20
5	62	8	124.00	2.00	1.50
6	68	6	136.00	2.00	2.00
7	72	4	144.00	2.00	3.00
8	74	2	148.00	2.00	6.00
9	72	−2	144.00		
10	68	−4	136.00		

Marginal Cost

Marginal cost (MC) is defined as *the change in cost or the additional cost incurred from producing another unit of output.* It is calculated from the equation

$$MC = \frac{\Delta \text{ total input cost}}{\Delta \text{ total physical product}}$$

where total input cost is the same as previously defined. In Table 5-3 marginal cost decreases slightly and then begins to increase as additional input is used. Notice the inverse relationship between MPP and MC. When MPP is declining (diminishing returns), MC is increasing, as it takes relatively more input to produce an additional unit of output. Therefore, the additional cost of another unit of output is increasing.

The Decision Rule

In a manner similar to analyzing MVP and MIC, MR and MC are compared to find the profit-maximizing output level. When MR is greater than MC, the additional unit of output increases profit as the additional income exceeds the additional cost. Conversely, if MR is less than MC, producing the additional unit of output will decrease profit. The profit-maximizing output level is therefore at the output level where

$$\text{Marginal revenue} = \text{marginal cost}$$
or $$MR = MC$$

In Table 5-3 this occurs at 68 units of output.

It should be no surprise that the optimum output level of 68 units is produced by the optimum input level of 6 units found in the previous section. The only real difference between MVP and MIC and MR and MC is that the former are expressed in units of input and the latter are expressed in units of output. There is only one profit-maximizing combination of input and output for a given production function and a given set of prices. Once either the optimum input or output level is found, the other value can be determined from the production function.

APPLYING THE MARGINAL PRINCIPLES

A common management decision is how much fertilizer, seed, or irrigation water to use per acre for a certain crop. Table 5-4 contains some information for a hypothetical irrigation problem where the manager must select the amount of water to apply to an acre of corn. Given prices for water and corn, the principles from the last two sections can be used to determine the amount of water and the corresponding amount of output that will maximize profit. The first two columns of Table 5-4 contain the production function data, which can often be obtained from experimental trials. Marginal physical product is computed from these data and represents the additional bushels of corn produced by an additional acre-inch of water within each 2-inch increment.

An examination of the MVP and MIC data shows they are never exactly equal.

TABLE 5-4 DETERMINING THE PROFIT-MAXIMIZING IRRIGATION LEVEL FOR CORN PRODUCTION
(Water at $3.00 per Acre-Inch and Corn at $2.50 per bu)

Irrigation water (acre-inch)	Corn yield per acre (bu)	Marginal physical product MPP (bu)	Marginal value product MVP ($)	Marginal input cost MIC ($)	Marginal revenue MR ($)	Marginal cost MC ($)
10	104.0					
12	116.8	6.40	16.00	3.00	2.50	0.47
14	128.6	5.90	14.75	3.00	2.50	0.51
16	138.2	4.80	12.00	3.00	2.50	0.63
18	144.8	3.30	8.25	3.00	2.50	0.91
20	149.0	2.10	5.25	3.00	2.50	1.43
22	151.8	1.40	3.50	3.00	2.50	2.14
24	153.6	0.90	2.25	3.00	2.50	3.33
26	154.2	0.30	0.75	3.00	2.50	10.00

This will often happen when using a tabular production function with an input level increasing by several units at each step. If the MVP = MIC rule cannot be met exactly, then the next best alternative is to get as close to this point as possible but never go past it. Never select an input level with the MVP less than MIC as that step results in less profit than a lower input level.

Table 5-4 assumes water can be applied only in increments of 2 acre-inches. This makes 22 acre-inches the profit-maximizing amount to apply. Up to and including that level, MVP is greater than MIC and each additional acre-inch causes profit to increase. The step to 24 inches would result in a MVP less than MIC and profit would be reduced. Actually profit would be reduced by $0.75 for *each* of the two additional inches or a total reduction of $1.50. The $0.75 comes from the difference between the additional cost of $3.00 per inch and the additional income of only $2.25.

The same logic can be applied to marginal revenue and marginal cost. Profit will be maximized at 151.8 bushels of output where MR is still greater than MC. Producing 153.6 bushels causes MR to be less than MC, which will reduce profit. As before, the actual reduction in profit can be found by taking the difference between MR and MC ($0.83), which is the reduction in profit *per bushel* of additional output, and multiplying it by the additional output of 1.8 bushels. The result of $1.50 is the same as that found in the previous paragraph. This helps to verify that 22 acre-inches and 151.8 bushels is the profit-maximizing combination of input and output. Moving to the combination of 24 acre-inches and 153.6 bushels will reduce profit by $1.50.

What would happen to the profit-maximizing levels if one or more of the prices changes? A change in the price of water would cause MIC and MC to change, and a change in the price of corn would cause MVP and MR to change. Such changes are very likely to change the profit-maximizing solution. Assuming corn price is constant at $2.50, the MVP and MIC columns indicate that the price of water would have to fall to $2.25 per inch to make 24 inches the new profit-maximizing level. It would take a price of $3.50 or greater (but not over $5.25) to make 20 inches the correct amount to use.

The same sort of analysis can be done using MR and MC. With the price of water constant at $3.00, the price of corn would have to increase to $3.33 per bushel or higher to make 153.6 bushels the new profit-maximizing amount of output. Similarly, if the price of corn dropped to $2.14 per bushel or less (but not less than $1.43), 149.0 bushels would then become the optimum output level.

Price changes are common in agriculture, and they affect the optimum input and output levels as shown in this example. An increase in the input price or a decrease in the output price tends to lower the profit-maximizing input and output levels. Price increases for the output or price decreases for the input tend to increase the profit-maximizing input and output levels. However, it is important to note that there must be a change in the *relative* prices. Either doubling both prices or decreasing both by 50 percent will change the numbers in Table 5-4 but will not change the choice of input and output levels. One price must move proportionately more than the other, remain constant or move in the opposite direction to cause a change in the relative prices.

EQUAL MARGINAL PRINCIPLE

The discussion so far in this chapter has assumed sufficient input is available or can be purchased to set MVP = MIC for each and every acre or head of livestock. Another and possibly more likely situation is a limited amount of input that prevents reaching the MVP = MIC point for all possible uses. Now the manager must decide how the limited input should be allocated or divided among several or many possible uses or alternatives. Decisions must be made on the best allocation of fertilizer between many acres or fields and different crops, irrigation water between fields and crops, and feed between different types and weights of livestock. If capital is limited, it first must be allocated somehow to the purchase of the fertilizer, water, feed, and any other inputs.

The *equal marginal principle* provides the guidelines and rules to ensure that the allocation is done in such a way that profit is maximized from the use of any limited input. This principle can be stated as follows:

> A limited input should be allocated among alternative uses in such a way that the marginal value products of the last unit used on each alternative are equal.

Table 5-5 is an application of this principle where irrigation water must be allocated among three crops in three fields of equal size. The MVPs are obtained from the production functions relating water use to the yield of each crop and from the respective crop prices.

Assume a maximum of 2,400 acre-inches of water is available and can be applied only in increments of 4 acre-inches. The limited supply of water would be allocated among the three crops in the following manner using the MVPs to make the decisions. The first 400 acre-inches (4 acre-inches on 100 acres) would be allocated to cotton as it has the highest MVP. The second 400 acre-inches would be allocated to grain

TABLE 5–5   APPLICATION OF THE EQUAL MARGINAL PRINCIPLE TO THE ALLOCATION OF IRRIGATION WATER*

Irrigation water (acre-inch)	Marginal value products ($)		
	Wheat (100 acres)	Grain sorghum (100 acres)	Cotton (100 acres)
4	1,200	1,600	1,800
8	800	1,200	1,500
12	600	800	1,200
16	300	500	800
20	50	200	400

*Each application of 4 acre-inches on a crop is a total use of 400 acre-inches (4 acre-inches times 100 acres).

sorghum as it has the second highest MVP. In a similar manner the third 400 acre-inches would be used on cotton and the fourth, fifth, and sixth 400-acre-inch increments on wheat, grain sorghum, and cotton, respectively. Each successive 400-acre-inch increment is allocated to the field that has the highest MVP remaining after the previous allocations.

The final allocation is 4 acre-inches on wheat, 8 on grain sorghum, and 12 on cotton. Each final 4-acre-inch increment on each crop has an MVP of $1,200, which satisfies the equal marginal principle. If more water was available, the final allocation would obviously be different. For example, if 3,600 acre-inches were available, it would be allocated 8 acre-inches to wheat, 12 to grain sorghum, and 16 to cotton. This equates the MVPs of the last 4 acre-inch increment on each crop at $800, which again satisfies the equal marginal principle. When inputs must be applied in fixed increments, it may not be possible to exactly equate the MVPs of the last units applied to all alternatives. However, the MVP of the last unit allocated should always be equal to or greater than the MVP available from any other alternative use.

When allocating an input that is *thought* to be limited, care must be taken to *not* use it past the point where MVP = MIC for any alternative. This would result in something less than maximum profit. The input is not really limited if a sufficient amount is available to reach the point where MVP = MIC for every alternative.

The profit-maximizing property of this principle can be demonstrated for the 2,400-acre-inch example above. If 4 acre-inches on wheat were allocated to grain sorghum or cotton, $1,200 of income would be lost and $800 gained for a net loss of $400. The same loss would be incurred if the last 4-acre-inch increment was removed from either grain sorghum or cotton and reallocated to another crop. When the MVPs are equal, profit cannot be increased by a different allocation of the limited input.

The equal marginal principle can also be presented in graphical form as in Figure 5-2, where there are only two alternative uses for the limited input. The problem is to allocate the input between the two uses, keeping the MVPs equal, so that input quantity 0a plus quantity 0b just equals the total amount of input available. If 0a plus 0b is less than the total input available, more should be allocated to each use, again

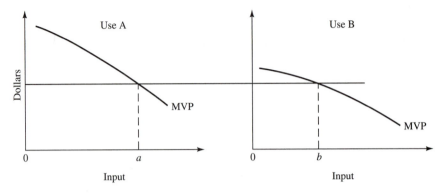

FIGURE 5-2   Illustration of the equal marginal principle.

keeping the MVPs equal, until the input is fully used. There would have to be a decrease in the input used on both alternatives if $0a$ plus $0b$ exceeded the total input available.

The equal marginal principle applies not only to purchased inputs but also to those that are already owned or available, such as land, the manager's labor, and machinery time. It also prevents making the mistake of maximizing profit from one enterprise by using the input until MVP = MIC and not having enough to use on other enterprises. Maximizing profit from the *total* business requires the proper allocation of limited inputs, which will not necessarily result in maximizing profit from any *single* enterprise.

SUMMARY

Economic principles using the concept of marginality provide useful guidelines for managerial decision making. They have direct application to the basic decisions of how much to produce, how to produce, and what to produce. The production function that describes the relationship between input levels and corresponding output levels provides some basic technical information. When this information is combined with price information, the profit-maximizing input and output levels can be found.

Marginal value product and marginal input cost are equated to find the proper input level, and marginal revenue and marginal cost are equated to find the profit-maximizing output level. These are all marginal concepts that measure the changes in revenue and cost resulting from small changes in either input or output. When a limited amount of input is available and there are several alternative uses for it, the equal marginal principle provides the rule for allocating the input and maximizing profit under these conditions.

A manager will seldom have sufficient information to use fully the economic principles discussed in this chapter. This does not detract from the importance of these principles, but their application and use are often hindered by insufficient physical and biological data. Prices must also be estimated before the product is available for sale, which adds more uncertainty to the decision-making process. However, a complete understanding of the economic principles permits making changes in the right direction in response to price and other changes. A manager should continually search for better information to use in refining the decisions made through the use of these basic economic principles.

QUESTIONS FOR REVIEW AND FURTHER THOUGHT

1 In Table 5-2 assume the price of both the input and output have doubled. Calculate the new MVPs and MICs and determine the profit-maximizing input level for the new prices. Now assume both prices have been cut in half and repeat the process. Explain your results.

2 Use several other price combinations for the production function data in Tables 5-2 and 5-3 and find the profit-maximizing input and output for each price combination. Pay close attention to what happens to these levels when the input price increases or decreases, or when the output price increases or decreases.

3 In Table 5-2 what would the profit-maximizing input level be if the input price was $0? Notice the TPP and MPP at this point. Can you think of any other situation where you would maximize profit by producing at this level?

4 Does the law of diminishing returns have anything to do with MVP declining and MC increasing? Why?

5 Find an extension service or experiment station publication for your state that shows the results of a seeding, fertilizer, or irrigation rate experiment. Do the results exhibit diminishing returns? Use the concepts of this chapter to find the profit-maximizing amount of the input to use.

6 Does the equal marginal principle apply to personal decisions when you have limited income and time? How do you allocate a limited amount of study time when faced with three exams on the same day?

7 Freda Farmer can invest capital in $100 increments and has three alternative uses for the capital, as shown in the following table. The values in the table are the marginal value products for each successive $100 of capital invested.

Capital invested	Fertilizer ($)	Seed ($)	Chemicals ($)
First $100	400	250	350
Second $100	300	200	300
Third $100	250	150	250
Fourth $100	150	105	200
Fifth $100	100	90	150

a If Freda has an unlimited amount of her own capital available and no other alternative uses for it, how much should she allocate to each alternative?

b If Freda can borrow all the capital she needs for 1 year at 10 percent interest, how much should she borrow and how should it be used?

c Assume Freda has only $700 available. How should this limited amount of capital be allocated among the three uses? Does your answer satisfy the equal marginal principle?

d Assume Freda has only $1,200 available. How should this amount be allocated? What is the total income from using the $1,200 this way? Would a different allocation increase the total income?

REFERENCES

Boehlje, Michael D., and Vernon R. Eidman: *Farm Management,* John Wiley & Sons, New York, 1984, chap. 5.

Buckett, Maurice: *An Introduction to Farm Organization and Management,* Pergamon Press, Oxford, 1981, chap. 3.

Castle, Emery N., Manning H. Becker, and A. Gene Nelson: *Farm Business Management,* 3d ed., Macmillan Publishing Co., New York, 1987, chap. 2.

Kadlec, John E.: *Farm Management,* Prentice-Hall, Englewood Cliffs, NJ, 1985, chap. 3.

McSweeny, William T.: *Producing at the Margin for Greater Farm Profits,* Extension Circular 342, Cooperative Extension Service, Pennsylvania State University, University Park, PA. (undated)

Workman, John P.: *Range Economics,* Macmillan Publishing Co., New York, 1986, chaps. 4, 5.

6

ECONOMIC PRINCIPLES—
CHOOSING INPUT AND
OUTPUT COMBINATIONS

CHAPTER OBJECTIVES

1 To explain the use of substitution in economics and decision making

2 To demonstrate how to compute a substitution ratio and a price ratio for two inputs

3 To use the preceding two ratios for finding the least-cost combination of two inputs to produce a given amount of output

4 To describe the characteristics of competitive, supplementary, and complementary enterprises

5 To show the use of a substitution ratio and price ratio to find the profit-maximizing combination of two enterprises

6 To understand the principle of comparative advantage and how it explains the location of agricultural production

Substitution takes place in the daily lives of most people. It occurs whenever one product is purchased or used in place of another, whenever personal income is spent for one type or class of product rather than another. Steak replaces hamburger on the dinner table, a large car replaces the economy model in the garage, and a new brand of soap or toothpaste is purchased instead of the old brand. Some substitutions or replacements are made due to an increase (or decrease) in one's personal income. The motivation for others comes from changes in relative prices or perceived differences in quality.

Substitution also takes place in the production of goods and services. Often there is more than one way to produce a product or provide a service. Machinery, computers, and robots can substitute for labor, and one ingredient can often be replaced by another. Prices, specifically relative prices, play a major role in determining if and how much substitution should take place. In many production activities including those on farms and ranches, the substitution is not an all or nothing decision. It is often a matter of making some relatively small changes in the ratio or mix of two or more inputs being used. Substitution is therefore another type of marginal analysis that considers the changes in costs and/or income when making the substitution decision.

INPUT SUBSTITUTION

One of the basic decisions a farm or ranch manager must make is *how to produce* a given product. Most products require two or more inputs in the production process, but the manager can often choose the input combination or ratio to be used. The problem is one of determining if more of one input can be profitably substituted for less of another and what is the least-cost combination of inputs to produce a given amount of output.

Substitution of one input for another occurs frequently in agricultural production. One grain can be substituted for another or forage for grain in a livestock ration. Herbicides substitute for mechanical and hand cultivation, and machinery can replace labor. The manager must select that combination of inputs that will produce a given amount of output or perform a certain task for the least cost. In other words the problem is to find the least-cost combination of inputs, as this will maximize the profit from producing a given amount of output. The alert manager will always be looking for a different input combination that will do the same job and do it for less cost.

Input Substitution Ratio

The first step in analyzing a substitution problem is to determine if it is physically possible to make a substitution and at what rate. Figure 6-1 illustrates the more common types of substitution between two inputs. In Figure 6-1a the line PP' is an *isoquant* (from isoquantity meaning the same quantity) and shows a number of possible combinations of corn and barley in a feed ration. This line is called an isoquant as *any* of these combinations will produce the same quantity of output or weight gain in this example.

The substitution ratio, or the rate at which one input will substitute for another, is determined from the equation

$$\text{Substitution ratio} = \frac{\text{amount of input replaced}}{\text{amount of input added}}$$

where both the numerator and denominator are the differences or changes in the amount of inputs being used between two different points on the isoquant PP'.

In Figure 6-1a moving from point A to point B means 4 pounds of corn are being replaced by 5 additional pounds of barley in order to maintain the same weight gain or output. The substitution ratio is 4 ÷ 5 = 0.8, which means 1 pound of barley will replace only 0.8 pound of corn. Since PP' is a straight line, the substitution ratio will al-

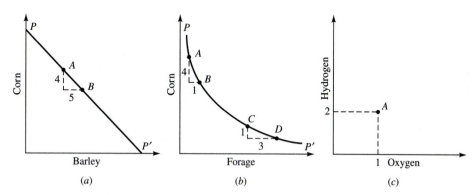

FIGURE 6-1   Three possible types of substitution.

ways be 0.8 between any two points on this isoquant. This is an example of a *constant rate of substitution* between two inputs. Whenever the substitution ratio is equal to the same numerical value over the full range of possible input combinations, the inputs exhibit a constant rate of substitution. This occurs most often when the two inputs contribute the same or nearly the same factor to the production process. Corn and barley, for example, both contribute energy to a feed ration.

Another and perhaps more common example of physical substitution is shown in Figure 6-1*b*. Here the isoquant *PP'* shows the different combinations of corn and forage that might produce the same weight gain on a steer or the same milk production from a dairy cow. This is an illustration of a *decreasing rate of substitution*. The substitution ratio is $4 \div 1 = 4$ when moving from point *A* to point *B* on isoquant *PP'*, but it is $1 \div 3 = \frac{1}{3}$ when moving from point *C* to point *D*. In this example the substitution ratio depends on the location on the isoquant and will decline with any movement down and to the right on it.

Many agricultural substitution problems will have a decreasing substitution ratio. This will occur when the two inputs are dissimilar or contribute different factors to the production process. As more of one input is substituted for another, it becomes increasingly difficult to make any substitution and still maintain the same level of output. More and more of the added input is needed to substitute for a unit of the input being replaced, which causes the substitution ratio to decrease. This is an indication the inputs work together best when combined at some combination containing a relatively large proportion of each. At low levels of one, there is a near surplus of the other, and it is not an efficient or productive combination.

One other possible substitution situation is where *no substitution* is possible. Many chemical reactions are an example, and Figure 6-1*c* illustrates that it takes two hydrogen atoms and one oxygen atom to produce a molecule of water. No other combination will work, at least not without ending up with an excess of either hydrogen or oxygen. This would be an inefficient use of inputs. While there are situations where no substitution is possible, managers should not be too quick to make that conclusion. For example, research has shown that even irrigation water and nitrogen fertilizer are substitutes but at a sharply decreasing rate and only over a limited range.

The Decision Rule

Identifying the type of physical substitution that exists and calculating the substitution ratio are necessary steps, but they alone do not permit a determination of the least-cost input combination. Input prices are needed, and the ratio of the input prices is compared to the substitution ratio. The price ratio is computed from the equation

$$\text{Price ratio} = \frac{\text{price of input being added}}{\text{price of input being replaced}}$$

With this ratio the least-cost combination can be found. The decision rule for finding this combination is where the

$$\text{Substitution ratio} = \text{price ratio}$$

Table 6-1 is an application of this procedure. Each of the feed rations is a combination of grain and forage that will put the same pounds of gain on a feeder steer. The problem is to select the ration that produces this gain for the least cost. The fourth and fifth columns of Table 6-1 contain the substitution ratio and price ratio for each feed ration. The substitution ratio is declining, and the price ratio is constant for the given prices.

If the substitution ratio is greater than the price ratio, the total cost of the feed ration can be reduced by moving to the next lower ration in the table. The converse is true if the substitution ratio is less than the price ratio. This can be verified by calculating the total cost of two adjacent rations for each situation. The least-cost ration in the example is ration F where the substitution ratio is equal to the price ratio.

It may not be apparent why this rule works. Using the substitution and price ratios is just a convenient method of comparing the additional cost of using more of one input versus the reduction in cost from using less of the other. Whenever the additional cost is less than the cost reduction, the total cost is lower and the substitution should be made. This is the case whenever the substitution ratio is greater than the price ratio.

In any substitution problem the least-cost combination depends on both the substitution ratio and the price ratio. The substitution ratios will remain the same over time

TABLE 6-1   SELECTING A LEAST-COST FEED RATION
(Price of Grain = 4.4¢ per lb; Price of Hay = 3.0¢ per lb)*

Feed ration	Grain (lb)	Hay (lb)	Substitution ratio	Price ratio
A	825	1,350		
			2.93	1.47
B	900	1,130		
			2.60	1.47
C	975	935		
			2.20	1.47
D	1,050	770		
			1.87	1.47
E	1,125	630		
			1.47	1.47
F	1,200	520		
			1.07	1.47
G	1,275	440		

*Each ration is assumed to put the same weight gain on a feeder steer with a given beginning weight.

provided the underlying physical or biological relationships do not change. However, the price ratio will change whenever the relative input prices change, which will result in a different input combination becoming the new least-cost combination. As the price of one input increases relative to the other, the new least-cost input combination will tend to have less of the higher-priced input and more of the now relatively less-expensive input.[1]

ENTERPRISE COMBINATIONS

The third basic decision to be made by a farm or ranch manager is *what to produce* or what combination of enterprises will maximize profit. A choice must be made from among all possible enterprises, which may include vegetables, wheat, soybeans, cotton, beef cattle, hogs, poultry, and others. Climate, soil, range vegetation, and limits on other inputs may restrict the list of possible enterprises to only a few on some farms and ranches. On others the manager may have a large number of possible enterprises from which to select the profit-maximizing combination. In all cases it is assumed that one or more inputs are limited, which places an upper limit on how much can be produced from a single enterprise or any combination.

Competitive Enterprises

The first step in determining profit-maximizing enterprise combinations is to determine the physical relationships between the enterprises. Given a limited amount of land, capital, or some other input, the production from one enterprise can often be increased only by decreasing the production from another enterprise. There is a trade-off, or substitution, to be considered when changing the enterprise combination. These are called *competitive* enterprises, as they are competing for the use of the same limited input at the same time.

Figure 6-2 illustrates two types of competitive enterprises. In the first graph corn and soybeans are competing for the use of the same 100 acres of land. Planting all corn would result in the production of 12,000 bushels of corn, and planting all soybeans would produce a total of 4,000 bushels. Other combinations of corn and soybeans totaling 100 acres would produce the combinations of corn and soybeans shown on the line connecting these two points. This line is called a *production possibility curve* (PPC), as it shows all possible combinations of corn and soybeans that can be produced from the given 100 acres.

Beginning with producing all corn, replacing an acre of corn with an acre of soybeans results in a loss of 120 bushels of corn and a gain of 40 bushels of soybeans. The trade-off, or substitution ratio, is 3, as 3 bushels of corn must be given up to gain 1 bushel of soybeans. With a straight-line production possibility curve, this substitution ratio is the same between any two points on it. This is an example of competitive enterprises with a *constant substitution ratio*.

---

[1]The reader should note that it is the price *ratio* and not the absolute prices that determine the least-cost combination. If both input prices double, the price ratio and least-cost combination are unchanged. However, costs have doubled and profit will be less.

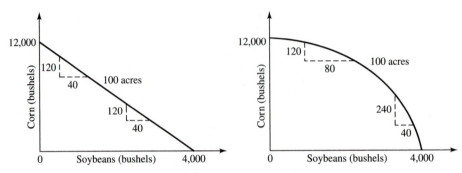

FIGURE 6-2   Production possibility curves for competitive enterprises.

Over a period of time a combination of crop enterprises may benefit each other because of better weed, disease, and insect control, erosion control, and timeliness in planting and harvesting a large acreage. This situation is shown in the second graph in Figure 6-2. The curved production possibility curve indicates that total corn production increases at a slower rate as a higher proportion of the land is used for corn production. A similar situation holds true for soybeans. This causes the substitution ratio to be different for different combinations of the two enterprises. The substitution ratio is 120 ÷ 80 = 1.5 near the top of the curve and increases to 240 ÷ 40 = 6.0 near the bottom of the curve. The enterprises are still competitive but have an *increasing substitution ratio.*

The most profitable combination of two competitive enterprises can be determined by comparing the substitution ratio and their price ratio. Each substitution ratio is calculated from the equation

$$\text{Substitution ratio} = \frac{\text{quantity of output lost}}{\text{quantity of output gained}}$$

where the quantities gained and lost are the *changes* in production between two points on the PPC. The price ratio is found from the equation

$$\text{Price ratio} = \frac{\text{unit price of the output gained}}{\text{unit price of the output lost}}$$

The decision rule[2] for finding the profit-maximizing combination of two competitive enterprises is the point where the

$$\text{Substitution ratio} = \text{price ratio}$$

Table 6-2 is an example of two competitive enterprises, corn and wheat, that have an increasing substitution ratio. The method for determining the profit-maximizing enter-

---

[2]This rule assumes total production costs are the same for any combination of the two enterprises. If this is not true, the ratio of the *profit or gross margin* per unit of each enterprise should be used instead of the price ratio.

TABLE 6-2   SELECTING A PROFIT-MAXIMIZING ENTERPRISE COMBINATION
WITH LAND THE FIXED INPUT
(Price of Corn = $2.80 per bu; Price of Wheat = $4.00 per bu)

Combination number	Corn (bu)	Wheat (bu)	Substitution ratio	Price ratio
1	0	6,000		
2	2,000	5,600	0.20	0.70
3	4,000	5,000	0.30	0.70
4	6,000	4,100	0.45	0.70
5	8,000	3,000	0.55	0.70
6	10,000	1,600	0.70	0.70
7	12,000	0	0.80	0.70

prise combination is basically the same as for determining the least-cost input combination but with one important exception. For enterprise combinations, when the price ratio is greater than the substitution ratio, substitution should continue by moving downward and to the right on the PPC and down to the next combination in Table 6-2. Conversely, a substitution ratio greater than the price ratio means too much substitution has taken place, and the adjustment should be upward and to the left on the PPC and up to the next combination in the table. Following this method, the profit-maximizing combination in Table 6-2 is combination number 6 (10,000 bushels of corn and 1,600 bushels of wheat) because the substitution ratio equals the price ratio at this point.

The decision rule of substitution ratio equals price ratio is a short-cut method for comparing the income gained from producing more of one enterprise versus the income lost from producing less of the other. Whenever the price ratio is greater than the substitution ratio, the additional income is greater than the income lost, and the substitution will increase total income. Profit will also increase as total cost is assumed to be constant for any enterprise combination.

When the enterprises have a constant substitution ratio, the profit-maximizing solution will be to produce all of one or all of the other enterprise and not a combination. This is because the price ratio will be either greater than or less than the substitution ratio for all combinations. An increasing substitution ratio will generally result in the production of a combination of the enterprises, with the combination dependent on the current price ratio. Any change in the price(s) of the output that changes the price ratio will affect the profit-maximizing enterprise combination when there is an increasing substitution ratio. As with input substitution, it is the price *ratio* that is important and not just the price level.

Supplementary Enterprises

While competitive enterprises are the most common, other types of enterprise relationships do exist. One of these is *supplementary,* and an example is shown in the left dia-

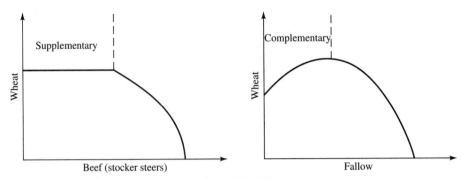

FIGURE 6-3   Supplementary and complementary enterprise relationships.

gram of Figure 6-3. Two enterprises are supplementary if the production from one can be increased without affecting the production level of the other.

The example in Figure 6-3 shows beef production from stocker steers run on winter wheat pasture can be increased over a range without affecting the amount of wheat produced. The PPC shows a relationship that eventually becomes competitive as beef production cannot be increased indefinitely without affecting wheat production. In states where landowners lease hunting rights to hunters, this leasing enterprise can be supplementary with livestock and crop production.

A manager should take advantage of supplementary relationships by increasing production from the supplementary enterprise. This should continue at least up to the point where the enterprises become competitive. If the supplementary enterprise shows any profit, however small, total profit will be increased by producing some of it since production from the primary enterprise does not change. Two general conclusions can be drawn about the profit-maximizing combination of supplementary enterprises. First, this combination will not be within the supplementary range. It will be at least at the point where the relationship changes from supplementary to competitive. Second, it will in all likelihood be in the competitive portion of the production possibility curve. The exact location will depend as it always does with competitive enterprises on the substitution ratio and the price ratio.

Complementary Enterprises

Another possible type of enterprise relationship is *complementary*. This type of relationship exists whenever increasing the production from one enterprise causes the production from the other to increase at the same time. The right-hand graph in Figure 6-3 illustrates a possible complementary relationship between wheat production and land left fallow.

In many dryland wheat production areas with limited rainfall, some land is left fallow, or idle, each year as a way to store part of one year's rainfall to be used by a wheat crop the following year. Leaving some acres fallow reduces the acres in wheat, but the per acre yield may increase because of the additional moisture available. This yield increase may be enough that total wheat production is actually greater than from planting all acres to wheat every year.

A complementary enterprise should be increased to at least the point where production from the primary enterprise (wheat in this example) is at a maximum. This is true even if the complementary enterprise has no value as production from the primary enterprise is increasing at the same time. Enterprises will generally be complementary only over some limited range and then become competitive. As with supplementary enterprises, two general conclusions can be drawn about the profit-maximizing combination of complementary enterprises. First, this combination will not be within the complementary range. It will be at least at the point where the relationship changes from complementary to competitive. Second, given positive prices, it will be somewhere in the competitive range. The correct combination can be found using the substitution ratio and price ratio decision rule for competitive enterprises.

COMPARATIVE ADVANTAGE

Certain crops can be grown in only limited areas because of specific soil and climatic requirements. However, even those crops and livestock that can be raised over a broad geographical area often have production concentrated in one region. For example, why do Iowa farmers specialize in corn and soybean production while Kansas farmers specialize in wheat production? All three crops can be grown in each state. Regional specialization in the production of agricultural commodities and other products is explained by the *principle of comparative advantage.*

While livestock and some crops can be raised over a broad geographical area, the yields, production costs, and profit may be different in each area. The principle of comparative advantage states that *individuals or regions will tend to specialize in the production of those commodities for which their resources give them a relative or comparative advantage.* It is *relative* yields, costs, and profits that are important for this principle.

An illustration of this principle using yields is shown below. Region A has an *absolute advantage* in the production of both crops because of the higher yields. However, region A must give up $2\frac{1}{2}$ bushels of corn for every bushel of wheat it grows while region B has to give up only 2 bushels of corn to get a bushel of wheat. Region B has an *absolute disadvantage* in the production of both crops but a relative or *comparative advantage* in the production of wheat, since it gives up less corn for a bushel of wheat than region A.

	Yield per acre (bushels)	
	Region A	Region B
Corn	100	60
Wheat	40	30

Wheat farmers in region B would be willing to give up 0.5 bushel of wheat to get a bushel of corn, and corn farmers in region A would be willing to take 0.4 bushel or more of wheat for a bushel of corn. Assume region A specializes in corn production and region B in wheat production. Farmers in both regions would be willing to trade if

0.4 to 0.5 bushel of wheat could be exchanged for a bushel of corn. Farmers in each region could obtain the product from the other region at less cost than raising it themselves and therefore will tend to specialize in producing the product for which they have the comparative advantage.

Specialization can also make the combined regions better off. Assume there are 100 acres in each region and each plants one-half corn and one-half wheat. Total production is 8,000 bushels of corn and 3,500 bushels of wheat. If region A specializes in corn and region B in wheat, total production is 10,000 bushels of corn and 3,000 bushels of wheat. As long as the income from an additional 2,000 bushels of corn is more than the income lost from 500 bushels less wheat, specialization increases the total value of the crops produced. Trade between the regions allows each to obtain the final desired combination of the two products.

SUMMARY

This chapter continued the discussion of economic principles begun in Chapter 5. Here the emphasis was on the use of substitution principles to provide a manager with a procedure to answer the questions *How to produce* and *What to produce.* The question of how to produce concerns finding the least-cost combination of inputs to produce a given amount of output. Calculation and use of the substitution and price ratios was shown as well as the decision rule of finding the point where these two values are equal. This decision rule determines the least-cost combination of two inputs. Marginality is still an important concept here as substitution ratios are computed from small or marginal changes in the two inputs.

The question of what to produce is one of finding the profit-maximizing combination of enterprises given some limited input. That combination will depend first on the type of enterprise relationship that exists—competitive, supplementary, or complementary. The profit-maximizing combination for competitive enterprises is found by computing substitution ratios and a price ratio. Finding the point where they are equal determines the correct combination. Supplementary and complementary enterprises have unique properties over a limited range but will eventually become competitive. The profit-maximizing combination for these two types of enterprise relationships will not be within these ranges but generally within their competitive range.

Comparative advantage was discussed as well as the difference between it and an absolute advantage. It was used to explain why certain regions, states, and areas tend to specialize in the production of certain commodities.

QUESTIONS FOR REVIEW AND FURTHER THOUGHT

1 First double and then halve both prices used in Table 6-1 and find the new least-cost input combination in each case. Why is there no change? But what would happen to profit in each case?

2 Is there always one "best" livestock ration? Or does it depend on prices? Would you guess that feed manufacturers know about input substitution and least-cost input combinations? Explain how they might use this knowledge.

3 Explain how the slope of the isoquant affects the substitution ratio.

4 Carefully explain the difference between an isoquant and a production possibility curve.

5 Increase the price of one of the enterprises in Table 6-2, calculate the new price ratio and find the new profit maximizing combination. Do you produce more or less of the enterprise with the now higher price? Why? What would be the effect of doubling both prices? Halving both prices?

6 Why is the profit-maximizing combination of even supplementary and complementary enterprises generally in the competitive range?

7 Except for some wheat farms in the Great Plains, most crop farms produce two or more different crops. Explain this observation in terms of the shape of the PPC found on these farms and the price ratio.

8 For two very similar inputs such as soybean oil meal and cottonseed oil meal, would you expect the substitution ratio to be nearly constant or to decline sharply as one input is substituted for another? Why?

9 Would you expect supplementary enterprises to use very similar or very different resources?

REFERENCES

Boehlje, Michael D., and Vernon R. Eidman: *Farm Management,* John Wiley & Sons, New York, 1984, chaps. 3, 5.

Buckett, Maurice: *An Introduction to Farm Organization and Management,* Pergamon Press, Oxford, 1981, chap. 3.

Casavant, Kenneth L., and Craig L. Infanger: *Economics & Agricultural Management,* Reston Publishing Co., Reston, VA, 1984, chaps. 3, 4.

Castle, Emery N., Manning H. Becker, and A. Gene Nelson: *Farm Business Management,* 3d ed., Macmillan Publishing Co., New York, 1987, chap. 2.

Workman, John P.: *Range Economics,* Macmillan Publishing Co., New York, 1986, chaps. 6, 7.

7

COST CONCEPTS IN ECONOMICS

CHAPTER OBJECTIVES

1 To explain the importance of opportunity cost and its use in managerial decision making

2 To clarify the difference between short run and long run

3 To discuss the difference between fixed and variable costs

4 To identify the fixed costs and show how to compute them

5 To show how to compute the different average costs

6 To demonstrate the use of fixed and variable costs in making short-run and long-run production decisions

7 To explore economies of size and how they help explain changes in farm size and profitability

A number of costs are important and useful in economics and for making management decisions. Costs can be classified a number of ways depending on whether they are fixed or variable, cash or noncash, included in record books as an accounting expense or not. Opportunity cost is a type of cost that is not included in the accounting expenses for a business but is an important economic cost. It will be widely used in later chapters and will be the first cost discussed in this chapter.

OPPORTUNITY COST

Opportunity cost is an economic concept and not a cost that can be found in an accountant's ledger. However, it is an important and basic concept that needs to be considered when making managerial decisions. Opportunity cost is based on the fact that every input or resource has an alternative use even if the alternative is nonuse. Once an input is committed to a particular use, it is no longer available for any other alternative, and the income from the alternative must be foregone.

Opportunity cost can be defined one of two ways:

1 The value of the product not produced because an input was used for another purpose *OR*

2 The income that would have been received if the input had been used in its most profitable alternative use

The latter definition is perhaps the more common. Either of these definitions of opportunity cost should be kept in mind as a manager makes decisions on input use. The *real* cost of an input may not be its purchase price. Its real cost, or its *opportunity cost,* in any one use is the income it would have earned in its next best alternative use. If this is greater than the income expected from the planned use of the input, the manager should reconsider the decision. The alternative appears to be a more profitable use of the input.

The concept can be illustrated by referring to Table 7-1, which is the same example used when discussing the equal marginal principle in Chapter 5. If the first 400 acre-inches of water is put on the 100 acres of wheat, the opportunity cost is $1,800. This income would be lost by not putting the water to its best alternative use, which is cotton. Since the expected return from wheat is $1,200, the higher opportunity cost indicates this first unit of water is not earning its greatest possible income. If the first 400 acre-inches was used on cotton, the opportunity cost is $1,600 (from grain sorghum), indicating there is no more profitable use for the water. The expected income exceeds the opportunity cost. This same line of reasoning can be used to determine the most profitable use of each successive 400-acre-inch increment of water. A little thought will reveal this to be just another way of applying the equal marginal principle. It also illustrates the relationship between these two concepts.

TABLE 7-1   THE OPPORTUNITY COST OF APPLYING IRRIGATION WATER AMONG THREE USES*

Irrigation water (acre-inch)	Marginal value products ($)		
	Wheat (100 acres)	Grain sorghum (100 acres)	Cotton (100 acres)
4	1,200	1,600	1,800
8	800	1,200	1,500
12	600	800	1,200
16	300	500	800
20	50	200	400

*Each application of 4 acre-inches on a crop is a total use of 400 acre-inches (4 acre-inches times 100 acres).

Opportunity costs are widely used in economic analysis. For example, the opportunity costs of a farm operator's labor, management, and capital are used in several types of budgeting and when analyzing farm profitability. The opportunity cost of a farm operator's labor (and perhaps that of other unpaid family labor) would be what that labor would earn in its next best alternative use. That alternative use could be non-farm employment but, depending on skills, training, and experience, it might also be employment on someone else's farm.

Opportunity cost of management is difficult to estimate. For example, what is the value of management per crop acre? Some percentage of the gross income per acre is often used but what is the proper percent? In other cases an annual opportunity cost of management is needed. Here and elsewhere the analyst must be careful to exclude labor from the estimate. For example, if the opportunity cost of labor is estimated at $20,000 per year and the individual could get a job in "management" paying $30,000 per year, the opportunity cost of management is estimated at the difference, or $10,000 per year. The opportunity cost of labor plus that of management cannot be greater than the total salary in the best alternative job. Because it is difficult to estimate the opportunity cost of management, the opportunity cost of labor and management are often left combined as one value.

Capital presents a different set of problems when estimating opportunity costs. There are many uses of capital and generally a wide range of possible returns. However, the higher the average return expected on an investment the higher the expected risk as measured by the variability of the return. To avoid the problem of identifying a comparable level of risk, the opportunity cost of farm capital is often set equal to the interest rate on savings accounts or the current cost of borrowed capital. This represents a minimum opportunity cost and is a somewhat conservative approach. If comparable levels of risk can be determined, it would be appropriate to use higher rates when analyzing farm investments with above average levels of risk.

A special problem is how to determine the opportunity cost on the annual service provided by long-lived inputs such as land, buildings, breeding stock, and machinery. It may be possible to use rental rates for land and custom rates for machinery services. However, this does not work well for all such items, and the opportunity cost of these inputs should often be determined by the most profitable alternative use outside the farm or ranch business. This is the real or true cost of using inputs to produce agricultural products.

It is difficult to evaluate directly the opportunity cost of these long-lived inputs in terms of nonagricultural alternative uses. Land, fences, buildings, and machinery may have a very low opportunity cost based on nonagricultural use, but they represent a large capital investment. This capital investment is generally used to estimate their opportunity cost. The total dollar value of the input is estimated and then multiplied by the opportunity cost of capital. This is an indirect procedure for estimating the annual opportunity cost. However, it has the advantages of converting everything into a common unit, its capital equivalent, and then using an opportunity cost that is relatively easy to estimate, that of capital.

COSTS

A number of cost concepts are used in economics. Marginal input cost was studied in Chapter 5. Seven additional cost concepts and their abbreviations are

1 Total fixed cost (TFC)
2 Average fixed cost (AFC)
3 Total variable cost (TVC)
4 Average variable cost (AVC)
5 Total cost (TC)
6 Average total cost (ATC)
7 Marginal cost (MC)

All these costs are output related. Marginal cost, which was also studied in Chapter 5 is the additional cost of producing an additional unit of output. The others are either the total or average costs for producing a given amount of output.

Short Run and Long Run

Before these costs are discussed in some detail, it is necessary to distinguish between what economists call the short run and the long run. These are time concepts, but they are not defined as fixed periods of calendar time. The short run is that period of time during which one or more of the production inputs is fixed in amount and cannot be changed. For example, at the beginning of the planting season it may be too late to increase or decrease the amount of cropland owned or rented. The current crop production cycle would be a short-run period, as land is fixed in amount.

Over a longer period of time, land may be purchased, sold, or leased or leases may expire and the amount of land available may be increased or decreased. The long run is defined as that period of time during which the quantity of all necessary productive inputs can be changed. In the long run a business can expand by acquiring additional inputs or go out of existence by selling all its inputs. The actual calendar length of the long run as well as the short run will vary considerably depending on the situation and circumstances. Depending on which input(s) are fixed, the short run may be anywhere from several days to several years. One year or one crop or livestock production cycle are common short-run periods in agriculture.

Fixed Costs

The costs associated with owning a fixed input or resource are called fixed costs. These are the costs that are incurred even if the input is not used, and there may be additional costs if it is actually used in producing some product. Fixed costs do not change as the level of production changes in the short run but can change in the long run as the quantity of the fixed input changes. Since there need not be any fixed inputs owned in the long run, fixed costs exist only in the short run and are equal to zero in the long run.

Another characteristic of fixed costs is that they are not under the control of the manager in the short run. They exist and at the same level regardless of how much or how little the resource is used. The only way they can be avoided is to sell the item, which can be done in the long run.

Total fixed cost (TFC) is simply the summation of the several types of fixed costs. Depreciation, insurance, repairs, taxes (property taxes, not income taxes), and interest are the usual components of TFC. They are easily remembered as the "DIRTI 5" costs, where DIRTI is derived from the first letter of each of the fixed cost items. Repairs may not always be included as a fixed cost, as they tend to increase with increased use of the resource. However, some minimum level of repairs may be needed to keep a resource in working order even if it is not used. For this reason some of the repair expenditure is sometimes included as a fixed cost, particularly for buildings. If repairs are not included, fixed costs can be remembered as being the "DITI 4."

Computing the average annual TFC for a fixed input requires some method of finding the average annual depreciation and interest. The straight-line depreciation method discussed in Chapter 2 provides the average annual depreciation from the equation

$$\text{Annual depreciation} = \frac{\text{cost} - \text{salvage value}}{\text{useful life}}$$

where cost is equal to the purchase price, useful life is the number of years the item is expected to be owned, and salvage value is its expected value at the end of that useful life.

Since capital invested in a fixed input has an opportunity cost, interest on that investment is included as part of the fixed cost. This interest is always included and always computed the same way regardless of whether or not any money was borrowed to finance the purchase. However, it is not correct to charge interest on the purchase price or original cost of a depreciable asset every year because its value is decreasing over time. Therefore, the interest component of total fixed cost is commonly computed from the formula

$$\text{Interest} = \frac{\text{cost} + \text{salvage value}}{2} \times \text{interest rate}$$

where the interest rate is the opportunity cost of capital.[1] This equation gives the interest charge for the *average* value of the item over its life (as shown by the first part of the equation) and reflects the fact that it is decreasing in value over time. Depreciation is being charged to account for this decline in value.

As an example, assume the purchase of a machine for $20,000 with a salvage value of $5,000 and a useful life of 5 years. Annual property taxes are estimated to be $25, annual insurance is $50, and the opportunity cost of capital is 12 percent. Using these values and the two preceding equations results in the following annual total fixed cost:

---

[1]This equation is widely used but only approximates the actual opportunity cost. The methods discussed in Chapter 15 can be used to find the true annual charge, which combines depreciation and interest into one value. This value recovers the investment in the asset plus compound interest over its useful life.

$$\text{Depreciation} \qquad \frac{\$20,000 - 5,000}{5 \text{ years}} = \$3,000$$

$$\text{Interest} \qquad \frac{\$20,000 + 5,000}{2} \times 12\% = \$1,500$$

Taxes	25
Insurance	50
Annual total fixed cost	$4,575

Repairs are not shown as they are commonly included as a variable cost. It is often difficult or impractical to divide total repair cost between fixed and variable cost, so repairs are generally included as a variable cost. Notice that annual total fixed costs are over 20 percent of the purchase price. It is not unusual for TFC to be 20 to 25 percent of the purchase price for a depreciable asset.

Fixed cost can be expressed as an average per unit of output. Average fixed cost (AFC) is found using the equation

$$AFC = \frac{TFC}{Output}$$

where output is measured in physical units such as bushels, bales, or hundredweights. Acres or hours are often used as the measure of output for machinery. Since by definition TFC is a fixed or constant value, regardless of the quantity of output, AFC will decline continuously as output is increased (see Figure 7-2). One way to lower the cost of producing a given commodity is to get more output from the fixed resource. This will always lower the AFC per unit of output.

Fixed costs can be either cash or noncash expenses. They can be easily overlooked or underestimated because a large part of total fixed cost can be noncash expenses as shown in the following chart.

Expense item	Cash expense	Noncash expense
Depreciation		X
Interest on the investment	X	X
Repairs	X	
Taxes, property	X	
Insurance	X	X

Depreciation is always a noncash expense, as there is no annual cash outlay for this fixed cost. Repairs and property taxes are always cash expenses, and interest and insurance may be either. If money is borrowed to purchase the asset, there will be some cash interest expense. When the item is purchased with the buyer's own capital, the interest charge would be opportunity cost on this capital, and there is no cash payment to a lender. Insurance would be a cash expense if insurance is carried with an insurance company or noncash if the risk of loss is assumed by the owner. In the latter case there would be no annual cash outlay, but the insurance charge should still be included in fixed costs to cover the possibility of damage to or loss of the item from fire, windstorm, and so forth.

This distinction between cash and noncash expenses does not imply that noncash expenses are any less important than cash expenses. In the short run noncash expenses do mean less cash is needed to meet current expenses. However, income must be sufficient to cover all expenses in the long run if the business is to survive and prosper.

Variable Costs

Variable costs are those that the manager has control over at a given point in time. They can be increased or decreased at the manager's discretion and will increase as production is increased. Items such as feed, fertilizer, seed, chemicals, fuel, and livestock health expenses are examples of variable costs. A manager has control over these expenses in the short run, and they will tend to change as total production is changed.

Total variable cost (TVC) can be found by summing each of the individual variable costs, each of which is equal to the quantity of the input purchased times its price. Average variable cost (AVC) is calculated from the equation

$$AVC = \frac{TVC}{output}$$

where output is again measured in physical units. Average variable cost may be either increasing or decreasing depending on the underlying production function and the output level. For the production function illustrated in Figure 5-1, AVC will initially decrease as output is increased and then will increase beginning at the point where average physical product starts to decline.

Variable costs exist in both the short run and the long run. All costs are considered to be variable costs in the long run as there are no fixed inputs. The distinction between a fixed and variable cost also depends on the exact point in time where the next decision is to be made. Fertilizer is generally a variable cost. Yet once it has been purchased and applied, the manager no longer has any control over the size of this expenditure. It must be considered a fixed cost for the remainder of the crop season, and future decisions must be made accordingly. Labor cost and cash rent for land are similar examples. After a labor or lease contract is signed, the manager cannot change the amount of money obligated, and the salary or rent must be considered a fixed cost for the duration of the contract.

Total and Marginal Costs

Total cost (TC) is the sum of total fixed cost and total variable cost (TC = TFC + TVC). In the short run, it will increase only as TVC increases, as TFC is a constant value.

Average total cost (ATC) can be found by one of two methods. For a given output level it is equal to AFC + AVC. It can also be calculated from the equation

$$ATC = \frac{TC}{output}$$

which will give the same result. Average total cost will typically be decreasing at low output levels because AFC is decreasing rapidly and AVC may be decreasing also. At higher output levels AFC will be decreasing less rapidly and AVC will eventually increase and be increasing at a faster rate than the rate of decrease in AFC. This combination causes ATC to increase.

Marginal cost (MC) is defined as the change in total cost divided by the change in output.

$$MC = \frac{\Delta\, TC}{\Delta\, output} \quad or \quad MC = \frac{\Delta\, TVC}{\Delta\, output}$$

It is also equal to the change in total variable cost divided by the change in output. Since TC = TFC + TVC and TFC is constant, the only way TC can change is from a change in TVC. Therefore, MC can be calculated either way with the same result.

Cost Curves

Relationships among the seven output-related cost concepts can be illustrated graphically by a series of curves. The shape of these cost curves depends on the characteristics of the underlying production function. Figure 7-1 contains cost curves representative of the general production function shown in Figure 5-1. Other types of production functions would have cost curves with different shapes.

The relationships among the three total costs are shown in Figure 7-1. Total fixed cost is constant and unaffected by output level. Total variable cost is always increasing, first at a decreasing rate and then at an increasing rate. Because it is the sum of total fixed cost and total variable cost, the total cost curve has the same shape as the

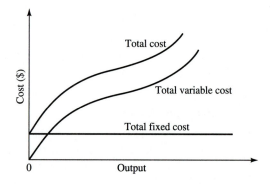

FIGURE 7-1   Typical total cost curves.

total variable cost curve. However, it is always higher by a vertical distance equal to total fixed cost.

The general shape and relationship of the average and marginal cost curves are shown in Figure 7-2. Average fixed cost is always declining but at a decreasing rate. The other two average curves are U-shaped, declining at first, reaching a minimum, and then increasing at higher levels of output. Notice that they are not an equal distance apart. The vertical distance between them is equal to average fixed cost, which changes with output level. This accounts for their slightly different shape and for the fact that their minimum points are at two different output levels.

The marginal cost curve will generally be increasing. However, for this particular production function it decreases over a short range before starting to increase. Notice that the marginal cost curve crosses both average curves at their minimum points. As discussed earlier, as long as the marginal value is below the average value, the latter will be decreasing and vice versa. For this reason the marginal cost curve will always cross the average cost curves at their minimum point if such a point exists.

Application of Cost Concepts

Table 7-2 is an example of some cost figures for the common problem of determining the profit-maximizing stocking rate for a pasture that is fixed in size. It is illustrative of many similar problems where an understanding of the different cost concepts and relationships will help a manager in planning and decision making.

Total fixed costs are assumed to be $4,000 in the example. This would cover the annual opportunity cost on the land and any improvements, depreciation on fences and water facilities, and insurance. Variable costs are assumed to be $395 per steer (steers are the variable input in this example). This includes the cost of the steer, transportation, health expenses, feed, interest on the investment in the steer, and any other expenses that increase along with the number of steers purchased.

Since the size of the pasture and amount of forage available are both limited, running more and more steers will eventually cause the average weight gain per steer to

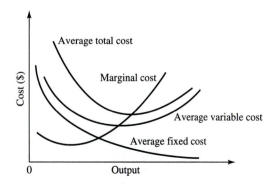

FIGURE 7-2 Average and marginal cost curves.

TABLE 7-2   ILLUSTRATION OF COST CONCEPTS APPLIED TO A STOCKING RATE PROBLEM*

Number of steers	Output (hundred weight of beef)	MPP	TFC ($)	TVC ($)	TC ($)	AFC ($)	AVC ($)	ATC ($)	MC ($)	MR ($)
0	0		4,000	0	4,000	—	—	—		
		7.2							54.86	70.00
10	72		4,000	3,950	7,950	55.56	54.86	110.42		
		7.6							51.97	70.00
20	148		4,000	7,900	11,900	27.03	53.38	80.41		
		7.7							51.30	70.00
30	225		4,000	11,850	15,850	17.78	52.66	70.44		
		7.0							56.43	70.00
40	295		4,000	15,800	19,800	13.56	53.56	67.12		
		6.5							60.77	70.00
50	360		4,000	19,750	23,750	11.11	54.86	65.97		
		6.0							65.83	70.00
60	420		4,000	23,700	27,700	9.52	56.43	65.95		
		5.5							71.82	70.00
70	475		4,000	27,650	31,650	8.42	58.21	66.63		
		5.0							79.00	70.00
80	525		4,000	31,600	35,600	7.62	60.19	67.81		
		4.5							87.78	70.00
90	570		4,000	35,550	39,550	7.02	62.37	69.39		
		4.0							98.75	70.00
100	610		4,000	39,500	43,500	6.56	64.75	71.31		

*Total fixed cost is $4,000 and variable costs are $395 per steer. Selling price of the steers is assumed to be $70 per hundredweight.

decline. This is reflected in diminishing returns and a declining MPP when more than 30 steers are placed in the pasture. Total beef sold off the pasture still increases but at a decreasing rate as more steers compete for the limited forage.

The cost figures in Table 7-2 have the usual or expected pattern as production or output is increased. Total fixed cost remains constant by definition while both TVC and TC are increasing. Average fixed cost declines rapidly at first and then continues to decline but at a slower rate. Average variable cost, average total cost, and marginal cost decline initially, reach a minimum at some output level, and then begin to increase.

The profit-maximizing output level was defined in Chapter 5 to be where MR = MC. A point of exact equality does not exist in Table 7-2, but 60 steers and 420 hundredweight of beef produced would be the profit-maximizing levels. Moving from 50 to 60 steers, MC is less than MR while MC is slightly greater than MR moving from 60 to 70 steers. When MC is greater than MR the additional cost per hundredweight of beef produced from the additional 10 steers is greater than the additional income per hundredweight. Therefore, placing 70 steers on the pasture would result in less profit than 60 steers.

The profit-maximizing point will depend on the selling price, or MR, and MC. Selling prices often change, and MC can change with changes in variable cost (primarily from a change in the cost of the steer in this example). The values in the MC column indicate the most profitable number of steers would be 70 if the selling price is $71.82 or higher. That number would drop to 50 steers if the selling price drops below $65.83. As was shown in Chapter 5, the profit-maximizing input and output levels will always depend on prices.

At the table price of $70.00 per hundredweight and running 60 steers, the TR is $29,400 and TC is $27,700 leaving a profit of $1,700. For a price of $60.77, the MR = MC rule indicates a profit-maximizing point of 50 steers and 360 hundredweight. However, the ATC per hundredweight at this point is $65.97 which means a *loss* of $5.20 per hundredweight or a total *loss* of $1,872. What should a manager do in this situation? Should any steers be purchased if the expected selling price is less than ATC and a loss will result?

The answer to this question is yes for some situations and no for others. Data in Table 7-2 indicate there would be a loss equal to the TFC of $4,000 if no steers were purchased. This loss would exist in the short run as long as the land is owned. It could be avoided in the long run by selling the land, which eliminates the fixed costs. However, the fixed costs cannot be avoided in the short run and the relevant question is: Can a profit be made *or* the loss reduced to less than $4,000 in the short run by purchasing some steers? Obviously steers should not be purchased if it would result in a loss greater than $4,000 since the loss can be minimized at $4,000 by not purchasing any.

Variable costs are under the control of the manager and can be reduced to zero by not purchasing any steers. Therefore, no variable costs should be incurred unless the expected selling price is at least equal to or greater than the minimum AVC. This will provide enough total revenue to cover total variable costs. If the selling price is greater than minimum AVC but less than minimum ATC, the income will cover all variable costs with some left over to pay *part* of the fixed costs. There would be a loss but it would be less than $4,000 in this example. To answer the question above, *yes,* steers should be purchased when the expected selling price is less than the minimum ATC *but* only if it is above the minimum AVC. This action will result in a loss, but it will be the minimum loss possible given these cost and price relationships.

If the expected selling price is less than the minimum AVC, total revenue will be less than TVC, there will be a loss, and it will be greater than $4,000. Under these conditions no steers should be purchased, which will minimize the loss at $4,000. In Table 7-2 the lowest AVC is $52.66 and the lowest ATC is $65.95. The loss would be minimized by not purchasing steers when the expected selling price is less than $52.66 and by purchasing steers when the expected selling price is between $52.66 and $65.95. In the last situation, the loss minimizing output level is where MR = MC.

Production Rules for the Short Run   The preceding discussion leads to three rules for making production decisions in the short run. They are:

1 *Expected selling price is greater than minimum ATC* (or TR greater than TC). A profit can be made and is maximized by producing where MR = MC.

2 *Expected selling price is less than minimum ATC but greater than minimum AVC* (or TR is greater than TVC but less than TC). A loss cannot be avoided but will be minimized by producing at the output level where MR = MC. The loss will be somewhere between zero and total fixed cost.

3 *Expected selling price is less than minimum AVC* (or TR less than TVC). A loss cannot be avoided but is minimized by *not* producing. The loss will be equal to TFC.

The application of these rules is illustrated graphically in Figure 7-3. With a selling price equal to MR_1, the intersection of MR and MC is well above ATC, and a profit is being made. When the selling price is equal to MR_2, the income will not be sufficient to cover total costs but will cover all variable costs with some left over to pay part of fixed costs. In this situation the loss is minimized by producing where MR = MC as the loss will be less than TFC. Should the selling price be as low as MR_3, income would not even cover variable costs and the loss would be minimized by stopping production. This would minimize the loss at an amount equal to TFC.

Production Rules for the Long Run The preceding discussion relates only to the short run where fixed costs exist. What about the long run where there are no fixed costs? Continual losses incurred by producing in the long run will eventually force the firm out of business. There are only two rules for making production decisions in the long run.

1 *Selling price is greater than ATC* (or TR greater than TC). Continue to produce as a profit is being made. This profit is maximized by producing at the point where MR = MC.

2 *Selling price is less than ATC* (or TR less than TC). There will be a continual loss. Stop production and sell the fixed asset(s), which eliminates the fixed costs. The money received should be invested in a more profitable alternative.

This does not mean assets should be sold the first time a loss is incurred. Short-run losses will occur when there is a temporary drop in the selling price. Long-run rule number 2 should be invoked only when the drop in price is expected to be long lasting or permanent.

ECONOMIES OF SIZE

Economists and managers are interested in farm size and the relationship between costs and size for a number of reasons. The following are examples of questions being

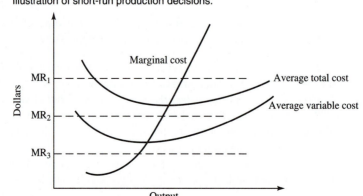

FIGURE 7-3   Illustration of short-run production decisions.

asked that relate to farm size and costs. What is the most profitable farm size? Can larger farms produce food and fiber cheaper? Are large farms more efficient? Will family farms disappear and be replaced by large corporate farms? Will the number of farms continue to decline? The answers to all these questions depend at least in part on what happens to costs and the cost per unit of output as farm size increases.

First, how is farm size measured? Number of livestock, number of acres, number of full-time workers, net worth, total assets, profit, and other factors have all been used to measure size, and all have some advantages and disadvantages. For example, number of acres is a common and convenient measure of farm size but should be used only to compare farm sizes in a limited geographical area where farm type, soil type, and climate are very similar. It is obvious that 100 acres of irrigated vegetables in California is not the same size operation as 100 acres of arid range land in neighboring Arizona or Nevada.

Gross farm income or total revenue is a common measure of farm size. It has the advantage of converting everything into the common denominator of the dollar. This and other measures that are in dollar terms are better than any physical measure for measuring and comparing farm size across different farming regions.

Size in the Short Run

In the short run, one or more inputs are fixed in amount, with land often being the fixed input. Given this fixed input, there will be a short-run average total cost curve as shown in Figure 7-4. Short-run average cost curves will typically be U-shaped, with the average cost increasing at higher levels of production because the limited fixed input makes additional production more and more difficult and therefore increases average cost per unit of output.

For simplicity, size is measured as the output of a single product in Figure 7-4. The product can be produced at the lowest average cost per unit by producing the quantity 0a. However, this may not be the profit-maximizing quantity, as profit is maximized at the output level where marginal revenue is equal to marginal cost. Since output price is equal to marginal revenue, a price of P' would maximize profit by producing the

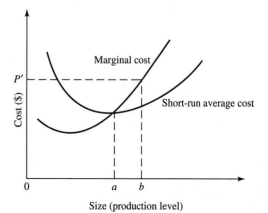

FIGURE 7-4   Farm size in the short run.

quantity 0*b*. A higher or lower price would cause output to increase or decrease to correspond with the point where the new price is equal to marginal cost.

Because of a fixed input such as land, output can be increased in the short run only by intensifying production. This means the use of more variable inputs such as fertilizer, chemicals, irrigation water, labor, and machinery time. However, the limited fixed input tends to increase average costs as production is increased past some point and a production limit is eventually reached. Additional production is possible only by acquiring more of the fixed input, which is a long-run problem.

Size in the Long Run

The economics of farm size is more interesting when analyzed in a long-run context. This gives the manager time to adjust all inputs to the level that will result in the desired farm size. One measure of the relationship between output and costs as farm size increases is expressed in the following ratio:

$$\frac{\text{Percent change in costs}}{\text{Percent change in output value}}$$

Both changes are calculated in monetary terms to allow combining the cost of the many inputs and the value of several outputs into one figure. This ratio can have three possible results called *decreasing costs, constant costs,* or *increasing costs.*

Ratio value	Type of costs
< 1	Decreasing
= 1	Constant
> 1	Increasing

These three possible results are also called, respectively, *increasing returns to size, constant returns to size,* and *decreasing returns to size.* Decreasing costs means increasing returns to size and vice versa. These relationships are shown in Figure 7-5 using a long-run average cost curve that is the average cost per unit of output. When decreasing costs exist, the average cost per unit of output is decreasing, so that the av-

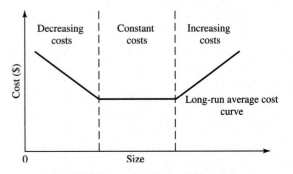

FIGURE 7-5   Possible size-cost relationships.

erage *profit* per unit of output is increasing. Therefore, increasing returns to size are said to exist. The same line of reasoning explains the relation between constant costs and constant returns and between increasing costs and decreasing returns.

Studies on the economics of size are concerned with identifying which type of costs and returns exist for a given farm type and hence the shape of the long-run average cost curve (LRAC). A short-run average cost curve (SRAC) exists for each and every possible farm size as defined by the amount of the fixed input available. Theoretically an infinite number of SRAC curves exist. Two such curves are shown in Figure 7-6, along with their relation to the LRAC curve. The latter can be thought of as the "envelope curve" containing a point on each of the infinite number of SRAC curves. Any point on the LRAC can be attained by selecting a given SRAC curve with its associated farm size and then operating at the point where it is just tangent to the LRAC curve.

The answers to the questions raised at the beginning of this section depend on the shape of the LRAC curve for actual farms and ranches. If it is falling, there is an incentive for farmers to increase the size of their business as average costs are decreasing and profit per unit of output is increasing given fixed output prices. Conversely, if the LRAC curve is rising after some point, as is commonly assumed, there will be little or no incentive to continue increasing farm size, as costs are increasing and profit per unit of output will begin to decrease.

Numerous studies have shown that the long-run average cost curve for farms declines rapidly at first, as shown in Figure 7-6. The LRAC curve then declines more slowly, with studies showing that most of the decline has taken place at a farm size equivalent to what two full-time workers can produce. This result is fairly consistent regardless of farm type. What is less clear is whether or not these economies of size continue for even larger and larger farms. In other words, does the LRAC curve continue to decrease because of economies of size, does it become flat, or does it begin to increase because of diseconomies of size?

FIGURE 7-6   Long-run average cost curve.

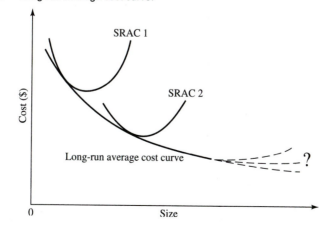

Economies of Size   Economies of size or increasing returns to size exist when the LRAC curve is decreasing. These economies can come from a number of sources, including the spreading of total fixed costs over a larger amount of output. (Remember the shape of the average fixed cost curve.) Full utilization of labor, machinery, and buildings is another factor. These two factors account for much of the initial decrease in the LRAC curve. Other possible economies of size include price discounts for volume purchasing of inputs, possible price advantages when selling large amounts of output, and added advantages of more specialized management. Larger tractors and buildings also tend to cost less per horsepower and per square foot than smaller ones. These factors all combine to exert a downward pressure on the LRAC curve.

Diseconomies of Size   Diseconomies of size exist when the LRAC curve is increasing, and this combination discourages further increases in farm size. Economists have long assumed an LRAC curve will eventually begin to rise as business size continues to increase. A lack of sufficient managerial skill to keep a large business running smoothly and efficiently is used to explain the rising LRAC curve. Related to this is the need to hire, train, supervise, and coordinate the activities of a larger labor force, which becomes more difficult as size increases. Another diseconomy of size in agriculture is the dispersion over a larger geographical area as size increases. This increases travel time and the cost of getting to and from work and from field to field. It also makes management and supervising labor more difficult. For livestock operations, increasing production may eventually lead to higher costs per head for disease control and manure disposal.

Figure 7-6 indicates some uncertainty about the shape of the LRAC curve for larger farm sizes. Both economies and diseconomies of size probably exist for large farms and ranches, with their relative strengths determining what happens to the LRAC curve. The studies mentioned previously generally show a rapidly decreasing LRAC curve, which then becomes relatively flat over a wide range of sizes. Insufficient data usually prevent arriving at a general conclusion about the shape of the LRAC curve for very large farms and ranches. The relatively poor success of some very large farming operations in the United States is cited as evidence of diseconomies of size. However, there are also some very successful large farm and ranch operations. Any analysis of economies of size is further clouded by the fact that successful managers can achieve lower average costs than others and tend to expand their operations over time. Thus, what appears to be economies of size may actually be due to superior management.

SUMMARY

This chapter discussed the different economic costs and their use in managerial decision making. Opportunity costs are often used in budgeting and farm financial analysis. This noncash cost stems from inputs having more than one use. Using an input one way means it cannot be used in any other alternative use, and the income from that alternative must be foregone. The income given up is the input's opportunity cost.

An analysis of costs is important for understanding and improving the profitability of a business. The distinction between fixed and variable costs is important and useful when making short-run production decisions. In the short run, production should take place only if the expected income will be more than variable costs. Otherwise losses will be minimized by not producing. Production should take place in the long run only if income is high enough to pay *all* costs. If all costs are not covered in the long run, the business will eventually fail or will be receiving less than opportunity cost on one or more inputs.

An understanding of costs is also necessary for analyzing economies of size. The relationship between cost per unit of output and size of the business determines if there are increasing, decreasing, or constant returns to size. If unit costs decrease as size increases, there are increasing returns to size, and the business would have an incentive to grow and vice versa. The type of returns that exist for an individual farm will determine in large part the success or failure of expanding farm size. Future trends in farm size, number of farms, and form of business ownership and control will be influenced by economies and diseconomies in farm and ranch businesses.

QUESTIONS FOR REVIEW AND FURTHER THOUGHT

1 How would you estimate the opportunity cost for each of the following items? What do you think the opportunity cost would be?
  a Capital invested in land
  b Your labor used in a farm business
  c Your management used in a farm business
  d One hour of tractor time
  e The hour you wasted instead of studying for your next exam
  f Your college education

2 For each of the following indicate if it is a fixed or variable cost and a cash or noncash expense. (Assume short run.)

	Fixed	Variable	Cash	Noncash
a Gas and oil	——	——	——	——
b Depreciation	——	——	——	——
c Property taxes	——	——	——	——
d Salt and minerals	——	——	——	——
e Labor hired on an hourly basis	——	——	——	——
f Labor contracted for 1 year in advance	——	——	——	——
g Insurance premiums	——	——	——	——
h Electricity	——	——	——	——

3 Assume Freda Farmer has just purchased a new combine. She has calculated total fixed cost to be $10,500 per year and estimates variable costs will be $8.50 per acre.
  a What will her average fixed cost per acre be if she combines 1,000 acres per year? 800 acres per year?
  b What is the additional cost of combining an additional acre?

    c Assume Freda plans to use the combine only for custom work on 900 acres per year. How much should she charge *per acre* to be sure all costs are covered? If she would custom harvest 1,200 acres per year?

4 Assume the purchase of a combine for $82,500 that is estimated to have a salvage value of $22,000 and a useful life of 8 years. The opportunity cost of capital is 12 percent. When computing annual total fixed cost, how much should be included for depreciation? For interest?

5 Using the data in Table 5-3 and a TFC of $50, compute the three total costs and the three average costs.

    a What is the maximum profit that can be made with the given prices?

    b To continue production in the long run, the output price must remain equal to or above $————————.

    c In the short run production should stop whenever the output price falls below $————————.

6 Why is interest included as a fixed cost even though no money was borrowed to purchase the item?

7 What will profit be if production takes place where MR = MC just at the point where ATC is minimum? At the point where MR = MC and AVC is at its minimum?

8 Explain why and under what conditions it is rational for a farmer to produce a product at a loss.

9 Select a typical farm or ranch in your local area. Assume it doubles in size to where it is producing twice as much of each product as before.

    a If total cost also doubles, is there increasing, decreasing, or constant returns to size? What if total cost increased by only 90 percent?

    b Which individual costs would you expect to exactly double? Which might increase by more than 100 percent? By less than 100 percent?

    c Would you expect this farm or ranch to have increasing, decreasing, or constant returns to size? Economies or diseconomies of size? Why?

REFERENCES

Boehlje, Michael D., and Vernon R. Eidman: *Farm Management,* John Wiley & Sons, New York, 1984, chap. 3.

Castle, Emery N., Manning H. Becker, and A. Gene Nelson: *Farm Business Management,* 3d ed., Macmillan Publishing Co., New York, 1987, chaps. 2, 10.

Osburn, Donald D., and Kenneth C. Schneeberger: *Modern Agricultural Management,* 2d ed., Reston Publishing Co., Reston, VA, 1983, chaps. 3, 4.

8

ENTERPRISE BUDGETING

CHAPTER OBJECTIVES

    1 To define an enterprise budget and discuss its purpose and use
    2 To illustrate the different sections of an enterprise budget
    3 To learn how to construct a crop enterprise budget
    4 To outline additional problems and steps to consider when constructing a livestock enterprise budget
    5 To show how data from an enterprise budget can be analyzed and used for computing cost of production and breakeven prices and yields

The previous three chapters have discussed some basic economic and cost concepts. One practical application of these concepts is their use in various types of budgets. Enterprise, partial, whole farm, and cash flow budgets are important and useful tools for the farm and ranch manager, and each provides a framework for the use of these concepts. These budgeting techniques are the subject of the next four chapters.

Budgeting is often described as "testing it out on paper" before committing resources to a plan or to a change in an existing plan. It is a way to estimate the profitability of a plan, a proposed change in a plan, or an enterprise before making the decision and implementing it. Paper, pencils, calculators, and computers are the tools of budgeting and, as such, may be the most important tools a manager uses.

The combination of budgeting and economic principles provides some powerful, practical, and useful techniques for the manager to use when "analyzing alternatives." This step is an important one in the decision-making process as only a proper and correct analysis will lead to making the right decision. The discussion of budgeting will begin with enterprise budgets as they often provide much of the data needed for developing the other types of budgets.

ENTERPRISE BUDGETS

An enterprise budget is an organization of revenue, expenses, and profit for a single enterprise. Each type of crop or livestock that can be grown is an enterprise. Therefore, there can be enterprise budgets for cotton, corn, wheat, beef cows, dairy cows, farrow-finish hogs, watermelons, soybeans, peanuts, and so forth. They can be created for different levels of production or types of technology so there can be more than one budget for a given enterprise. The base unit for enterprise budgets is typically one acre for crops and one head for livestock. Using these common units permits an easy and fair comparison across different enterprises.

The primary purpose of enterprise budgets is to estimate costs, returns, and profit per acre or per head for the enterprises. Once this is done, the budgets have many uses. They identify the more profitable enterprises to be included in the whole farm plan. Since a whole farm plan often consists of several enterprises, enterprise budgets are often called the "building blocks" in a whole farm plan. (See Chapter 10 for more on whole farm planning.) Although constructing an enterprise budget requires a large amount of data, once completed, it is a source of data for other types of budgeting. A manager will often refer to enterprise budgets for information and data when making many types of decisions. An illustration of other uses for enterprise budgets will be postponed until after discussing their organization and construction.

Table 8-1 is an example that will be used to discuss the content, organization, and structure of a crop enterprise budget. Although there is no single organization or structure used by everyone, most budgets will contain the sections or parts included in Table 8-1. This example does not include all the detail on physical quantities and prices usually found on an enterprise budget, but it will serve to illustrate the basic organization and content.

The name of the enterprise being budgeted and the budgeting unit are shown first. For some long-term enterprises it is useful to state the time period being covered as well. Income or revenue from the enterprise is typically shown next. Quantity, unit, and price should all be included to provide full information to the user. The cost section comes next and is generally divided into two parts: variable or operating costs and fixed or ownership costs. Some budgets further divide variable costs into preharvest variable costs, or those that occur prior to harvest, and harvest costs, or those that are a direct result of harvesting. Income or revenue above variable costs is an intermediate calculation and shows the revenue remaining to be applied to fixed costs.

Fixed costs on a crop enterprise budget include the fixed costs for the machinery used in producing the crop and a charge for land use. These costs are probably the most difficult to estimate and discussion of the procedures to be used will be postponed until the next section.

The estimated profit per unit is the final value and is found by subtracting total costs from total revenue. Everyone is interested in this value, and it is important that it be interpreted correctly. Most enterprise budgets such as the one in Table 8-1 are eco-

CHAPTER 8: ENTERPRISE BUDGETING **141**

TABLE 8-1 EXAMPLE ENTERPRISE BUDGET FOR CORN PRODUCTION (1 ACRE)

Item	Value per acre	
Revenue:		
125 bushels @ $2.50 per bu		$312.50
Variable Costs		
Seed	$22.00	
Fertilizer	38.00	
Chemicals	20.00	
Machinery expenses	23.75	
Drying	12.50	
Hauling	7.50	
Labor	18.00	
Interest @ 10% for 6 months	7.09	
Total variable cost		$148.84
Income above variable cost		$163.66
Fixed Costs		
Machinery depreciation, interest, taxes, and insurance	$ 48.00	
Land charge	100.00	
Total fixed costs		$148.00
Total costs		$296.84
Estimated profit (return to management)		$ 15.66

nomic budgets. This means that, in addition to cash expenses and depreciation, opportunity costs are also included. Typically there would be opportunity costs for operator labor, capital used for variable costs, capital invested in machinery, and possibly for that invested in land. Therefore, the profit or return shown on an enterprise budget is an estimated *economic* profit. This is different than the concept of profit in accounting where opportunity costs are not recognized.

The one opportunity cost often omitted from an enterprise budget is the one for management. While it would certainly be appropriate to include it, this is a difficult opportunity cost to estimate, which explains its frequent omission. If there is no opportunity cost for management shown on the budget, the estimated profit should be interpreted as "estimated return to management and profit." This terminology is used on some enterprise budgets.

On enterprise and other types of budgets a slightly different terminology is often used to describe the different types of costs. Economic variable costs may be called operating or direct costs, emphasizing they arise from the actual operation of the enterprise. These costs would not exist except for the production from this enterprise.

Fixed costs may be called ownership or indirect costs. The term ownership costs refers to the fixed costs arising from *owning* machinery, buildings, or land. They result from owning the asset and would exist even if they were not used for this enterprise. The usual economic fixed costs and other farm expenses such as property and liability insurance, legal and accounting fees, pickup expenses, and subscriptions to agricultural publications are necessary and appropriate expenses but exist not because of any single enterprise. It is difficult to directly assign these expenses to any particular enterprise(s). They are often prorated on some basis to all enterprises to make sure all farm expenses, direct and indirect, are charged to the farm enterprises.

CONSTRUCTING A CROP ENTERPRISE BUDGET

The first step in constructing a crop enterprise budget is to determine tillage and agronomic practices, input levels, type of inputs, and so forth. What seeding rate? Fertilizer levels? Type and quantity of herbicide? Number and type of tillage operations? These and similar questions must be answered before beginning work on the enterprise budget.

Table 8-2 is an enterprise budget for wheat that will be used to discuss the steps in constructing a crop enterprise budget. This budget assumes an owner/operator who is receiving all of the crop and paying all of the expenses.

Revenue

The revenue section should include all cash and noncash revenue from the crop. Some crops have two sources of revenue such as cotton lint and cottonseed, oat grain and oat straw, and wheat grain and winter grazing of the growing wheat. All cash revenue

TABLE 8-2   ENTERPRISE BUDGET FOR WHEAT (ONE ACRE)

Item	Unit	Quantity	Price	Amount
Revenue:				
Wheat grain	bu	48	$3.00	$144.00
Total revenue				$144.00
Operating expenses:				
Seed	lb	80	$0.10	$ 8.00
Fertilizer: N	lb	60	0.20	12.00
P$_2$O$_5$	lb	30	0.22	6.60
K$_2$O	lb	30	0.15	4.50
Chemicals	acre	1	5.50	5.50
Fuel, oil, lub.	acre	1	7.60	7.60
Machinery repairs	acre	1	12.40	12.40
Labor	hr	2.1	8.00	16.80
Interest (operating expenses for 6 months)	$	36.70	10%	3.67
Total operating expenses				$ 77.07
Return above operating expenses				$ 66.93
Ownership expenses:				
Mach. depreciation	acre	1	14.20	14.20
Mach. interest	acre	1	10.60	10.60
Mach. taxes & insur.	acre	1	2.00	2.00
Land charge	acre	1	50.00	50.00
Misc. overhead	acre	1	5.00	5.00
Total ownership expenses				$ 81.80
Total expenses				$158.87
Profit (return to management)				$ (14.87)

should be included and the concept of opportunity cost used to value any noncash sources of revenue.

The accuracy of the projected profit for the enterprise may depend more on the estimates made in this section than in any other. It is important that both yield and price estimates be as accurate as possible. Projected yield should be based on historical yields, yield trends, and the type and amount of inputs to be used. The appropriate selling price will depend somewhat on whether the budget is for only the next year or for long-run planning purposes. In either case a review of historical price levels, price trends, and outlook for the future should be conducted before selecting a price.

Operating or Variable Costs

This section includes those costs that will be incurred only if this crop is produced. The amount to be spent in each case is under the control of the decision maker and can be reduced to $0 by not producing this crop.

Seed, Fertilizer, and Chemicals Costs for these items are relatively easy to determine once the levels of these inputs have been selected. Prices can be found by contacting input suppliers, and the total per acre cost for these items is found by simply multiplying quantity times price.

Fuel, Oil, and Lubrication Fuel, oil, and lubrication expense is related to the type and size of machinery used and to the number and type of machinery operations performed for this crop. A quick and simple way to obtain this value is to divide the total farm expense for fuel, oil, and lubrication by the number of crop acres. However, this method is not accurate if some machinery is also used for livestock production and if some crops take more machinery time than others.

A more accurate method is to determine fuel consumption per acre for each machine operation and then simply sum the fuel usage for all the operations scheduled for this crop. The result can be multiplied by the price of fuel to find the per acre cost. Another method is to compute fuel consumption per hour of tractor use and then determine how many hours will be needed to perform the machine operations. This method is used by many of the computer programs written to calculate enterprise budgets.

Machinery Repairs Estimating machinery repairs per acre has some of the same problems as estimating fuel use. A method must be devised that allocates repair expense relative to the type of machinery used and the amount of use. Any of the methods discussed for estimating fuel expense can also be used to estimate machinery repair expense. Chapter 20 contains more detailed methods for estimating repair costs for all types of machinery.

Labor Some enterprise budgets go into enough detail to divide labor requirements into that provided by the farm operator and that provided by hired labor. However, most use one estimate for labor with no indication of the source. Total labor hours needed for crop production are heavily influenced by the size of the machinery used and the number of machine operations. In addition to the labor needed to operate machinery in the field, care should be taken to include time needed to get to and from fields, adjust and repair machinery, and perform any other tasks related directly to the crop being budgeted.

The opportunity cost of farm operator labor is generally used to value labor. If some hired labor is used, this value should certainly be at least equal to the cost of hired labor including any fringe benefits. When estimating the opportunity cost of labor, it is important to include only the cost of labor, not management. A management charge should be shown as a separate item or included in the estimated net return.

Interest This interest is on the capital tied up in operating expenses. Since it is generally less than a year from the time of the expenditure until harvest when income is or can be received, interest is charged for some time less than a year. This time should be the average length of time from when the operating costs are incurred until harvest. Interest is charged on operating expenses without regard to how much is borrowed or even if any is borrowed. Even if no capital is borrowed, there is an opportunity cost on the farm operator's capital. If the amount of borrowed capital and equity capital that will be used is known, a weighted average of the interest rate on borrowed money and the opportunity cost of equity capital can be used.

Return above Operating Expenses This value shows how much an acre of this enterprise will contribute to payment of fixed or ownership expenses. It also shows how much revenue could decrease, and this enterprise would still cover its variable or operating expenses.

Machinery Depreciation The amount of machinery depreciation to charge to a crop enterprise will depend on the size and type of machinery used and the number and type of machine operations. As with the other machinery expenses discussed, the problem is one of properly allocating the total machinery depreciation to a specific enterprise. The first step is to find the average annual depreciation on each machine. This can be quickly and easily determined using the straight-line depreciation method. Annual depreciation on each machine can then be converted to a per acre or per hour value based on acres or hours used per year. Next, it can be prorated to a specific crop enterprise based on the amount of use.

Machinery Interest Interest on machinery is based on the average investment in the machine over its life and is computed the same way regardless of how much, if any, money was borrowed to purchase it. The equation used in Chapter 7 to compute the interest component of total fixed costs should be used to get the average annual interest charge. Then the annual interest should be prorated to this enterprise using the same method as was used for depreciation.

Machinery Taxes and Insurance Machinery is subject to personal property taxes in some states, and most farmers carry some type of insurance on their machinery. The annual expense for these items should be computed and then allocated again using the same method as for the other machinery ownership costs.

Land Charge There are several ways to calculate a land charge: (1) What it would cost to cash rent similar land, (2) the net cost of a share rent lease for this crop on similar land and, (3) for owned land, the opportunity cost of the capital invested, that is, the value of an acre multiplied by the opportunity cost of the owner's capital. The three methods can give widely different values. This is particularly true for the third method if it is used during periods of rapidly increasing land values. During inflationary periods, land values reflect the appreciation potential of the land as well as its value for crop production.

Most enterprise budgets use one of the rental charges even if the land is owned. Assuming a short-run enterprise budget, the land owner/operator could not sell the acre and invest the resulting capital. As long as the land is owned, if it is not farmed by the owner, the alternative is to rent it to another farm operator. The rental amount then becomes the short-run opportunity cost for the land charge.

Miscellaneous Overhead Many enterprise budgets will contain an entry for miscellaneous overhead expense. This can be used to cover many expenses such as a share of pickup expenses, farm liability insurance, farm shop expenses, and other farm expenses that cannot be directly associated with a single enterprise but are necessary and important farm expenses.

Profit or Return to Management The estimated profit is found by subtracting total expenses from total revenue. If a charge for management has not been included in the budget, this value should be considered the return to management. Management is an economic cost and should be recognized on an economic budget either as a specific expense or as part of the residual net return or loss.

Other Considerations on Crop Enterprise Budgets

The wheat example used to describe the process of constructing a crop enterprise budget is a fairly simple example. Other crops may have many more variable inputs or revenue and expenses specific to that crop. Special problems that may be encountered when constructing crop enterprise budgets include:

1 Double cropping where two crops are grown on the same land in the same year. In this case budgets should be developed for each crop with the annual ownership costs for land divided equally between the two crops.

2 Storage, transportation, and marketing expenses may be important for some crops. Most enterprise budgets assume sale at harvest and do not include storage costs. They are assumed to be part of a marketing decision and not a production decision. However, there may still be transportation and marketing expenses even with a harvest sale. If storage costs are included in the budget, then the selling price should be that expected at the end of the storage period and not the harvest price.

3 Establishment costs for perennial crops, orchards, and vineyards present another problem. These and other crops may take a year or more to begin production, but their enterprise budget is typically for a year of full production. It is often useful to develop separate budgets for the establishment phase and for the production phase. The latter budget must include a cost representing the establishment costs and any other costs incurred prior to receiving any income. This is done by accumulating costs for all years prior to the onset of production, determining the present value of these costs and then amortizing this present value over the life of the crop. Chapter 15 discusses the principles and methods for determining present value and amortization.

4 The methods for computing machinery depreciation and interest discussed earlier are easy to use and widely used. However, they do not result in the exact amounts needed to cover depreciation and provide interest on the machine's remaining value each year. The capital recovery method, while more complex, does provide the correct amount. Capital recovery combines depreciation and interest into one value.

The general equation for the annual capital recovery amount is

$$(\text{TD} \times \text{amortization factor}) + (\text{salvage value} \times \text{interest rate})$$

where TD is the total depreciation over the life of the machine (i.e., the amount of capital that must be recovered), and the amortization factor is the value from Table 1 in the Appendix corresponding to the interest rate used and the life of the machine. (See the discussion of amortization in Chapter 15 for more information.) The capital recovery amount obtained from this equation is the amount of money required at the end of each year to pay interest on the remaining value of the machine *and* recover the capital lost through depreciation. Since the salvage value will be recovered at the end of the machine's useful life, only interest is charged on this amount.

CONSTRUCTING A LIVESTOCK ENTERPRISE BUDGET

Livestock enterprise budgets include many of the same entries and problems of a crop enterprise budget. However, livestock budgets can have some unique and particular problems and only these will be discussed in this section. The cow/calf budget in Table 8-3 will be used as the basis for the discussion.

Unit   The budgeting unit for livestock is usually one head, but units such as one litter for swine, one cow unit, or 100 birds for poultry can also be used.

Time Period   Although many livestock enterprises are budgeted for one year, some feeding and finishing enterprises require less than a year. Some types of breeding livestock, such as swine, produce offspring more than once a year. Regardless of the time period chosen, it is important that all costs and revenue in the budget be calculated for the same time period.

Multiple Products   Many livestock enterprises will have more than one product producing revenue. For example, dairy would have revenue from cull cows, calves,

TABLE 8-3 EXAMPLE OF A COW/CALF ENTERPRISE BUDGET FOR ONE COW UNIT*
(1 cow unit = 1 cow, 0.04 bull, 0.9 calf, 0.12 replacement heifer)

Item	Unit	Quantity	Price	Amount
Revenue:				
Cull cow (0.10 head)	cwt	10.0	$50.00	$ 50.00
Heifer calves (0.33 head)	cwt	5.20	80.00	137.28
Steer calves (0.45 head)	cwt	5.50	88.00	217.80
Total revenue				$405.08
Operating expenses:				
Hay	ton	1.50	60.00	90.00
Grain & supplement	cwt	8.00	7.00	56.00
Salt, minerals	cwt	0.40	6.00	2.40
Pasture maintenance	acre	3.00	7.50	22.50
Vet. & health expense	head	1.0	10.00	10.00
Livestock facilities	head	1.0	8.00	8.00
Machinery & equipment	head	1.0	5.00	5.00
Breeding expenses	head	1.0	5.00	5.00
Labor	head	5.0	6.00	30.00
Miscellaneous	head	1.0	10.00	10.00
Interest (on half of	$	119.45	10%	11.95
operating expenses)				
Total operating expenses				$250.85
Ownership expenses:				
Interest on breeding herd	$	750.00	10%	75.00
Livestock facilities				
Deprec. & interest	head	1.0	10.00	10.00
Mach. & equipment				
Deprec. & interest	head	1.0	6.50	6.50
Land charge	acre	3.0	35.00	105.00
Total ownership expenses				$196.50
Total expenses				$447.35
Profit (return to management)				($42.27)

*Budget assumes raised replacements and a 90% calf crop.

and milk while a sheep flock would have revenue from cull breeding stock, lambs, and wool. All sources of revenue must be identified and then prorated correctly to an average individual animal in the enterprise.

The cow/calf budget in Table 8-3 shows average revenue from cull cows, heifer calves, and steer calves *per cow unit* in the producing herd. A cow unit includes the cow and a portion of the calves, bulls, and replacement heifers. There is an implied 10 percent replacement rate each year based on the 0.10 cull cow sold per producing cow. Less than one calf is sold per cow unit for two reasons: (1) The calving percentage is less than 100 percent and (2) some heifer calves must be retained for replacing cull cows.

Breeding Herd Replacement An important consideration in enterprise budgets for breeding herds is properly accounting for replacements for the producing animals.

The example in Table 8-3 assumes replacements are raised rather than purchased. With a 90 percent calving rate, there are 0.45 steer calves available for sale per cow, and there should also be 0.45 heifer calves. However, at least 0.10 heifer calves per cow must be retained as replacements. This example shows 0.12 (0.45 available *less* 0.33 actually sold) are actually retained. The additional heifer calves retained would cover death loss among the replacements and the producing herd. Whenever replacement animals are raised, the number of female offspring sold must be adjusted to reflect the percentage retained for replacing animals culled from the breeding herd.

If replacements are purchased, the enterprise budget would show revenue from all female offspring. Including the net cost of the replacements can be done one of two ways. First, a pro rata annual depreciation charge can be included to account for the decline in value from purchase price to market value at time of culling. No revenue from cull animals nor expense for purchasing replacements would be shown. Second, a pro rata share of the purchase cost of replacements can be shown as an expense and a pro rata share of revenue from culled animals can be shown as revenue. The *difference* between these two values should be the same value as the annual depreciation in the first method.

Feed and Pasture Many livestock enterprises will consume both purchased feed and farm-raised feed. Purchased feed is easily valued at cost. Farm-raised feed should be valued at its opportunity cost or what it would sell for if marketed off the farm. Care should be taken to include expenses for salt, minerals, any annual charges for establishment costs, fertilizer, spraying, or mowing of pastures and for the feed needed to maintain the replacement herd if replacements are being raised. Pasture costs would also include a charge for the land used based either on the cost of renting the pasture or, if owned, its opportunity cost.

Livestock Facilities Livestock facilities include buildings, fences, pens, working chutes, feeders, waterers, wells, windmills, and other specialized items used for livestock production. Operating expenses for these items include repairs and any fuel or electricity required to operate them.

These items also incur fixed costs and present some of the same computational and allocation problems outlined when discussing crop machinery. Annual depreciation, interest, taxes, and insurance should be computed for each livestock facility. For specialized items used only by one livestock enterprise, the total annual fixed costs can simply be divided by the number of head to get the per head charge. When more than one livestock enterprise uses the item, some method must be used to allocate the fixed costs among the enterprises.

Machinery and Equipment Tractors, pickups, and other machinery and equipment may be used for both crop and livestock production. Both operating and ownership expenses should be divided between crop and livestock enterprises according to the proportional use of the item.

COMPUTERIZED ENTERPRISE BUDGETS

Because of the many computations and details involved in constructing enterprise budgets, computers are often used. Once the computer program is written to do the repetitive calculations and organize the data, it then becomes relatively easy to produce a large number of enterprise budgets. Computerized budgets are often available through state land grant universities, local or state extension offices, and other sources. Different states and organizations often use slightly different formats for organizing and presenting enterprise budget data. Tables 8-4, 8-5, and 8-6 are examples of computerized enterprise budgets with different formats. A close examination will show Tables 8-5 and 8-6 are for the exact same enterprise, only the budget format and the amount of detail information provided are different.

It is often convenient to use enterprise budgets that have been prepared by someone else. However, no budget prepared by a third party is likely to fit any individual farm situation exactly. Most prepared budgets use typical or average values for a given geographical area, and there are differences in yields, input levels, and other management practices from farm to farm. It is also important to know the assumptions and equations used in the computer program as they may not fit all situations. Prepared budgets should only be considered a guide that will need adjustments to fit any individual farm or ranch. Some prepared budgets, such as the one in Table 8-4, include a column for recording individual adjustments.

GENERAL COMMENTS ON ENTERPRISE BUDGETS

Several factors need to be considered when constructing and using enterprise budgets. The economic principles of marginal value product (MVP) equals marginal input cost (MIC) and least-cost input combinations should be considered when selecting input levels for a budget. However, it should *not* be assumed that all prepared budgets were constructed using these principles and, even if they were, these levels may not be correct for a specific, individual situation. The typical or average input levels used in many prepared budgets may not be the profit-maximizing levels for any individual farm.

Because it is possible to use a large number of different input levels and input combinations even on a single farm, there is not a single budget for each enterprise. There are as many potential budgets as there are possible input levels and combinations. This again emphasizes the importance of selecting the profit-maximizing input levels for the individual situation or, if capital is limited, the input levels that will satisfy the equal marginal principle.

The fixed-cost estimates for enterprise budgets are usually based on an assumed farm size or level of use. Different estimates may be needed if fixed costs are spread over significantly more or less output units than the level assumed in the budget.

Enterprise budgets require a large amount of data. Past farm or ranch records are an excellent source if available in sufficient detail and for the enterprise being budgeted. Many states publish an annual summary of the average income and expenses for farms participating in a statewide record-keeping service. These summaries may contain suf-

TABLE 8-4   EXAMPLE OF A COMPUTER-GENERATED CROP ENTERPRISE BUDGET
(Does not include a land charge)

Summary of estimated costs and returns per acre,
Soybeans, clay soil, 6-row equipment, (38 inch rows),
owner-operators, Ouachita Area, Louisiana, 1992.

ITEM	UNIT	PRICE	QUANTITY	AMOUNT	YOUR FARM
		DOLLARS		DOLLARS	
INCOME					
Soybean	bu	5.50	25.000	137.50	_____
TOTAL INCOME				137.50	_____
DIRECT EXPENSES					
CUSTOM					
Airplane lo-vol	acre	2.00	0.500	1.00	
HERBICIDES					
2,4-DB	pt	3.00	0.500	1.50	_____
Scepter	pt	23.50	0.333	7.83	_____
Treflan 4L	pt	3.63	2.000	7.26	_____
HIRED LABOR					
Other labor	hour	5.50	0.400	2.20	_____
INSECTICIDES					
Ambush 2EC	pt	12.38	0.200	2.48	_____
SEED					
Soybean seed	lb	0.27	45.000	12.15	_____
OPERATOR LABOR					
Tractors	hour	5.50	1.001	5.51	_____
Self-propelled Eq.	hour	5.50	0.555	3.05	_____
DIESEL FUEL					
Tractors	gal	0.80	7.101	5.68	_____
Self-propelled Eq.	gal	0.80	1.775	1.42	_____
GASOLINE					
Self-propelled Eq.	gal	1.11	1.400	1.55	_____
REPAIR & MAINTENANCE					
Tractors	acre	4.75	1.000	4.75	_____
Self-propelled Eq.	acre	11.03	1.000	11.03	_____
Implements	acre	3.71	1.000	3.71	_____
INTEREST ON OP. CAP.	acre	2.38	1.000	2.38	_____
TOTAL DIRECT EXPENSES				73.50	_____
RETURNS ABOVE DIRECT EXPENSES				64.00	_____
FIXED EXPENSES					
Tractors	acre	8.90	1.000	8.90	_____
Self-propelled Eq.	acre	19.80	1.000	19.80	_____
Implements	acre	6.26	1.000	6.26	_____
TOTAL FIXED EXPENSES				34.96	_____
TOTAL SPECIFIED EXPENSES				108.46	_____
RETURNS ABOVE SPECIFIED EXPENSES				29.04	_____
RESIDUALS					
Overhead (owner)				55.71	_____
RESIDUAL RETURNS				-26.67	_____

Source: Projected Costs and Returns and Cash Flows for Major Agricultural Enterprises Louisiana, 1992, Department of Agricultural Economics and Agribusiness, Louisiana State University, A. E. A. Information Series Nos. 99-105, January 1992.

TABLE 8-5 EXAMPLE OF A COMPUTER-GENERATED LIVESTOCK ENTERPRISE BUDGET

```
                          STOCKER CALF PRODUCTION
                    Central Texas District (8), Eastern*
                  1992 Projected Costs and Returns per Head
================================================================      Your
PRODUCTION Description          Quantity    Unit   $ / Unit   Return  Estimate
  FEEDER STEERS           0.98Hd   6.500    cwt.    84.0000   535.08  _____
                                                             ===========
Total GROSS Income                                            535.08  _____
================================================================
OPERATING INPUT or CUSTOM OPERATION
          Description          Input Use   Unit   $ / Unit    Cost
  GRAIN SUPPL.    STOCKER         1.000    cwt.     7.500      7.50   _____
  HAY             STOCKER         0.840    cwt.     3.000      2.52   _____
  PASTURE         NATIVE          0.100    acre     1.400      0.14   _____
  SALES COMMISSIONSTOCKER         0.980    head    14.250     13.97   _____
  SALT & MINERALS STOCKER         0.200    cwt.    15.300      3.06   _____
  SMALL GRAINS    PASTURE         1.000    acre    60.000     60.00   _____
  STOCKER STEERS                  3.750    cwt.   111.000    416.25   _____
  VET. MEDICINE   STOCKER         1.000    head     5.000      5.00   _____
  Fuel                                                         1.39   _____
  Lube                                                         0.14   _____
  Repair                                                       0.34   _____
                                                             ===========
Total OPERATING INPUT and CUSTOM OPERATION Costs             510.30  _____
================================================================
Residual returns to capital, ownership
  labor, land, management, and profit                         24.78  _____
================================================================
CAPITAL INVESTMENT Description  Quantity    Unit   Rate of     Cost
                                Invested           Return
  Interest - IT Borrowed         57.771    Dol.    0.113       6.53   _____
  Interest - OC Borrowed        317.793    Dol.    0.100      31.78   _____
                                                             ===========
Total CAPITAL INVESTMENT Costs                                38.31  _____
================================================================
Residual returns to ownership, labor,
  land, management, and profit                               -13.53  _____
================================================================
OWNERSHIP COST Description (Depreciation, Taxes, and Insurance)  Cost
  Machinery and Equipment                                      8.00  _____
                                                             ===========
Total OWNERSHIP Costs                                          8.00  _____
================================================================
Residual returns to labor, land, management, and profit      -21.53  _____
================================================================
LABOR COST Description         Input Use   Unit   Average     Cost
                                                  Rate
  Machinery and Equipment        0.667    Hr.     5.500       3.67   _____
  Other                          1.300    Hr.     5.500       7.15   _____
                                                             ===========
Total LABOR Costs                                             10.82  _____
================================================================
Residual returns to land, management, and profit            -32.35  _____
================================================================
LAND COST Description          Input Use   Unit   Rate of     Cost
                                                  Return
  PASTURE RENT    NATIVE
    Annual Lease                 0.100    Acre     8.000       0.80   _____
  SMALL GRAINS    MACH. FC
    Annual Lease                 1.000    Acre    35.000      35.00   _____
                                                             ===========
Total LAND Costs                                              35.80  _____
================================================================
Residual returns to management and profit                   -68.15  _____
================================================================

  -WARNING- No Management Cost Specified

================================================================
Residual returns to profit                                  -68.15  _____
================================================================
Total Projected Cost of Production                          603.23  _____

*  Intended for eastern portion of Central Texas Extension district.
   50 steer unit, 300 pounds gain/stocker, stocking rate: 1.0 acres/head,
   small grain winter pasture, December-May, 2% death loss.
```

Source: Lovell, Ashley C.: *Texas Livestock Enterprise Budgets Central Texas District,* Texas Agricultural Extension Service, B-1241(L08).

TABLE 8-6 EXAMPLE OF A COMPUTER-GENERATED LIVESTOCK ENTERPRISE BUDGET

```
                      Stocker Calf Production
                 Central Texas District (8), Eastern*
                 1992 Projected Costs and Returns per Head
                                                                      Your
GROSS INCOME Description        Quantity   Unit    $ / Unit    Total   Estimate
============================    ========   ====   ==========  =======  ========
   FEEDER STEERS             0.98Hd   6.500   cwt.   84.0000   535.08  _____
                                                             ==========
Total GROSS Income                                            535.08  _____

VARIABLE COST Description                                      Total
================================                             ==========
   GRAIN SUPPL.    STOCKER                                       7.50  _____
   HAY            STOCKER                                        2.52  _____
   HAY RACKS                                                    0.04  _____
   Interest - OC Borrowed                                      31.78  _____
   LIVESTOCK LABOR                                              7.15  _____
   PASTURE        NATIVE                                        0.14  _____
   PICKUP TRUCK   3/4 TON                                       5.49  _____
   SALES COMMISSIONSTOCKER                                     13.97  _____
   SALT & MINERALS STOCKER                                      3.06  _____
   SMALL GRAINS    PASTURE                                     60.00  _____
   STOCKER STEERS                                             416.25  _____
   VET. MEDICINE   STOCKER                                      5.00  _____
                                                             ==========
Total VARIABLE COST                                           552.90

  Break-Even Price, Total Variable Cost  $   86.79 per cwt. of FEEDER STEERS

GROSS INCOME minus VARIABLE COST                              -17.82  _____

FIXED COST Description                    Unit               Total
================================          ====              ==========
   Machinery and Equipment               Acre                14.53  _____
   Land                                  Acre                35.80  _____
                                                             ==========
Total FIXED Cost                                             50.33  _____

  Break-Even Price, Total Cost $   94.69 per cwt. of FEEDER STEERS

Total of ALL Cost                                            603.23  _____

NET PROJECTED RETURNS                                        -68.15  _____
```

* Intended for eastern portion of Central Texas Extension district.
  50 steer unit, 300 pounds gain/stocker, stocking rate: 1.0 acres/head,
  small grain winter pasture, December-May, 2% death loss.

Source: Lovell, Ashley C.: *Texas Livestock Enterprise Budgets Central Texas District,* Texas Agricultural Extension Service, B-1241(L08).

ficient detail to be a useful source of data for enterprise budgeting. Research studies conducted by agricultural colleges, U.S. Department of Agriculture, and agribusiness firms are reported in bulletins, pamphlets, special reports, and farm magazines. This information often includes average yields and input requirements for individual enterprises.

The appropriate price and yield data may depend on the purpose of the budget. A budget to be used only for making adjustments in next year's crop and livestock plan should contain estimates of next year's price and expected yield. Estimates of long-run prices and yields should be used in budgets constructed to assist in developing a long-run plan for the business. The appropriate yield for a particular enterprise budget will also depend on the types of inputs included in the budget and the input levels. Higher input levels should be reflected in higher yields and vice versa.

INTERPRETING AND ANALYZING ENTERPRISE BUDGETS

Any economic enterprise budget must be interpreted correctly. An "economic" budget means opportunity costs on labor, capital, land, and perhaps management were included as expenses. The resulting profit (or loss) is the revenue remaining after covering all expenses including opportunity costs. This can be thought of as an economic profit, which will not be the same as accounting profit. The latter would not include any opportunity costs as operating expenses. A projected economic profit of zero is not as bad as it might seem. This result simply means all labor, capital, and land are just earning their opportunity costs—no more, no less. A positive projected profit means one or more of these factors is earning more than its opportunity cost.

The data in an enterprise budget can be used to perform several types of analyses. These include calculating cost of production and computing break-even prices and yields.

Cost of Production   Cost of production is a term used to describe the average cost of producing one unit of the commodity. It is the same as average total cost discussed in Chapter 7, provided the same costs and production level are used to compute each. The cost of production equation for crops is

$$\text{Cost of production} = \frac{\text{total cost}}{\text{yield}}$$

which is the same as for average total cost with "output" and "yield" being interchangeable terms. For the example in Table 8-4, the cost of production for soybeans is $108.46 divided by 25 bushels or $4.34 per bushel. Notice that cost of production will change if either costs or yield change.[1]

Cost of production is a useful concept particularly when marketing the product. Any time the product can be sold for more than its cost of production, a profit is being made. If opportunity costs are in the expenses, the profit is an economic profit. The resulting accounting profit will be even higher.

Break-even Analysis   The data contained in an enterprise budget can be used to perform a break-even analysis for prices and yields. The formula for computing the break-even yield is

---

[1]Note that the enterprise budget in Table 8-4 does not include a land charge. Therefore, the cost of production and break-even values computed in this section cover only the "Total Specified Expenses" shown in the budget and do not include any return to land.

$$\text{Break-even yield} = \frac{\text{total costs}}{\text{output price}}$$

This is the yield necessary to cover all costs at a given output price. For the example in Table 8-4 it would be $108.46 divided by $5.50 or 19.72 bushels per acre. Since the output price is only an estimate, it is often useful to compute the break-even yield for a range of possible prices as shown below:

Price per bushel ($)	Break-even yield (bu)
4.50	24.1
5.00	21.7
5.50	19.7
6.00	18.1
6.50	16.7

This often provides some insight into how sensitive the break-even yield is to changes in the output price. As shown in this table, the break-even yield is often very sensitive to changes in the output price.

The break-even price is the output price needed to just cover all costs at a given output level and can be found from the equation

$$\text{Break-even price} = \frac{\text{total costs}}{\text{expected yield}}$$

Again using the example in Table 8-4, the break-even price would be $108.46 divided by 25.0 bushels, or $4.34. Notice that the break-even price is the same as the cost of production. They are only two different ways of looking at the same value.

The break-even price can also be computed for a range of possible yields as in the following table. Different yields cause different break-even prices (and cost of production), and these prices can vary widely depending on the yield level.

Yield (bu)	Break-even price ($)
20.0	5.42
22.5	4.82
25.0	4.34
27.5	3.94
30.0	3.62

Since both the yield and output price in an enterprise budget are estimated rather than actual values, the calculation of break-even yields and prices can aid managerial decision making. By studying the various combinations of break-even yields and prices, managers can form their own expectations about the probability of obtaining a price and yield combination that would just cover total costs. Break-even prices and yields can also be calculated from total variable costs rather than total costs. These results can help managers make the decisions discussed in Chapter 7 concerning continuing or stopping production to minimize losses in the short run.

SUMMARY

Enterprise budgets are an organization of projected income and expenses for a single enterprise. They are usually for a single small unit of the enterprise such as one acre for a crop and one head for a livestock enterprise. Most enterprise budgets are economic budgets and, as such, include all variable or operating expenses, all fixed or ownership expenses, as well as opportunity costs on factors such as operator labor, capital, and management.

Enterprise budgets can be used to compare the profitability of alternative enterprises and are particularly useful when developing a whole farm plan. They can also be used to make minor year-to-year adjustments in the farm plan in response to short-run price and yield changes. Once completed, an enterprise budget contains the data needed to compute cost of production, the break-even price, and the break-even yield.

QUESTIONS FOR REVIEW AND FURTHER THOUGHT

1 Suggest how the "return above variable costs" value found on most enterprise budgets can be used to make some short-run production decisions.
2 Should the economic principles for determining profit-maximizing input levels be applied before or after completing an enterprise budget? Why?
3 An enterprise budget for soybeans shows a yield of 36 bushels, a selling price of $5.85 per bushel, and total costs of $220.00 per acre. What is the cost of production? The break-even yield? The break-even price?
4 Why should the opportunity costs of a farmer's labor, capital, and management be included on an enterprise budget?
5 If the land is owned, should a land charge be included on an enterprise budget? Why?
6 Would you expect two farms of widely different size to have the same costs on their enterprise budgets for the same enterprise? Might economies or diseconomies of size explain any differences?
7 "There are potentially a very large number of enterprise budgets for a single enterprise." Defend or refute that statement.
8 How might an agricultural loan officer use enterprise budgets? A property tax appraiser? A farmer when ordering input supplies for the coming year?

REFERENCES

Boehlje, Michael D., and Vernon R. Eidman: *Farm Management,* John Wiley & Sons, New York, 1984, chaps. 3, 4.

Castle, Emery N., Manning H. Becker, and A. Gene Nelson, *Farm Business Management,* 3d ed., Macmillan Publishing Company, New York, 1987, chaps. 15, 17.

Enterprise Budgets—A Guide for Development, Texas Agricultural Extension Service, Department of Agricultural Economics, Texas A&M University, 1990.

McSweeny, William T.: *Crop Budgeting: How to Use Crop Budgets for Greater Farm Profits,* Extension Circular 343, Cooperative Extension Service, The Pennsylvania State University, (undated).

Projected Costs and Returns and Cash Flows for Major Agricultural Enterprises Louisiana, 1992, A. E. A. Information Series Nos. 99–105, Department of Agricultural Economics and Agribusiness, Louisiana State University, 1992.

9

PARTIAL BUDGETING

CHAPTER OBJECTIVES

1 To discuss the purpose and use of a partial budget

2 To illustrate the format of a partial budget

3 To show what types of entries are made on a partial budget

4 To emphasize the importance of including only changes in revenue and expenses on a partial budget

5 To demonstrate the use of partial budgeting with several examples

Enterprise budgets are certainly useful, but they do have limitations. Analyzing a possible change involving several enterprises and any interactions between them may require a different budgeting technique. A partial budget will often be an appropriate way to analyze these types of problems using information obtained from enterprise budgets.

Many of the day-to-day management decisions made by farmers and ranchers are really adjustments to, or "fine-tuning" of, an existing farm plan. Even the best farm plan will need some occasional fine-tuning as changes occur and new information becomes available. These adjustment decisions often affect revenue and expenses. A convenient and practical method for analyzing the profit potential of these partial changes in the overall whole farm plan is the use of a partial budget.

Examples of decisions that can be analyzed with a partial budget are whether to participate in government farm programs, to own harvesting equipment or custom hire harvesting, and to plant more barley and less wheat. Most of these decisions could be evaluated by comparing two whole farm budgets, but much effort would be wasted collecting and organizing information that will not change and therefore does not affect the decision.

A partial budget provides a formal and consistent method for calculating the expected change in profit from a proposed change in the farm business. It compares the profitability of one alternative, typically what is now being done, with a proposed change or new alternative. Throughout the discussion of partial budgeting, the emphasis will be on *change*. The analysis is for a relatively small *change* in the farm plan, so only *changes* in revenue and expenses are included on a partial budget, and the result is the expected *change* in profit.

Because it is designed to analyze relatively small changes in the farm business, partial budgeting is really a form of marginal analysis. Figure 9-1 illustrates this point by showing how typical changes analyzed by partial budgeting relate to a production function, an isoquant and a production possibility curve. The production function in the first panel of Figure 9-1 shows, assuming the current input/output combination is point *A,* possible increases or decreases in that combination. Examples would be using more or less fertilizer, irrigation water, labor, or capital.

The second panel of Figure 9-1 shows possible movements up or down an isoquant or different combinations of two inputs to produce a given amount of output. Possible changes in input combinations can be easily analyzed with a partial budget. Substituting larger machinery for less labor would be an example. Another typical use of a partial budget is to analyze the change in profit from substituting more of one enterprise for another. This adjustment is shown in the third panel of Figure 9-1 by possible movements up or down the production possibility curve from the current combination at point *A*. A fourth general type of alternative adapted to partial budget analysis is expanding or contracting a single enterprise. This would be illustrated by moving to a higher or lower isoquant or a higher or lower production possibility curve.

PARTIAL BUDGETING PROCEDURE

Steps in the decision-making process discussed in Chapter 1 included identifying and defining the problem, gathering information, and identifying and analyzing alterna-

FIGURE 9-1   Partial budgeting and marginal analysis.

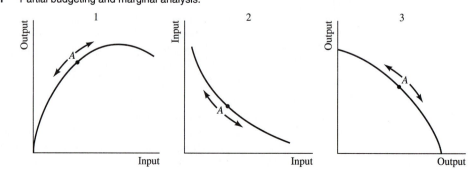

tives. Partial budgeting fits this process with one modification. It is capable of analyzing only two alternatives at a time, the current situation and a single, proposed alternative. Of course, several partial budgets can be used to evaluate a number of alternatives. Identifying the alternative to be analyzed before gathering any information reduces the amount of information needed. The only information required is that related to changes in costs and revenues if the proposed alternative is implemented. There is no need for information on any other alternative or for information about costs and returns that will not be affected by the proposed change.

The changes in costs and revenues needed for a partial budget can be identified by considering the following four questions. They should be answered on the basis of what would happen *if* the proposed alternative was implemented.

1 What new or additional costs will be incurred?
2 What current costs will be reduced or eliminated?
3 What new or additional revenue will be received?
4 What current revenue will be lost or reduced?

For many problems it will be easier to first identify all physical changes that would result from the alternative being analyzed and then put an economic value on each.

THE PARTIAL BUDGET FORMAT

The answers to the preceding questions are organized under one of the four categories shown on the partial budget form in Table 9-1. There are different partial budgeting forms but all will have these four categories arranged in some manner. It is important to remember that, for each category, only the *changes* are included, not any total costs or total revenues.

Additional Costs   These are costs that do not exist at the current time with the current plan. A proposed change may cause additional costs because of a new or expanded enterprise that requires the purchase of additional inputs. Other causes would be increasing the current level of input use or substituting more of one input for another. Additional costs may be either variable or fixed as there will be additional fixed costs whenever the proposed alternative requires additional capital investment.

Reduced Revenue   This is revenue currently being received but that will be lost should the alternative be adopted. Revenue may be reduced if an enterprise is eliminated or reduced in size, if the change causes a reduction in yields or production levels or if the selling price will decrease. Estimating reduced revenue requires careful attention to information about yields, livestock birth and growth rates, and output selling prices.

Additional Revenue   This is revenue to be received only if the alternative is adopted. It is not being received under the current plan. Additional revenue can be received if a new enterprise is added, if there is an increase in the size of a current enter-

TABLE 9-1 PARTIAL BUDGETING FORM

Partial Budget

Problem:

Additional Costs:	**Additional Revenue:**
Reduced Revenue:	**Reduced Costs:**
A. Total additional costs and reduced revenue $ _____	B. Total additional revenue and reduced costs $ _____
	$ _____

Net Change in Profit (B minus A) $ _____

prise, or if the change will cause yields, production levels, or selling price to increase. As with reduced revenue, accurate estimates of yields and prices are important.

Reduced Costs   Reduced costs are those now being incurred that would no longer exist under the alternative being considered. Cost reduction can be due to eliminating an enterprise, reducing the size of an enterprise, reducing input use, substituting more of one input for another, or being able to purchase inputs at a lower price. Reduced costs may be either fixed or variable. A reduction in fixed costs will occur if the proposed alternative will reduce or eliminate the current investment in machinery, equipment, breeding livestock, land, or buildings.

The categories on the left-hand side of the partial budget are the two that reduce profit—additional costs and reduced revenue. On the right-hand side of the budget are the two that increase profit—additional revenue and reduced costs. Entries on the two sides of the form are summed and then compared to find the net change in profit. Whenever opportunity costs are included on a partial budget, the result is the estimated change in "economic profit." This will not be the same as the change in "accounting profit."

PARTIAL BUDGETING EXAMPLES

Two examples will illustrate the procedure and possible uses of partial budgeting. Table 9-2 is a relatively simple budget analyzing the profitability of purchasing a combine to replace the current practice of hiring a custom combine operator to harvest 1,000 acres of wheat. All values in the budget are average annual values, and the result is net change in annual profit. Costs or revenue for wheat production are not included in the budget as these will be the same under either alternative.

Purchasing the combine will increase fixed costs as well as incur the additional variable costs associated with operating the combine. Depreciation and interest are computed from the equations used to compute fixed costs in Chapter 7. The values used were a cost of $100,000, a salvage value of $20,000, an 8-year useful life, and a 10 percent opportunity cost for capital. The additional fixed costs total $16,150. Additional variable costs such as repairs, fuel and oil, and labor to operate the combine were estimated to be $4,100. No opportunity cost on capital tied up in these additional variable costs was included because little time elapses between their occurrence and when income from the wheat can or will be received. There is no reduced revenue so the total profit reducing effects will be $20,250.

No additional revenue is shown in the example, although some might be expected if timeliness of harvesting would be improved or field losses reduced. These factors are difficult to estimate but should be included if a significant change is expected. The only reduced cost is the elimination of the payment for custom combining, which is $17,500. This example shows purchasing the combine instead of continuing to hire the custom operator would reduce annual economic profit by $2,750.

Once a partial budget is completed, further analysis is possible. For example, how much would the charge for custom combining have to increase to make purchasing the combine the better alternative? The total charge would have to increase by $2,750, or

TABLE 9-2 PARTIAL BUDGET FOR OWNING COMBINE VERSUS CUSTOM HIRING

Partial Budget

Problem: Purchase combine to replace custom hiring (1000 acres of wheat)

Additional Costs:		Additional Revenue:
Fixed costs		None
Depreciation	$10,000	
Interest	6,000	
Taxes	50	
Insurance	100	
Variable costs		
Repairs	2,400	
Fuel and oil	1,200	
Labor	500	

Reduced Revenue:		Reduced Costs:	
None		Custom combining charge	
		1000 acres @ $17.50	$17,500

A. Total additional costs
and reduced revenue $ 20,250

B. Total additional revenue
and reduced costs $ 17,500

$ 20,250

Net Change in Profit (B minus A) $ (2,750)

$2.75 per acre. Therefore, a custom rate of $17.50 + $2.75 = $20.25 per acre is the break-even value. Any charge above this value would make it profitable to purchase the combine.

Another question might be: What if the purchased combine was used to do custom work for others? Then the partial budget would need to show additional revenue from this source and additional variable costs that would increase directly with the number of acres custom harvested. The fixed costs would not change unless the additional use decreased useful life or salvage value. Because repair costs might increase more than proportionally, a variable cost of $5.00 per acre was assumed. With a custom charge of $17.50 per acre, this would leave a net profit of $12.50 for each acre of custom work done. Dividing this value into the $2,750 reduction in profit indicates that custom combining 220 acres would be the break-even amount. If more than 220 acres can be custom harvested, purchasing the combine becomes the more profitable alternative.

A second and more detailed example of a partial budget is shown in Table 9-3. The proposed change is the addition of 50 beef cows to an existing herd. However, not enough forage is available, and 100 acres currently in grain production must be converted to forage production.

This proposed change will cause additional fixed ownership costs. There will be additional interest on the increased investment in beef cows, depreciation on additional bulls, and additional property taxes on the new animals. Herd replacements are assumed to be raised, so there is no depreciation included on the cows. Variable or operating costs will also increase as shown, including labor, feed, and health costs for the additional livestock as well as fertilizer for the new 100 acres of pasture and opportunity cost interest on all additional variable costs. Income from the grain now being produced on the 100 acres will no longer be received, and this reduced revenue is estimated to be $15,000. The sum of additional costs and reduced revenue equals $23,220.

Additional revenue will be received from the sale of cull cows, steer calves, and heifer calves from the additional 50 cows. Several items are important in estimating this revenue in addition to carefully estimating prices and weights. First, it is unrealistic to assume every cow will wean a calf every year. This example assumes 46 calves from the 50 cows, or a 92 percent weaned calf crop. Second, herd replacements are raised rather than purchased, so 5 heifer calves must be retained each year to replace the 5 cull cows that are sold. This assumption results in the sale of only 18 heifer calves each year compared with 23 steer calves.

The reduced costs include all expenses from planting the 100 acres of grain that would no longer exist. Opportunity cost interest on these expenses are also included as a reduced cost. No reduction in machinery fixed costs is included, as the machinery complement is assumed to be no different after the proposed change than before. However, should the reduction in crop acres be large enough that some machinery could be sold or a smaller size used, reduced costs should include the change in related fixed costs.

The total additional revenue and reduced costs are $27,053, or $3,833 higher than the total additional costs and reduced revenue. This positive difference indicates the proposed change would increase profit.

TABLE 9-3 PARTIAL BUDGET FOR ADDING 50 BEEF COWS

Partial Budget

Problem: Add 50 beef cows and convert 100 acres to forage production

Additional Costs:		Additional Revenue:	
Fixed costs		5 cull cows	$2,500
Interest on cows/bulls	$2,500		
Bull depreciation	200	23 steer calves	
Taxes	100	500 lb @ 85¢	9,775
Variable cost		18 heifer calves	
Labor	600	460 lb @ 78¢	6,458
Vet and health	500		
Feed and hay	2,000		
Hauling	300		
Miscellaneous	200		
Pasture fertilizer	1,500		
Interest on variable costs	320		
Reduced Revenue:		**Reduced Costs:**	
Grain production		Fertilizer	$2,750
5000 bu @ $3.00	$15,000	Seed	1,400
		Chemicals	1,200
		Labor	1,500
		Machinery	1,000
		Interest on variable costs	470
A. Total additional costs		B. Total additional revenue	
and reduced revenue $ 23,220		and reduced costs $ 27,053	
		$ 23,220	
Net Change in Profit (B minus A)		$ 3,833	

FACTORS TO CONSIDER WHEN COMPUTING CHANGES IN REVENUE AND COSTS

In addition to the usual problem of acquiring relevant and accurate information, there are several other factors to consider. The first is nonproportional changes in costs and revenue. This problem will occur more often with costs but it is possible for revenues.

Assume the proposed change is a 20 percent increase (decrease) in the size of an enterprise. It would be easy to take the totals for each existing expense and revenue and assume each will be 20 percent higher (lower). This could be wrong for two reasons. Fixed costs would not change unless the 20 percent change caused an increase or decrease in capital investment. Many relatively small changes will not. Even variable costs may not change proportionally. For example, adding 20 cows to an existing beef or dairy herd of 100 cows will increase labor requirements but probably by something less than 20 percent. Economies and diseconomies of size must be considered when estimating cost and revenue changes.

Opportunity costs can be another easily overlooked item. They should be included on a partial budget to permit a fair comparison of the alternatives. This is particularly true if the difference in capital or labor requirements is large. Additional variable costs represent capital that could be invested elsewhere, so opportunity cost on them should be included as another additional cost. The reverse of this argument holds true for reduced variable costs, so an opportunity cost on these should be included as a reduced cost. Opportunity cost on any additional capital investment becomes part of additional fixed costs and likewise should be a part of reduced fixed costs if the capital investment will be reduced.

Opportunity cost on the farm operator's labor may also be needed on a partial budget. However, many things should be considered when estimating this opportunity cost. Is there really an opportunity cost for using additional labor? Free or leisure time would be given up, and there may be an opportunity cost on it. Alternatively, the farm operator may desire some minimum return before using any excess labor in a new alternative. The same question exists in reverse if the alternative will reduce labor requirements. Is there a productive use for an additional 50 or 100 hours of available labor or will it just be additional leisure time? What will it earn in the alternative use or what is the value of an additional hour of leisure time? Answers to these questions help determine the appropriate opportunity cost of operator labor on a partial budget.

FINAL CONSIDERATIONS

The partial budget in Table 9-3 can be used to illustrate two additional factors to be considered in the decision. Before adopting a proposed change that appears to increase profit, any additional risk and capital requirements should be carefully evaluated. If risk is measured in terms of annual variability in profit, is the profit from the additional 50 cows more variable than the profit from the 100 acres of grain? If the answer is "Yes," then the decision maker must evaluate the potential effects of additional risk on the financial stability of the business. Is the additional average profit worth the additional risk or variability of profit?

Adding 50 cows to the herd will require an additional capital investment. Is the capital for acquiring the cows available or can it be borrowed? If it is borrowed, how will this affect the financial structure of the business, risk, cash flow requirements, and repayment ability? Will this additional investment cause a capital shortage in some other part of the business? These questions need to be carefully evaluated before making the final decision to adopt the change. A profitable change may not be adopted if the increase in profit is relatively small, it increases risk, or an additional capital investment is required.[1]

SUMMARY

A partial budget is an extremely useful type of budgeting. It can be used to analyze many of the common, everyday problems and opportunities confronting the farm and ranch manager. Partial budgets are intended to analyze the profitability of proposed changes in the operation of the business where the change affects only part of the plan or organization of the farm. The current situation is compared to the expected situation after implementing a proposed change.

Data requirements are rather small as only *changes* in costs and revenues are included on a partial budget. The sum of additional costs and reduced revenue is subtracted from the sum of additional revenue and reduced costs to find the estimated change in profit. A positive result indicates the proposed change would increase profit. However, any additional risk and capital requirements should be considered before making the final decision.

QUESTIONS FOR REVIEW AND FURTHER THOUGHT

1 Will all partial budgets contain some fixed ownership costs? Why or why not?
2 List the types of changes that would appear in a partial budget for determining the profitability of participating in a government farm program. The program requires 10 percent of your cropland be left idle in exchange for a lump-sum payment.
3 Why are opportunity costs included in partial budgets?
4 Assume a proposed change would reduce labor requirements by 200 hours. If this was the farm operator's labor rather than hired hourly labor, would you include a reduced cost for labor? What factors would determine the value to use?
5 True or false and explain. Partial budgeting can be used to develop a whole farm plan.
6 Besides additional profit, what other factors should a farm operator take into account when evaluating a proposed change?

[1]Potential changes requiring additional capital investment can also be analyzed other ways. See Chapter 15 for capital budgeting and other more comprehensive investment analysis methods.

REFERENCES

Boehlje, Michael D., and Vernon R. Eidman: *Farm Management,* John Wiley & Sons, New York, 1984, chap. 6.

Castle, Emery N., Manning H. Becker, and A. Gene Nelson: *Farm Business Management,* 3d ed., Macmillan Publishing Company, New York, 1987, chap. 6.

Luening, Robert A.: "Partial Budgeting," *Journal of the American Society of Farm Managers and Rural Appraisers,* vol. 38, no. 1, pp. 33–38, April 1974.

Ott, Gene: *Easier Decisions with Partial Budgeting,* New Mexico Cooperative Extension Service Circular 452, 1973.

Retzlaff, Robert E. J.: *Budgeting a Change,* Nebraska Cooperative Extension Service Publication G 77-333, 1977.

10

WHOLE FARM PLANNING

CHAPTER OBJECTIVES

    1 To show how whole farm planning summarizes the resources available and the enterprises to be carried out on a farm or ranch

    2 To outline the need and procedure for taking a resource inventory

    3 To learn the steps and procedures to follow in developing a whole farm budget

    4 To understand the various uses for a whole farm budget

    5 To compare the assumptions used for short-run and long-run budgeting

    6 To introduce the use of simplified programming and linear programming as tools for choosing the most profitable combination of enterprises

Once a manager has defined the goals for a farm or ranch business, the next logical step is to develop a plan to attain these goals. Every manager has a plan of some kind—about what to produce, how to produce, and how much to produce—even if it has not been fully developed on paper. However, a systematic procedure for developing a whole farm plan may well result in one that will increase farm profit or come closer to attaining other goals.

DEFINITION AND DESCRIPTION

As the name implies, a *whole farm plan* is an outline or summary of the resources available and the type and volume of production to be carried out. It may have sufficient detail to include fertilizer, seed, and chemical application rates and actual feed rations for livestock, or it may simply list the enterprises to be carried out and their levels of production. When the expected costs and returns for the plan are organized into a detailed projection, the result is a *whole farm budget.*

The last two chapters discussed enterprise and partial budgeting. Enterprise budgets can be used as building blocks for the development of a whole farm plan and its associated budget. Partial budgets are particularly useful for making minor adjustments or fine-tuning a whole farm plan. However, a major change that affects several enterprises may require a new whole farm plan and budget to adequately compare the profitability and feasibility of the proposed change with the current plan.

A whole farm plan should be completed upon taking over the operation of a new farm business. A new plan may also be needed whenever a major change is planned for an existing farm business, such as adding more land or taking in a partner. The plan may be for the current or upcoming year, or it may reflect a typical or average year. In some instances a transitional plan may be needed for the period it takes to implement a major change in the operation. The effects on the financial position and risk exposure of the business may also need to be considered.

THE PLANNING PROCEDURE

The development of a whole farm plan can be divided into six steps, as shown in Figure 10-1: (1) formulate goals and objectives, (2) take inventory of the resources

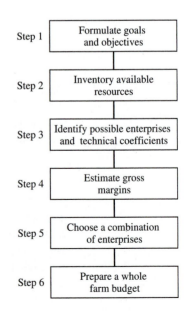

Step 1	Formulate goals and objectives
Step 2	Inventory available resources
Step 3	Identify possible enterprises and technical coefficients
Step 4	Estimate gross margins
Step 5	Choose a combination of enterprises
Step 6	Prepare a whole farm budget

FIGURE 10-1   Procedure for developing a whole farm plan.

available, (3) identify possible enterprises and their technical coefficients, (4) estimate gross margins for each enterprise, (5) choose a plan—the feasible enterprise combination that best meets the specified goals, and (6) develop a whole farm budget that projects the profit potential of the plan. Each step will be discussed and illustrated by an example.

Usually the plan is for one year, and resources such as land, full-time labor, machinery, and buildings are considered to be fixed in quantity. However, alternatives that assume an increase in the supply of one or more of these key resources can also be considered. Some techniques for analyzing the profitability of new investments are discussed in Chapter 15.

Formulate Goals and Objectives

Most planning techniques assume profit maximization is the primary goal, but this is often subject to a number of both personal and societal restrictions. Maintaining the long-term productivity of the land, protecting the environment, guarding the health of the operator and workers, maintaining financial independence, and allowing time for leisure activities are all goals that affect the total farm plan. Certain enterprises may be included in the plan simply because of the satisfaction the operator receives from them regardless of their economic results.

Inventory Resources

The second step in the development of a whole farm plan is to complete an accurate inventory of available resources. The type, quality, and quantity of resources available determine which enterprises can be considered in the whole farm plan and which are not feasible.

Land   The land resource is generally the most valuable resource and one of the most difficult to alter. Land is also a complex resource with many characteristics that influence the type and number of enterprises to be considered. The following are some of the important items to be included in the land inventory.

1 Total number of acres available in cropland, pasture, timber, and wasteland.

2 Climatic factors including annual rainfall and length of the growing season.

3 Soil types and such factors as slope, texture, and depth.

4 Soil fertility levels and needs. A soil testing program may be needed as part of the inventory.

5 The current irrigation system and water supply or the potential for developing an irrigation system.

6 Surface and subsoil drainage problems and possible corrective measures.

7 Soil conservation practices and structures including any current and future needs for improvement.

8 The current soil conservation plan, and any limitations it may place on land use or technology.

9 Crop bases, established yields, long-term contracts, or other characteristics related to government programs or regulations.

10 Existing and potential pest and weed problems that might affect enterprise selection and crop yields.

11 Tenure arrangements to determine if the land is owned or leased and, if leased, what are the lease terms.

This is also a good time to draw up a map of the farm showing field sizes, field layout, fences, drainage ways and ditches, tile lines, and other physical features. This map can assist in planning changes or documenting past practices. If available, the cropping history of each field including crops grown, yields, fertilizers applied, and chemicals used can be recorded on a copy of the field map. This information is very useful for developing a crop program where a crop rotation is desirable or herbicide carryover may be a problem.

Buildings   The building inventory should include a listing of all buildings along with their capacity and potential uses. Livestock enterprises and crop storage may be severely limited in both number and size by the buildings available. Feed-handling equipment, grain storage, water supply, and arrangement and size of livestock pens should all be noted in the inventory.

Information from the building inventory can be used to draw a map of the farmstead. This map can be used to plan new buildings and additions to livestock facilities. Any new construction should be done in such a way that it improves farmstead appearance, livestock disease control, manure runoff, and the efficiency of handling grain and livestock.

Labor   The labor resource should be analyzed for both quantity and quality. Quantity can be measured in months of labor currently available from the operator, family members, and hired labor, including its seasonal distribution. The availability and cost of additional full- or part-time labor should also be noted in the inventory, as the final farm plan might profitably use additional labor. Labor quality is more difficult to measure, but any special skills, training, and experience that would affect the possible success of certain enterprises should be noted.

Machinery   Machinery can be a short-run, fixed resource, and the number, size, and capacity of the available machinery should be included in the inventory. Special attention should be given to any specialized, single-purpose machines. The capacity limit of a specialized machine such as a cotton picker will often determine the maximum size of the enterprise where it is needed.

Capital   Capital for both short-run and long-run purposes can be another limiting resource. The lack of ready cash or limited access to operating credit can affect the enterprises that are chosen. Reluctance to tie up funds in fixed assets or to leverage the business through long-term borrowing may also prevent expansion of the farming operation or the adoption of labor-saving technology.

Management   The last part of a resource inventory is an assessment of the management skills available for the business. What is the age and experience of the manager? What is the past performance of the manager and his or her capacity for making management decisions? What special skills or critical weaknesses are present? If a manager has no training, experience, or interest in a certain enterprise, that enterprise is likely to be inefficient and unprofitable. The quality of the management resource should be reflected in the technical coefficients incorporated into the farm budgets. Past experience and records are the best indicators to use.

Other Resources   The availability of local markets, transportation, or specialized inputs may also be important resources to consider when developing the whole farm plan.

Identify Enterprises and Technical Coefficients

The resource inventory will show certain crop and livestock enterprises are feasible as the resources needed for their production are available. Those requiring a resource that is not available can be eliminated from consideration unless purchasing this resource is a feasible alternative. Custom and tradition should not be allowed to restrict the list of potential enterprises, only the resource limitations. Many farms have incorporated alternative or nontraditional enterprises into their farm plan as some can be quite profitable.

The budgeting unit for each enterprise should be defined, which would typically be 1 acre for crops and 1 head for livestock. Next, the resource requirements per unit of each enterprise, or the *technical coefficients,* must be estimated. For example, the technical coefficients for an acre of corn might be 1 acre of land, 4 hours of labor, 3.5 hours of tractor time, and $85 of operating capital, while one beef cow might require 4 acres of pasture, 3 hours of labor, and $120 of operating capital. These technical coefficients, or resource requirements, become very important in determining the maximum size of enterprises and the final enterprise combination. Enterprise budgets, which were discussed in Chapter 8, are very useful for obtaining and organizing this information.

Estimate Gross Margin per Unit

The next step is to estimate the income and variable costs per unit for each feasible enterprise. The gross margin per unit, or the difference between total income and total variable costs, can then be computed for each enterprise. Gross margin is the enterprise's contribution to fixed costs and profit after the variable costs have been paid.

If profit maximization is the assumed goal, the need to estimate gross margins may seem strange. However, the plan is for the short run where fixed costs are constant regardless of the farm plan selected. Any positive gross margin represents a contribution toward paying those fixed costs. Therefore, in the short run, maximizing gross margin is equivalent to maximizing profit (or minimizing losses) because the fixed costs are constant.

Calculating gross margins requires the manager's best estimates of yields or output for each enterprise and expected prices. The calculation of total variable costs requires identifying each variable input needed, the amount required, and its purchase price. Optimum input levels should be determined according to the economic principles of Chapter 5. Yield estimates should be consistent with the input levels selected and the management ability present.

Choosing the Enterprises

For many farmers and ranchers the decision about what enterprises to include in the farm plan is already determined by personal experience and preferences, fixed investments in specialized equipment and facilities, or the regional comparative advantages for certain products. For them the whole farm planning process focuses on preparing the whole farm budget for their plan.

Other managers will want to experiment with different enterprise combinations by developing a series of budgets and comparing them. However, choosing the most profitable enterprises can be a tedious project unless a systematic procedure is used. Two such procedures will be illustrated later in this chapter.

Preparing the Whole Farm Budget

The last step in the planning process is to prepare the *whole farm budget.* A whole farm budget can be used for the following purposes:

1 To estimate the expected income, expenses, and profit for a given farm plan
2 To estimate the cash inflows, cash outflows, and liquidity of a given farm plan
3 To compare the effects of alternative farm plans on profitability and liquidity
4 To evaluate the effects of expanding or otherwise changing the present farm plan
5 To estimate the need for and availability of resources such as capital, labor, livestock feed, or irrigation water
6 To communicate the farm plan to a lender, landowner, partner, or stockholder.

The starting point for a whole farm budget is the income and variable expenses used for computing enterprise gross margins. These values can be multiplied by the number of units of each enterprise in the plan to get a first estimate of total plan income and variable expenses. Other farm income that does not come directly from the budgeted enterprises, such as custom work income and gas tax refunds, should then be added. Past records are a good source of information for estimating these.

Any costs not included in the enterprise variable costs also need to be added. In practice, expenses such as building repairs, auto and pickup expenses, interest, utilities, and similar farm overhead expenses are very difficult to allocate to specific enterprises and are affected very little by the final enterprise combination. Other expenses, such as cash rent or property taxes, may apply to part of the acres in the plan but not to others. Although these and other fixed costs do not affect the selection of a profit-maximizing plan, they may be a major portion of total expenses and should be included in the whole farm budget. Budgets based on farm plans that include investments in additional fixed assets should have their fixed costs revised accordingly.

Forms for organizing and recording whole farm budgets are often available in farm record books. The Extension Service in each state may also have publications, forms, or computer software available. Use of such forms or computer programs will save time and improve the accuracy of the budget estimates.

EXAMPLE OF WHOLE FARM PLANNING

The procedure used in whole farm planning is best learned by working through an example. A relatively small problem will be used to simplify the discussion and concentrate on the planning procedure itself. Once the planning procedure is mastered, it becomes relatively easy to move on to larger problems.

Goals and Objectives

It will be assumed that the manager of the example farm wishes to choose a combination of crop and livestock activities that will maximize total gross margin for the farm for the coming year. For agronomic reasons the operator wishes to carry out a crop rotation that does not include more than 50 percent cotton.

Resource Inventory

Table 10-1 contains the resource inventory for the example farm. The land resource has been divided into three types: Class A cropland, which can be planted entirely in row crops (cotton and milo), Class B cropland, which is subject to erosion and limited to 50 percent row crops, and land that is in permanent pasture. Labor is the only other resource with an effective limit with only 2,000 hours available per year. Capital and machinery are both available in adequate amounts, and the few buildings available serve mostly to limit the number of livestock enterprises that can be considered.

TABLE 10-1   RESOURCE INVENTORY FOR EXAMPLE FARM

Resource	Amount and comments
Class A cropland	400 acres (adapted to continuous row crop production)
Class B cropland	200 acres (conservation needs limit row crops to 50% of total)
Pasture	200 acres
Buildings	Only hay shed and cattle shed are available
Labor	2,000 hours available annually
Capital	Adequate for any farm plan
Machinery	Adequate for any potential crop plan but all harvesting will be custom hired
Management	Manager appears capable and has experience with crops and beef cattle
Other limitations	Any hay produced must be fed on farm, not sold

Identifying Enterprises and Technical Coefficients

Potential crop and livestock enterprises are identified and listed in Table 10-2. Any enterprise that is obviously unprofitable or that requires a resource which is not available should be eliminated. However, if there is any question about an enterprise's profitability, it should be considered. The availability of surplus resources will sometimes cause one of the apparently less profitable enterprises to be included in the final plan.

As shown in Table 10-2, the example farm has three potential crop enterprises on Class A cropland (cotton, wheat, and milo) and two on Class B cropland (wheat and milo). The livestock enterprises are limited to beef cows or stocker steers because of the few buildings available. The beef cow enterprise also requires some Class B land for hay production. Resource requirements per unit of each enterprise, or technical coefficients, are also shown in Table 10-2. For example, an acre of cotton requires 1 acre of Class A cropland, 4 hours of labor, and $115 of operating capital. Even though operating capital is not limited in this example, it is included for illustration.

Estimating Gross Margins

Table 10-3 contains the estimated income, variable costs, and gross margins for the seven potential enterprises to be considered in the whole farm plan. Detailed breakdowns of the variable costs have been omitted to save space but should be done on a separate worksheet. The accuracy of the whole farm plan depends heavily on the estimated gross margins. Careful attention should be paid to estimating yields, output prices, input levels, and input prices.

Choosing the Enterprise Combination

For this example, assume the farm plan is to produce 200 acres of cotton and 200 acres of milo on Class A land, 150 acres of wheat and 50 acres of milo on Class B land, and 100 head of stocker steers. Procedures for choosing the most profitable combination of enterprises will be illustrated later in this chapter.

TABLE 10-2   POTENTIAL ENTERPRISES AND TECHNICAL COEFFICIENTS OR RESOURCE REQUIREMENTS

| | Crops (per acre) | | | | | Livestock (per head) | |
| | Class A cropland | | | Class B cropland | | | |
Resource	Cotton*	Milo	Wheat	Milo[†]	Wheat	Beef cows	Stocker steers
Class A cropland (acres)	1	1	1	—	—	—	—
Class B cropland (acres)	—	—	—	1	1	0.5	—
Pasture (acres)	—	—	—	—	—	4	2
Labor (hours)	4	3	2	3	2	3	1.5
Operating capital ($)	115	60	30	65	30	250	510

*Limited to one-half of the Class A cropland for crop rotation needs.
[†]Limited to one-half of the Class B cropland for conservation purposes.

TABLE 10-3 ESTIMATING GROSS MARGINS

	Class A cropland			Class B cropland		Livestock	
	Cotton (acre)	Milo (acre)	Wheat (acre)	Milo (acre)	Wheat (acre)	Beef cows (head)	Stocker steers (head)
Yield	450 lb	45 cwt	40 bu	35 cwt	30 bu	—	—
Price ($)	0.60	4.00	3.00	4.00	3.00	—	—
Total income ($)	270	180	120	140	90	400	600
Total variable costs ($)	115	60	30	65	30	250	510
Gross margin ($)	155	120	90	75	60	150	90

Completing the Budget

The whole farm budget for the plan is shown in Table 10-4 under the Plan 1 column. Estimated net farm income is $40,500. Income and variable costs for individual enterprises were calculated by multiplying the income and expenses per unit by the number of units to be produced.

Note that an estimate of income from other sources, such as custom work done for neighbors, is included in the budget. Also shown are other expenses that do not vary directly with the number of acres of crops produced or number of livestock raised. Included are items such as property taxes, insurance, hired labor, depreciation, and interest on fixed debt. Interest on borrowed operating capital may be shown here if a general operating loan or credit line for the entire farm is used. However, interest on operating loans that is tied to individual enterprises, such as for the purchase of feeder cattle, should be included in the enterprise budget data.

Alternative Plans

Table 10-4 also contains whole farm budgets for two other farm plans. Plan 2 involves dropping the stocker steer enterprise and cash renting 400 acres of Class A cropland to be equally divided between cotton and milo. This plan has a projected net farm income of $49,500. There are additional fixed expenses for hired labor, cash rent, and miscellaneous.

Plan 3 involves purchasing 400 acres of pasture and adding 100 beef cows to the plan. A loan of $130,000 at 10 percent interest to be repaid over 10 years is used to finance the purchase. Fifty acres of Class B land is shifted from wheat to produce hay needed by the cows. Additional fixed expenses for some part-time hired labor and interest on the new loan are included. The profitability of this expansion should be analyzed over a period of more than one year. Therefore, the interest expense was calculated on the average principal amount that would be owed over the period of the loan, $65,000, rather than the initial debt of $130,000. For the first few years of the plan, the interest expense would be considerably higher, but it would also be lower in the later years. Plan 3 has a projected net farm income of $43,200, falling between that for Plan 1 and Plan 2.

TABLE 10-4 EXAMPLE OF A WHOLE FARM BUDGET

		Plan 1		Plan 2		Plan 3	
	$/Unit	Units	Total	Units	Total	Units	Total
Income							
Cotton-A (acre)	$270	200	$ 54,000	400	$108,000	200	$ 54,000
Milo-A (acre)	180	200	36,000	400	72,000	200	36,000
Milo-B (acre)	140	50	7,000	50	7,000	50	7,000
Wheat-B (acre)	90	150	13,500	150	13,500	100	9,000
Stocker steers (head)	600	100	60,000	0	0	100	60,000
Beef cows (head)	400	0	0	0	0	100	40,000
Gross income			$170,500		$200,500		$206,000
Variable costs							
Cotton-A (acre)	$115	200	$ 23,000	400	$ 46,000	200	$ 23,000
Milo-A (acre)	60	200	12,000	400	24,000	200	12,000
Milo-B (acre)	65	50	3,250	50	3,250	50	3,250
Wheat-B (acre)	30	150	4,500	150	4,500	100	3,000
Stocker steers (head)	510	100	51,000	0	0	100	51,000
Beef cows (head)	250	0	0	0	0	100	25,000
Total variable costs			$ 93,750		$ 77,750		$117,250
Total gross margin			$ 76,750		$122,750		$ 88,750
Other income			$ 5,000		$ 5,000		$ 5,000
Other expenses							
Property taxes			$ 2,600		$ 2,600		$ 3,800
Insurance			1,250		1,250		1,250
Interest on debt			22,000		22,000		28,500
Hired labor			0		10,000		1,600
Depreciation			10,400		10,400		10,400
Cash rent			0		24,000		0
Miscellaneous			5,000		8,000		5,000
Total other expenses			$ 41,250		$ 78,250		$ 50,550
Net farm income			$ 40,500		$ 49,500		$ 43,200
Net farm income with 10% less gross income			$ 23,450		$ 29,450		$ 22,600

Sensitivity Analysis

Although the whole farm budget may project a positive net income, unexpected changes in prices or production levels may quickly turn that into a loss. Analyzing how changes in key budgeting assumptions affect income and cost projections is called *sensitivity analysis.*

In the example in Table 10-4 a very simple sensitivity analysis was performed by reducing the anticipated gross farm income by 10 percent and recalculating the net farm income. Although Plan 3 has the second highest projected net income in a typical year, it would have the lowest net farm income if there was a 10 percent drop in gross income. Additional sensitivity analysis could be carried out by constructing several entire budgets using different values for key prices and production rates.

Note also that Plans 2 and 3, and particularly Plan 2, have higher fixed costs than Plan 1 due to increased labor, cash rent, and interest expenses. These higher fixed cash expenses would make it more difficult to meet cash flow requirements in years of reduced income.

Analyzing Liquidity

Liquidity refers to the ability of the business to meet cash flow obligations as they come due. A whole farm budget can be used to analyze liquidity as well as profitability. This is especially important when major investments in fixed assets and/or major changes in intermediate or long-term debts are being considered. Table 10-5 shows how net cash flow can be estimated from the whole farm budget. Besides cash farm income, income from nonfarm work and investments can be added to total cash inflows. Cash outflows include cash farm expenses (but not noncash expenses such as depreciation), cash outlays to replace capital assets, principal payments on intermediate and long-term debts (interest has already been included in cash farm expenses), and nonfarm cash expenditures for family living and income taxes.

Profitable farm plans will not always have a positive cash flow, particularly when they include a heavy debt load. Interest payments will be especially large during the first few years of the loan period, so it is wise to analyze liquidity for both the first year or two of a plan and for an average year. A negative cash flow projection indicates that some adjustments to the plan are needed if the business is going to be able to meet all of its obligations on time. All three plans in the example have a projected net cash flow that is positive, but Plan 3 would have a $12,000 negative cash flow if prices or yields fall 10 percent below expectations. In addition, the interest expense for the new loan projected in Plan 3 would be $13,000 the first year instead of the average amount of $6,500 that was included in the budget. Therefore, severe liquidity problems could be encountered in the early years of Plan 3.

TABLE 10-5 EXAMPLE OF LIQUIDITY ANALYSIS FOR A WHOLE FARM BUDGET

	Plan 1	Plan 2	Plan 3
Cash inflows:			
Cash farm income	$175,500	$205,500	$211,000
Nonfarm income	12,000	12,000	12,000
	$187,500	$217,500	$223,000
Cash outflows:			
Cash farm expenses	$124,600	$145,600	$157,400
Term debt principal	8,500	8,500	21,500
Equipment replacement	10,000	10,000	10,000
Nonfarm expenses	25,000	25,000	25,000
	$168,100	$189,100	$213,900
Net cash flow	$ 19,400	$ 28,400	$ 9,100
Net cash flow with 10% less gross farm income	$ 1,850	7,850	(12,000)

Long-Run versus Short-Run Budgeting

Short-run budgets that assume some fixed resources should generally incorporate assumptions about prices, costs, and other factors that are expected to hold true over the next year or so. However, when major changes in the supply of land, labor, or other assets, or in the way they are financed are being contemplated a longer run perspective is needed. Very few farms or ranches are profitable every year, but a plan that involves a long-term investment or financing decision should project a positive net income in an "average" or "typical" year.

The following procedures should be used for developing a typical year budget:

1 Use average or long-term planning prices for products and for inputs, not prices that are expected for the next production cycle. In particular, use prices that accurately reflect the long-term price relationships among various products.

2 Use average or long-term crop yields and livestock production levels. Use past records as a guide. For new enterprises use conservative performance rates.

3 Carryover inventories of crops and livestock, accounts payable and receivable, or cash balances do not need to be considered when estimating income, expenses, and cash flows. In the long run these will cancel out from year to year. Assume that all production is sold in a typical year.

4 The borrowing or repayment of operating loans can be ignored when projecting cash flows in a typical year because they will offset each other. If significant short-term borrowing is anticipated, however, the interest cost that results should be incorporated into the estimate of cash expenses.

5 Assume enough capital investment each year to maintain depreciable assets at their current level.

6 Assume that the operation is neither increasing nor decreasing in size. This is especially critical when projecting liquidity for a typical year.

In some cases the farm business may require several years to move from the current plan to a future plan. Profits and cash flows may be temporarily reduced due to inventory buildup, start-up costs, low production levels while learning new technology, and rapid debt repayment. Several "transitional" budgets may be needed to reflect conditions that will exist until the new farm plan is fully implemented.

The actual profitability of the operation may fall below levels projected by the whole farm budget in some years. If this is due to unfavorable weather or low price cycles, however, the whole farm plan chosen may still be the most profitable one for the long run.

SIMPLIFIED PROGRAMMING

The most profitable combination of enterprises in the whole farm plan depends on the gross margin realized from each enterprise and the size at which the supply of available resources allows each enterprise to be carried out. The *simplified programming* process can be used to approximate the most profitable enterprise combination.

The procedure for simplified programming can be summarized in the following steps:

1 Calculate the maximum units of each enterprise permitted by each limited resource.

2 Identify the smallest of these limits for each enterprise. This is the maximum possible level of production for each one.

3 Multiply the gross margin per unit by the maximum possible number of units.

4 Select the enterprise that has the greatest total gross margin and introduce it into the plan up to the limit.

5 Calculate the quantity of each limited resource that is still unused.

6 Repeat steps 1 to 5 until all the resources are exhausted, or no other enterprises are possible.

7 Check to see if the total gross margin can be increased by substituting one enterprise for another.

Example of Simplified Programming

This example uses the same information shown in Tables 10-1, 10-2, and 10-3. The values in step 1 of Table 10-6 indicate the maximum units of each enterprise that are possible without using more than the available supply of each resource. The smallest value for each enterprise is underlined.

For cotton grown on Class A cropland, there is enough labor for 500 acres but only enough land for 400 acres. Milo and wheat are also limited to 400 acres each by the Class A land resource, even though 666 and 1,000 acres, respectively, are permitted by the labor supply. In a similar manner the number of units of milo or wheat on Class B land is limited to 200 acres by the supply of land, even though there is enough labor for more acres. The beef cows are restricted to 50 head and the stocker steers to 100 head by the amount of pasture land available.

Step 2 in Table 10-6 shows the total gross margin possible from each enterprise, found by multiplying the gross margin per unit by the number of units possible. Cotton at $62,000 has the largest potential gross margin. However, for rotational reasons the operator does not want more than 200 acres planted to cotton in any year. This reduces the cotton gross margin to $31,000. The other 200 acres can be planted to milo, which is the crop with the next largest gross margin. This adds $24,000 to the gross margin, for a total of $55,000.

All Class A land has now been utilized, but there are other possible enterprises. Therefore, the process used in steps 1 and 2 is repeated in steps 3 and 4. There are only 600 hours of labor remaining because 800 hours will be used for cotton production and 600 hours for milo production. As shown by the underlined values in step 3 of Table 10-6, however, milo, wheat, beef cows, and stocker steers are still limited by the supply of Class B land or pasture land, not by the labor supply.

Of the enterprises not yet in the plan, step 4 shows milo on Class B land has the largest maximum gross margin at $15,000. However, to control erosion the operator does not wish to plant more than half of the Class B land to milo, so it is limited to 100 acres. This 100 acres of milo increases the plan's gross margin by $7,500, to a total of $62,500.

TABLE 10-6 SIMPLIFIED PROGRAMMING EXAMPLE

		Class A cropland			Class B cropland		Livestock		
	Resource available	Cotton (acre)	Milo (acre)	Wheat (acre)	Milo (acre)	Wheat (acre)	Beef cows (head)	Stocker steers (head)	Total
Step 1									
Class A cropland (acre)	400	400	400	400					
Class B cropland (acre)	200				200	200			
Pasture (acre)	200						50	100	
Labor (hour)	2,000	500	666	1,000	666	1,000	666	1,333	
Gross margin per unit ($)		155	120	90	75	60	150	90	
Step 2									
Maximum units		400	400	400	200	200	50	100	
Maximum gross margin ($)		62,000	48,000	36,000	15,000	12,000	7,500	9,000	
Optimum plan		200	200						
Gross margin ($)		31,000	24,000						55,000
Step 3									
Class A cropland (acre)	0	0	0	0					
Class B cropland (acre)	200				200	200			
Pasture (acre)	200						50	100	
Labor (hour)	600	150	200	300	200	300	200	400	
Step 4									
Maximum units		0	0	0	200	200	50	100	
Maximum gross margin ($)		0	0	0	15,000	12,000	7,500	9,000	
Optimum plan		200	200		100				
Gross margin ($)		31,000	24,000		7,500				62,500
Step 5									
Class A cropland (acre)	0	0	0	0					
Class B cropland (acre)	100				200	100			
Pasture (acre)	200						50	100	
Labor (hour)	300	75	100	150	100	150	100	200	
Step 6									
Maximum units		0	0	0	100	100	50	100	
Maximum gross margin ($)		0			7,500	6,000	7,500	9,000	
Optimum plan		200	200		100			100	
Gross margin ($)		31,000	24,000		7,500			9,000	71,500
Step 7									
Class A cropland (acre)	0	0	0	0					
Class B cropland (acre)	100				0	100			
Pasture (acre)	0						0	0	
Labor (hour)	150	37.5	50	75	50	75	50	100	
Step 8									
Optimum plan		200	200		50	150		100	
Gross margin ($)		31,000	24,000		3,750	9,000		9,000	76,750

In step 5 the calculation of maximum levels of additional enterprises is repeated. There are 100 acres of Class B land, 200 acres of pasture, and 300 hours of labor still available. The maximum possible enterprise levels are 100 acres of wheat, 50 head of beef cows, or 100 head of stocker steers. The stocker steers give the highest total gross margin of $9,000 as shown in step 6, so they are now added to the plan. Gross margin for the farm plan is increased to $71,500.

Beef cows are no longer feasible because all the pasture has been used for the stocker steers. There are still 100 acres of Class B land available, but the remaining 150 hours of labor permit only 75 acres of wheat to be grown as shown in step 7. Adding 75 acres of wheat to the plan would increase total gross margin by $4,500, to a total of $76,000. However, 25 acres of Class B land would remain unused. Comparing the gross margin per hour of labor (now the most limiting resource), milo returns $25 per hour and wheat $30 per hour. Thus, it would pay to shift labor from milo to wheat as long as unused land is available. Decreasing milo by one acre frees 3 hours of labor, and 1.5 more acres of wheat can be grown. Fifty acres of milo can be shifted to wheat in this manner, leaving 50 acres of milo and 150 acres of wheat on Class B land. Total gross margin for the plan is now $76,750. No further enterprises can be added unless the supply of at least one key resource is increased or the level of one of the current enterprises is reduced.

A large planning problem with many fixed resources and many alternative enterprises becomes difficult to solve by the simplified programming method. A computer can be used to save time and increase the accuracy of the planning process. Many different plans can be tested by changing a few coefficients and comparing the results.

LINEAR PROGRAMMING

Linear programming is a mathematical technique that uses a systematic procedure to find the best possible enterprise combination. A computer can solve large linear programming problems and find the "best" plan provided the data are accurate. In mathematical terms linear programming maximizes an objective function subject to specified restraints. The objective function in whole farm planning is usually maximization of total gross margin, and the restraints are the amounts of the fixed resources available. However, it is possible to place other limits or restraints on the various enterprises or activities.

Linear programming requires the same type of information needed for the earlier farm planning example. The resource inventory must be completed, potential crop and livestock enterprises identified, the technical coefficients or the resource requirements per unit of each enterprise computed, and gross margins estimated.

Example of Linear Programming

The basic logic for solving a linear programming problem can be illustrated in graphic form for a small problem involving two enterprises, corn and soybeans, and three limited resources. The necessary information is shown in Table 10-7, with land, labor, and operating capital as the limiting resources. Gross margins and the technical coefficients are also shown in the table.

TABLE 10-7 INFORMATION FOR LINEAR PROGRAMMING EXAMPLE

		Resource requirements (per acre)	
Resource	Resource limit	Corn	Soybeans
Land (acres)	120	1	1
Labor (hours)	500	5	3
Operating capital ($)	15,000	100	80
Gross margin ($)		120	96

The resource limits and technical coefficients are used to graph the possible enterprise combinations shown in Figure 10-2. Land limits corn and soybeans each to a maximum of 120 acres. These points, A and A', are found on the graph and connected with a straight line. Any point on the line AA' is a possible combination of corn and soybeans given only the land restriction. Labor, however, restricts corn to a maximum of 100 acres (500 hours divided by 5 hours per acre) and soybeans to 166.7 acres (500 hours divided by 3 hours per acre). These points on the graph are connected by the line BB'. Any point on the line BB' is a possible combination of corn and soybeans permitted by the labor restriction. In a similar manner the line CC' connects the maximum corn acres permitted by the operating capital restriction ($15,000 ÷ $100 per acre = 150 acres) with the maximum soybean acres ($15,000 ÷ $80 per acre = 187.5 acres). Line CC' identifies all the possible combinations based only on the operating capital restriction.

FIGURE 10-2 Graphical illustration of resource restrictions in a linear programming problem.

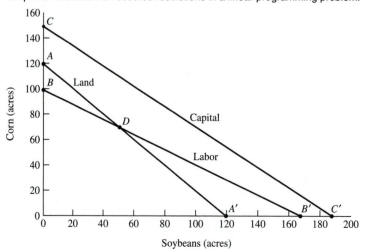

The sections of lines *AA'*, *BB'*, and *CC'* that are closest to the origin of the graph, or line *BDA'*, represent the maximum possible combinations of corn and soybeans when all limited resources are considered together. Line *BDA'* is actually a segmented production possibility curve similar to those for competitive enterprises discussed in Chapter 6. The graph reveals that operating capital is not actually a limiting resource. It is fixed in amount, but $15,000 is more than sufficient for any combination of corn and soybeans permitted by the land and labor resources.

The next step is to find which of the possible combinations of corn and soybeans will maximize total gross margin. The total gross margin from planting 100 acres of corn, the maximum possible amount if no soybeans are grown (represented by point *B*), is $12,000. Decreasing corn by one acre reduces the gross margin by $120 but frees up 5 hours of labor, which in turn allows 1.667 more acres of soybeans to be grown (remember, extra land is available). This adds (1.667 × $96) = $160 to total gross revenue, for a net increase of $40.

This substitution can be continued until no more unused land is available (point *D* in Figure 10-2). At this point reducing corn by one more acre allows only one more acre of soybeans to be grown, resulting in a net decrease in gross margin of $120 − $96 = $24. Thus, point *D* represents the combination of crops that maximizes gross margin. This combination is 70 acres of corn and 50 acres of soybeans with a total gross margin of $13,200.

Figure 10-3 shows the graphical solution to this example. Only the relevant production possibility curve or line segment *BDA'* is shown. The graphical solution to a profit-maximizing linear programming problem is the point where a line containing points of equal total gross margin just touches or is tangent to the production possibility curve on its upper side. This is point *D* in Figure 10-3, which is just tangent to a line representing all possible combinations of corn and soybeans producing a total gross margin of $13,200. Higher gross margins are not possible because they require combi-

FIGURE 10-3   Graphical solution for finding the profit-maximizing plan using linear programming.

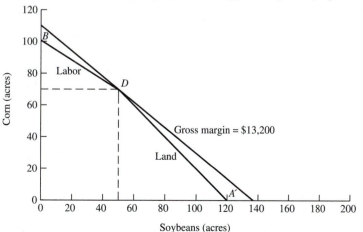

nations of enterprises or enterprise levels not permitted by the limited resources. Combinations other than 70 acres of corn and 50 acres of soybeans are possible, but they would have a total gross margin less than $13,200.

The solution at point D was found in a manner similar to that used to find the profit-maximizing enterprise combination in Chapter 6, where the substitution ratio was equated with the price ratio. One basic difference is that linear programming generates a production possibility curve with linear segments rather than a smooth, continuous curve, as in Figure 6-2. The solution will generally be at one of the corners or points on the production possibility curve, so the substitution ratio will not exactly equal the price (or gross margin) ratio. Instead, the gross margin ratio will fall in between the substitution ratios for the two most limiting resources. For the example, the substitution ratio of soybeans for corn is 1.667 along segment BD and 1.0 along segment DA'. The gross margin ratio is ($120 \div $96) = 1.25, which falls between the two substitution ratios.

Other Features of Linear Programming

Since computerized linear programming can quickly solve large problems, more detail can be included about the resource restrictions and the number of enterprises to be considered. For example, the labor resources can be defined on a monthly basis instead of annually, resulting in 12 labor restrictions instead of 1. This refinement ensures that potential labor bottlenecks during peak seasons are not overlooked. Limits on operating capital can be defined in a similar manner, and the land resource can be divided into several classes. Additional restraints that reflect desirable land use practices or subjective preferences of the manager can also be incorporated. Several enterprises can be defined for each crop or type of livestock, representing different types of technology or levels of production.

In addition to limitations imposed by fixed resources such as owned land and permanent labor, linear programming can incorporate activities that increase the supplies of certain resources. These activities could include renting land, hiring additional labor, or borrowing money. The cost of acquiring these resources (rent, wages, or interest) are included in the calculation of total gross margin. The effect is that the optimum solution to the planning problem will include the acquisition of additional resources only as long as total gross margin can be increased by doing so. It is also possible to transfer production from one enterprise to a resource needed by another enterprise and hence increase the amount of that resource available. For example, corn can be transferred from the corn enterprise to the feed resource to increase the feed available for livestock production.

The linear programming solution also provides values called *dual values,* or *shadow prices.* For resources the dual value represents the amount by which total gross margin would be increased if one more unit of that resource were available. For resources that are not completely used, this value would be zero. For enterprises the dual value shows by how much the gross margin would be reduced by forcing into the plan one unit of an enterprise that was not included in the optimal plan.

Table 10-8 shows the linear programming solution to the farm planning problem analyzed earlier with simplified programming. The level of each enterprise and total

TABLE 10-8 LINEAR PROGRAMMING SOLUTION TO THE FARM PLANNING EXAMPLE

Resource	Units	Quantity available	Class A cotton (acre)	Class A milo (acre)	Class A wheat (acre)	Class B milo (acre)	Class B wheat (acre)	Beef cows (head)	Stocker steers (head)	Total used	Shadow price ($)
						Resources required per unit					
Class A land	acre	400	1	1	1	0	0	0	0	400	75
Class B land	acre	200	0	0	0	1	1	0.5	0	200	30
Pasture	acre	200	0	0	0	0	0	4	2	200	34
Labor	hour	2,000	4	3	2	3	2	3	1.5	2,000	15
Gross margin ($)			155	120	90	75	60	150	90		
No. of units in optimum plan			200	200	0	50	150	0	100		
Shadow price ($)			0	0	15	0	0	45	0		
Total gross margin											$76,750

gross margin are the same. The shadow prices shown in the last column of Table 10-8 indicate that having one more acre of Class A land would increase gross margin by $75. Even though the gross margin for one acre of cotton, the most profitable crop on Class A land, is $155, the limited labor supply would cause a reduction in some other enterprise if one more acre of land were utilized for cotton, so the net benefit would be only $75. Similar interpretations can be placed on the other resource shadow prices. This information is useful for deciding how much the manager can afford to pay for additional resources. For example, it would be profitable to hire extra labor if the cost were less than $15 per hour.

The shadow prices for wheat on Class A land and for beef cows indicate that forcing one acre of Class A wheat or one head of beef cows into the solution would reduce total gross margin by $15 and $45, respectively. Before they would compete for a place in the farm plan, the income from these enterprises would have to increase by these amounts or more.

Both simplified and linear programming are mathematical tools for finding the best enterprise combination. They do not consider goals other than maximization of gross margin, although some of the more sophisticated programming techniques can take into account such concerns as reduction of risk, conservation of resources, interactions among enterprises, and optimal growth strategies over time.

SUMMARY

Whole farm planning and the resulting budget analyze the combined profitability of all enterprises in the farming operation. Whole farm planning starts with formulating goals and objectives and taking an inventory of the resources available. Feasible enterprises must be identified and their per unit costs, income, and gross margins computed. The combination of enterprises chosen can then be used to prepare a whole farm budget. Income and variable costs per unit are multiplied by the number of units to be produced and then combined with other farm income, fixed costs, and any additional variable costs. The completed whole farm budget is an organized presentation of the sources and amounts of income and expenses. Whole farm budgets can be based on either short-run or long-run planning assumptions.

Either simplified or linear programming can be used to select the combination of enterprises that maximizes gross margin without exceeding the supply of resources available. Linear programming can handle large, complex planning problems quickly and accurately, and provides information such as the value of additional resources.

QUESTIONS FOR REVIEW AND FURTHER THOUGHT

1 Why is a resource inventory needed for whole farm planning?
2 Why is land such an important resource in whole farm planning?
3 How does a whole farm budget differ from a partial budget and an enterprise budget?
4 Why can fixed costs be ignored when developing the whole farm plan but must be included in the whole farm budget?

5 Give some examples in which "fixed" costs would change when comparing whole farm budgets for alternative farm plans.

6 What values would you change in a whole farm budget to perform a sensitivity analysis?

7 What is the difference between analyzing profitability and analyzing liquidity?

8 What prices and yields would you use when developing a long-run whole farm plan? A short-run plan?

9 Use the simplified programming procedure explained in this chapter to find the profit-maximizing whole farm plan using the following information:

| | | Resource requirements per acre | |
Resource	Resource limit	Cotton	Grain sorghum
Land (acre)	400	1	1
Capital ($)	28,000	100	40
Labor (hour)	3,000	10	6
Gross margin ($)		100	80

10 Compare your answer to Question 9 to your results from solving the same problem using a piece of graph paper and the linear programming technique.

REFERENCES

Beneke, Raymond R., and Ronald D. Winterboer: *Linear Programming Applications to Agriculture,* Iowa State University Press, Ames, IA, 1973, chap. 3.

Boehlje, Michael D., and Vernon R. Eidman: *Farm Management,* John Wiley & Sons, New York, 1984, chaps. 6, 10.

Boutwell, J. L., and E. W. McCoy: *Simplified Programming as a Farm Management Tool,* Alabama Agricultural Experiment Station, Circular 232, 1977.

Hawkins, Richard O., Dale W. Nordquist, Robert H. Craven, James A. Yates, and Kevin S. Klair: *FINPACK USERS MANUAL,* Minnesota Extension Service, St. Paul, MN, 1987.

Osburn, Donald D., and Kenneth C. Schneeberger: *Modern Agricultural Management,* 2d ed., Reston Publishing Co., Reston, VA, 1983, chaps. 10, 12, 13.

Scott, John T., Jr.: *The Basics of Linear Programming and Their Use in Farm Management,* AER-3-70, Department of Agricultural Economics, University of Illinois, Urbana, IL, 1970.

Thomas, Kenneth, Richard O. Hawkins, Robert A. Luening, and Richard N. Weigle: *Managing Your Farm Financial Future,* North Central Regional Extension Publication 34, University of Minnesota, St. Paul, MN, 1980.

11

CASH FLOW BUDGETING

CHAPTER OBJECTIVES

1 To identify cash flow budgeting as a tool for financial decision making and business analysis

2 To understand the structure and components of a cash flow budget

3 To illustrate the procedure for completing a cash flow budget

4 To describe both the similarities and differences between a cash flow budget and an income statement

5 To discuss the advantages and potential uses of a cash flow budget

6 To show how to use a cash flow budget when analyzing a possible new investment

Even the most profitable farms and ranches occasionally find themselves short of cash. Anticipating these shortages and having a plan to deal with them is an important management activity. A cash flow budget is a financial analysis tool with applications in both forward planning and the ongoing analysis of a farm or ranch business. Preparing one is the next logical step after finishing the whole farm plan and budget. The cash flow budget will provide answers to some remaining questions. Is the plan financially feasible? Will there be sufficient capital available at the specific times it will be needed? If not, how much will need to be borrowed? Will the plan generate the cash needed to repay any new loans? These types of questions can be answered by preparing and analyzing a cash flow budget.

CHARACTERISTICS OF A CASH FLOW BUDGET

A cash flow budget is a summary of the projected cash inflows and cash outflows for a business over a given period of time. This period of time is typically a future accounting period and is divided into quarters or months. As a forward planning tool, its primary purpose is to estimate the amount and timing of future borrowing needs and the ability of the business to repay these and other loans on time. Given the large amount of capital today's commercial farms and ranches require and often must borrow, cash flow budgeting is an important budgeting and financial management tool.

A discussion of cash flow budgeting must continually emphasize the word *cash. All* cash flows must be identified and included on the budget. Cash flows into a farm business from many sources throughout the year, and cash is used to pay business expenses and to meet other needs for cash. Identifying and measuring these *sources* and *uses* of cash is the important first step in constructing a cash flow budget. The concept of cash flows is shown graphically in Figure 11-1. It assumes all cash moves through the business checking account making it the central point for identifying and measuring the cash flows.

Two things make a cash flow budget substantially different from a whole farm budget. First, a cash flow budget contains all *cash* flows, not just revenue and expenses, and it does not include any *noncash* items. For example, cash inflows would include cash from the sale of capital items and proceeds from new loans, but not inventory changes. Principal payments on debt and the full cost of new capital assets would be included as cash outflows but depreciation would not. The emphasis is on *cash* flows regardless of source or use and whether or not they are business revenue or expenses. For this reason, personal and nonfarm cash revenue and expenses are also included on a cash flow budget as they affect the amount of cash available for farm business use.

The second major difference between a whole farm budget and a cash flow budget is the latter's concern with the *timing* of revenue and expenses. A cash flow bud-

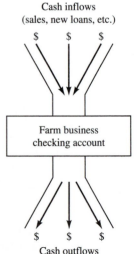

Cash inflows
(sales, new loans, etc.)

$ $ $

Farm business
checking account

$ $ $

Cash outflows
(expenses, debt payments, taxes, etc.)

FIGURE 11-1   Illustration of cash flows.

get also includes "when" cash will be received and paid out as well as "for what" and "how much." This timing is shown by preparing a cash flow budget on a quarterly or monthly basis. Because farming and ranching are seasonal businesses, most agricultural cash flow budgets are done on a monthly basis to permit a detailed analysis of the relationship between time and cash flows.

The unique characteristics of a cash flow budget do not allow it to replace any other type of budget or any of the records discussed in Part Two. It fills other needs and is used for different purposes. However, it will be shown shortly that much of the information needed for a cash flow budget can be found in the whole farm budget and farm records.

ACTUAL VERSUS ESTIMATED CASH FLOWS

By definition, a cash flow budget contains estimates of cash flows for a future time period. However, it is possible to record and organize the actual cash flows for some past time period. The Farm Financial Standards Task Force recommends a "Statement of Cash Flows" be developed as part of the standard set of financial statements for the farm business. This statement is a condensed version of a typical cash flow budget which contains actual data for the accounting period.

A record of actual cash flows has several uses. First, if done monthly, it can be compared with monthly budgeted values at the end of each month. This comparison can provide an early warning of any substantial deviations while there is still time to determine causes and make corrections. Second, the Statement of Cash Flows can provide useful insight into the financial structure of the business and show how the operating, financing, and investing activities combine and interact as sources and uses of cash. Third, actual cash flows provide a good starting point for developing the next cash flow budget. With the totals and timing of past cash flows, it is relatively easy to make the adjustments needed to project these into the future. Reviewing past cash flows also prevents some important items from being overlooked on the new budget.

STRUCTURE OF A CASH FLOW BUDGET

The structure and format of a cash flow budget are shown in Table 11-1 in condensed form. This condensed budget illustrates the sources and uses of cash which need to be included on any cash flow form. Note that there are five potential sources of cash:

1 The beginning cash balance or cash on hand at the beginning of the period
2 Farm product sales or cash revenue from the operation of the farm business
3 Capital sales, the cash received from the sale of capital assets such as land, machinery, breeding livestock, and dairy cattle
4 Nonbusiness cash receipts, which would include nonfarm cash income, cash gifts, and other sources of cash
5 New borrowed capital or loans received

The last source cannot be included in the cash inflow section because borrowing requirements are not known until the cash outflows are matched against the cash inflows.

TABLE 11-1   SIMPLIFIED CASH FLOW BUDGET

	Time period 1	Time period 2
1. Beginning cash balance	$1,000	$ 500
Cash inflow:		
2. Farm product sales	$2,000	$12,000
3. Capital sales	0	5,000
4. Miscellaneous cash income	0	500
5. Total cash inflow	$3,000	$18,000
Cash outflow:		
6. Farm operating expenses	$ 3,500	$ 1,800
7. Capital purchases	10,000	0
8. Miscellaneous expenses	500	200
9. Total cash outflow	$14,000	$ 2,000
10. Cash balance (line 5 – line 9)	–11,000	16,000
11. Borrowed funds needed	$11,500	0
12. Loan repayments (principal and interest)	0	11,700
13. Ending cash balance (line 10 + line 11 – line 12)	500	4,300
14. Debt outstanding	$11,500	$ 0

Table 11-1 also shows the four general uses of cash. They are:

1 Farm operating expenses, which are the normal and usual cash expenses incurred in producing the farm revenue

2 Capital purchases, which would be the full purchase price of any new capital assets such as land, machinery, and dairy or breeding livestock

3 Nonbusiness and other expenses, which would include cash used for living expenses, income and social security taxes, and any uses of cash not covered elsewhere

4 Principal payments on debt. Interest payments should also be included here unless they were included as part of the operating expenses.

The difference between total cash inflows and total cash outflows for any time period is shown as the ending cash balance for that period.

Only two time periods are shown in Table 11-1, but once the basic procedure is understood it can be extended to any number of time periods. In the first time period, the total cash inflow of $3,000 includes the beginning cash balance. The total cash outflow is $14,000 leaving a projected cash balance of –$11,000. This deficit will require borrowing $11,500 to provide a $500 minimum ending cash balance.

Total cash inflow for the second time period is estimated to be $18,000 resulting in a cash balance of $16,000 after subtracting total cash outflow of $2,000. This large cash balance permits paying off the debt incurred in the first time period, which is estimated at $11,700 when interest of $200 is included. The final result is an estimated $4,300 cash balance at the end of the second time period. Following this same procedure for all subsequent time periods traces out the projected level and timing of borrowing and debt repayment potential.

CONSTRUCTING A CASH FLOW BUDGET

Considerable information is needed to construct a cash flow budget. For simplicity and to ensure accuracy, a logical, systematic approach should be followed. The following steps summarize the process and information needs.

1 Develop a whole farm plan. It is impossible to estimate cash revenues and expenses without knowing what crops and livestock will be produced.

2 Estimate crop production and livestock feed requirements. This step projects crops available for sale after meeting livestock feed requirements and may show a need to purchase feed if production is less than what is needed for livestock. Most, if not all, of this information should be found in the whole farm plan.

3 Estimate cash receipts from livestock enterprises. This estimate would include sales of market livestock as well as livestock products such as milk and wool.

4 Estimate cash crop sales. An inventory of feed may be required at the end of the year to provide livestock feed until the next harvest.

5 Estimate other cash income. In this estimate, include interest and dividends on investments and nonfarm sources of cash revenue.

6 Estimate cash farm operating expenses. Reviewing actual cash flows for the last year will prevent overlooking items such as property taxes, insurance, repairs, and other cash expenses not directly related to crop and livestock production.

7 Estimate personal and nonfarm cash expenses. Included here would be cash needed for living expenses, income and social security taxes, and any other nonfarm cash expenses.

8 Estimate purchases and sales of capital assets. Include the full purchase price of any planned purchases of machinery, buildings, breeding livestock, and land, as well as the total cash to be received from the sale of any capital assets.

9 Find and record the scheduled principal and interest payments on existing debt. The amounts and dates of these payments are known and entered directly on the cash flow budget as another required use of cash.

The data estimated and organized in these steps can now be entered on a cash flow form. With a few more calculations, the end result will be an estimate of borrowing needs for the year, the ability of the business to repay these loans, and the timing of each. Key assumptions about selling prices, input costs, and production levels should be well documented before the plan is presented to a lender.

A CASH FLOW BUDGETING FORM

Printed forms for completing a cash flow budget are available from many sources including lending agencies and the agricultural extension service in most states. These sources as well as commercial software firms may have computer programs which can be used. A cash flow budget can also be constructed by anyone familiar with one of the commercial spreadsheet software programs. Table 11-2 illustrates one type of form for a cash flow budget. To save space, only the headings for the first three columns are shown. Other forms may differ in organization, headings, and details, but all provide the same basic information.

TABLE 11-2   FORM FOR CASH FLOW BUDGET

Name _____   Year _____

		Total	January	February	
1	Beginning cash balance				
	Operating receipts:				
2	Grain and feed				
3	Feeder livestock				
4	Livestock products				
5	Other				
6					
	Capital receipts:				
7	Breeding livestock				
8	Machinery and equipment				
9	Other				
	Nonfarm income:				
10	Wages and salary				
11	Investments				
12	Other				
13	Total cash inflow (add lines 1–12)				
	Operating expenses:				
14	Seed				
15	Fertilizer and lime				
16	Chemicals				
17	Other crop expenses				
18	Gas, oil, lubricants				
19	Hired labor				
20	Machine hire				
21	Feed and grain				
22	Feeder livestock				
23	Livestock expenses				
24	Repairs—machinery				
25	Repairs—buildings and improvements				
26	Cash rent				
27	Supplies				
28	Property taxes				
29	Insurance				
30	Utilities				
31	Auto and pickup (farm share)				
32	Other farm expenses				
33					
34					
35	Total cash operating expenses				

The first 13 lines of the form are for recording cash inflows projected from both farm and nonfarm sources. Since nonfarm cash income and expenses will both affect the cash available for farm business use, these items are included in the budget even though they are not directly related to the farm business. The total annual amount expected for each cash inflow is recorded in the Total column. This amount is then allocated to the month or months when it will be received.

TABLE 11-2 (continued)

		Total	January	February	
	Capital expenditures:				
36	Machinery and equipment				
37	Breeding livestock				
38					
	Other expenditures:				
39	Family living expenses				
40	Income tax and social security				
41	Other nonfarm expenses				
42					
43					
	Scheduled debt payments:				
44	Short-term —principal				
45	—interest				
46	Intermediate —principal				
47	—interest				
48	Long-term —principal				
49	—interest				
50	Total cash outflow (add lines 35–49)				
51	Cash available (line 13 – line 50)				
	New borrowing				
52	Short term				
53	Intermediate				
54	Long term				
55	Total new borrowing (add lines 52–54)				
	Payments on new short-term debt:				
56	Principal				
57	Interest				
58	Total debt payments (line 56 + line 57)				
59	Ending cash balance (51 + 55 – 58)				
	Summary of debt outstanding:				
60	Short term				
61	Intermediate				
62	Long term				
63	Total debt outstanding				

Cash farm operating expenses are listed on lines 14 through 34, with the total on line 35. As with all entries on a cash flow budget, the projected total for each expense item is put in the Total column, and this amount is then allocated to the month or months when the cash will actually be needed. The sum of total expenses for the individual months should always be compared with the sum in the Total column on line 35. This cross check will show if any errors were made when allocating individual cash expenses to specific months.

Several other possible cash outflows are shown on lines 36 through 49. Capital expenditures for replacement or expansion of machinery and equipment, breeding livestock, land, and buildings require cash. The full purchase price for any capital expenditures should be recorded even when borrowing will be used to make the purchase. One purpose of the budget is to estimate the amount of such borrowing. Family living expenses, income and social security taxes, and other nonfarm cash expenses should be entered on lines 39 through 43. All personal expenses such as automobile expenses and insurance premiums should be included in family living expenses.

Lines 44 through 49 are used to enter the scheduled principal and interest payments on debt incurred in past years. Entries here would be payments on intermediate and long-term debt, or noncurrent debt, as well as on any current debt carried over from the past year. Only scheduled payments on *old* debt are entered in this section because payments on any *new* debt incurred during the coming year are computed and entered in a later section.

Total annual cash outflows and totals for each month are entered on line 50 by adding lines 35 through 49. Next, the total estimated cash available at the end of the month is calculated by subtracting line 50 from line 13, with the result entered on line 51. If the total cash outflow is greater than the total cash inflow, the cash available will be negative, and new borrowing or other adjustments will be needed to obtain a positive ending cash balance. Any new borrowing is entered on lines 52, 53, or 54 depending on the type of loan. The total new borrowing is added to the cash available on line 51 to find the ending cash balance for the month. (Line 58 will be zero.)

If the cash available on line 51 is greater than zero, the total cash inflow for the month was greater than the total cash outflow. This amount can be used to pay off part or all of any *new* short-term debt incurred earlier in the year or carried over to a future time period. The principal and interest are entered on lines 56 and 57 with these amounts adjusted to permit a positive ending cash balance. After finding the ending cash balance by subtracting line 58 from line 51 (line 55 is zero), the result is entered on line 59 and on line 1 as the beginning cash balance for the next month.

Lines 60 through 63 are not a necessary part of a cash flow budget. However, they summarize the debt situation for the business and provide some useful information. For each type of debt, the amount outstanding at the end of any month will equal the amount at the end of the previous month plus any new debt minus any principal payments during the month. Lines 60 through 62 should be added for each month and the total debt outstanding entered on line 63. This figure shows the pattern of total debt and its changes throughout the annual production cycle. Because of the seasonal nature of agricultural production, income, and expenses, this pattern will often repeat itself each year. Some lenders specify a maximum amount of debt the operator can have at any one time. This part of the budget helps project if this borrowing limit can be met.

EXAMPLE OF A CASH FLOW BUDGET

A completed cash flow budget is shown in Table 11-3. It will be used to review the budgeting steps and to point out several special calculations. The first estimates needed are the beginning cash balance on January 1 and all sources and amounts of cash inflows for the year. These amounts are entered in the Total column and then allocated to the month or months when the cash will actually be received. This example shows an estimated total annual cash inflow of $256,900, including the beginning cash balance of $2,500. Notice that the total cash inflow for any month other than January cannot be determined until its beginning cash balance is known. The ending cash balance for each month becomes the beginning cash balance for the next month. This requires completing the budget for one month before beginning the next month.

The next step is to estimate total cash operating expenses by type, placing each amount in the Total column. Each expense estimate is then allocated to the appropriate month(s), and the total expenses for each month entered on line 35. In the example, the total annual cash operating expenses are projected to be $108,500 with the largest expenditures in November and March. The same procedure is followed for capital expenditures, family living expenses, income taxes, and social security payments. Another important requirement for cash is the scheduled principal and interest payments on debt outstanding at the beginning of each year. These amounts are shown on lines 44 through 49. Total cash outflow, found by summing lines 35 through 49, is $245,500 in the example. All the above steps should be completed before any of the calculations on lines 51 to 63 are attempted.

Beginning with the month of January, note that total cash inflow is $23,300 and total cash outflow is $2,650, leaving $20,650 of cash available at the end of January (line 51). There is no need to borrow any money in January, and there are no new loans that can be paid off. Therefore, lines 52 through 58 are all zero and the $20,650 becomes the ending cash balance on line 59. Lines 60 through 63 (short column on far left) indicate $40,000 intermediate debt and $300,000 long-term debt on January 1, and there are no changes in the balances during the month. These figures are transferred to the January column, showing $340,000 of total debt outstanding at the end of January.

The next step is to transfer the January ending cash balance of $20,650 to the beginning cash balance for February. Calculations for February are similar to those for January with one exception. There is a $6,000 principal payment on the long-term loan, which reduces its balance to $294,000 as shown on line 62, and the total debt outstanding is reduced to $334,000.

February's ending cash balance of $19,250 is transferred to the beginning balance for March where the first new borrowing occurs. Cash outflow exceeds cash inflow by $67,100 in March, requiring the borrowing of $68,000 to cover this amount and leave a small ending balance of $900. New machinery will be purchased in March at a cost of $60,000; so a *new* intermediate loan of $40,000 is included as part of the new borrowing, and the remaining $28,000 is shown as a *new* short-term loan. After the appropriate adjustments in the loan balances on lines 60 and 61, March concludes with $402,000 of debt outstanding.

This example shows additional new borrowing will also be needed in April, May, June, July, and September, and payments can be made in August when the feeder livestock are sold and in October when some crop income is received. Repayments are handled in the following manner, using August as the example. A total of $66,050 in cash (line 51) is available for repaying new debt incurred earlier in the year. This full amount cannot be used for principal payments, as there will be some interest due, and a positive ending balance is needed. A principal payment of $63,000 plus $2,000 interest on this amount will leave an ending balance of $1,050. The short-term debt outstanding must be reduced by the $63,000 and the long-term debt by the $6,000 principal payment (line 48). Total debt outstanding at the end of August is $360,500.

TABLE 11-3 EXAMPLE OF A CASH FLOW BUDGET

Name I. M. Farmer

Cash Flow Budget

#	Item	Total	Jan	Feb	March	April	May	June	July	Aug	Sept	Oct	Nov	Dec
1	Beginning cash balance	2,500	2,500	20,650	19,250	900	450	550	500	450	1,050	500	39,650	52,850
	Operating receipts:													
2	Grain and feed	160,000	20,000	20,000								60,000	60,000	
3	Feeder livestock	82,000								82,000				
4	Livestock products	1,200									1,200			
5	Other													
6														
	Capital receipts:													
7	Breeding livestock	3,800								3,800				
8	Machinery and equipment													
9														
	Nonfarm income:													
10	Wages and salary	7,200	600	600	600	600	600	600	600	600	600	600	600	600
11	Investments	200	200											
12														
13	Total cash inflow (add lines 1–12)	256,900	23,300	41,250	19,850	1,500	1,050	1,150	1,100	86,850	2,850	61,100	100,250	53,450
	Operating expenses:													
14	Seed	5,800			5,800									
15	Fertilizer and lime	26,000			8,000	10,000							8,000	
16	Chemicals	4,000			4,000									
17	Other crop expenses	2,600										2,600		
18	Gas, oil, lubricants	3,000					1,500							1,500
19	Hired labor	6,000			500	1,500	1,500	500				1,000	1,000	
20	Machine hire	800						400			400			
21	Feed and grain	4,000		800		800		800		800				800
22	Feeder livestock	36,000											36,000	
23	Livestock expenses	7,500	1,000	500	500	1,000	500	1,000	500	500	500	500	1,000	
24	Repairs—machinery	3,600	500	500	500	1,500	600							
25	Repairs—buildings & improvements	1,800									1,800			
26	Cash rent	1,000		250			250			250			250	
27	Supplies	3,400			1,700							1,700		
28	Property taxes	700		700										
29	Insurance	600												
30	Utilities	600	50	50	50	50	50	50	50	50	50	50	50	50
31	Auto and pickup (farm share)	1,200	100	100	100	100	100	100	100	100	100	100	100	100
32	Other farm expenses	500		100		100		100			100		100	
33														
34														

TABLE 11-3 *(continued)*

#		Total	Jan	Feb	March	April	May	June	July	Aug	Sept	Oct	Nov	Dec
35	Total cash operating expenses	108,500	1,650	3,000	21,150	14,050	4,500	2,950	650	1,800	2,850	6,050	46,400	3,450
	Capital expenditures:													
36	Machinery and equipment	60,000			60,000									
37	Breeding livestock	1,000						1,000						
38														
	Other expenditures:													
39	Family living expenses	12,000	1,000	1,000	1,000	1,000	1,000	1,000	1,000	1,000	1,000	1,000	1,000	1,000
40	Income tax and social security	4,800			4,800									
41	Other nonfarm expenses													
42														
43														
	Scheduled debt payments:													
44	Short term —principal	0												
45	—interest	0												
46	Intermediate—principal	20,000						20,000						
47	—interest	3,200						3,200						
48	Long term —principal	12,000		6,000						6,000				
49	—interest	24,000		12,000						12,000				
50	Total cash outflow (add lines 35–49)	245,500	2,650	22,000	86,950	15,050	5,500	28,150	1,650	20,800	3,850	7,050	47,400	4,450
51	Cash available (line 13–line 50)	11,400	20,650	19,250	(67,100)	(13,550)	(4,450)	(27,000)	(550)	66,050	(1,000)	54,050	52,850	49,000
	New borrowing													
52	Short term	77,000			28,000	14,000	5,000	27,500	1,000		1,500			
53	Intermediate	40,000			40,000									
54	Long term	0												
55	Total new borrowing (add lines 52–54)	117,000			68,000	14,000	5,000	27,500	1,000		1,500			
	Payments on new short-term debt:													
56	Principal	77,000								63,000		14,000		
57	Interest	2,400								2,000		400		
58	Total debt payments (line 56 + line 57)	79,400								65,000		14,400		
59	Ending cash balance (51 + 55 – 58)	49,000	20,650	19,250	900	450	550	500	450	1,050	500	39,650	52,850	49,000
	Summary of debt outstanding:													
60	Short term	0			28,000	42,000	47,000	74,500	75,500	12,500	14,000			
61	Intermediate	40,000	40,000	40,000	80,000	80,000	80,000	60,000	60,000	60,000	60,000	60,000	60,000	60,000
62	Long term	300,000	300,000	294,000	294,000	294,000	294,000	294,000	294,000	288,000	288,000	288,000	288,000	288,000
63	Total debt outstanding	340,000	340,000	334,000	402,000	416,000	421,000	428,500	429,500	360,500	362,000	348,000	348,000	348,000

The principal repaid in August assumes the oldest of the new short-term borrowing was repaid first. It included the $28,000 borrowed in March, the $14,000 borrowed in April, the $5,000 borrowed in May, and $16,000 of the June borrowing for a total of $63,000. The interest rate was assumed to be 10 percent and the interest due was calculated as below using the equation interest = principal × rate × time.

March borrowing:	$28,000 × 10% × 5/12 of a year =	$1,166
April borrowing:	14,000 × 10% × 4/12 of a year =	466
May borrowing:	5,000 × 10% × 3/12 of a year =	125
June borrowing:	16,000 × 10% × 2/12 of a year =	266
Totals	$63,000	$2,023

For convenience, the interest was rounded to $2,000.

Sufficient cash was available in October to repay all remaining new short-term debt consisting of $11,500 left from June borrowing, the $1,000 borrowed in July, and the $1,500 borrowed in September. Interest due was calculated as above and rounded to $400 for convenience.

Large cash balances are projected for October, November, and December after all new short-term debt has been repaid. However, the budget shows that new borrowing will peak in July with $75,500 of new short-term debt and $40,000 of new intermediate debt for the new machinery. With the repayment capacity exhibited in this cash flow budget, a borrower should have no trouble obtaining necessary funds from a lender.

The large cash balances toward the end of the year suggest several alternatives that should be considered. First, arrangements might be made to invest this money for several months rather than have it sit idle in a checking account. Any interest earned would be additional income. Second, the excess cash could be used to make early payments on the intermediate or long-term loans to reduce the interest charges. A third alternative relates to the $40,000 new intermediate loan used to purchase machinery in March. If this amount had been included as part of the new short-term borrowing instead of a new intermediate loan, it could have been paid off by November. This alternative would reduce the ending intermediate debt to $20,000, the total debt outstanding at year-end to $308,000, and the ending cash balance to slightly less than $9,000 after paying interest on the additional $40,000 of short-term debt.

This last alternative would require paying for all the new machinery in the year of purchase and would add considerable risk to the operation. If prices and yields were below expectations, there might not be sufficient cash available to repay all the new short-term debt. For this reason, using short-term debt to finance the purchase of long-lived items such as machinery and buildings is not recommended. A better practice would be to obtain a longer loan and then make additional loan payments in years when extra cash is available.

This example has assumed all monthly cash shortages would be met by new short-term borrowing. However, other potential solutions to monthly cash shortages should be investigated first. For example, could some products be sold earlier than projected to cover negative cash balances? Could some expenditures, particularly large capital expenditures, be delayed until later in the year? Would it be possible to change the due date of some expenditures such as loan payments, insurance premiums, and family expenditures to a later date? These adjustments may reduce or eliminate the amount of short-term borrowing needed for some months and hence save interest expense.

USES FOR A CASH FLOW BUDGET

The primary use of a cash flow budget is to project the timing and amount of new borrowing the business will need during the year and the timing and amount of loan repayments. Discussion in the last paragraph suggests other ways a cash flow budget can be used to improve the financial management of the business. Other uses and advantages are

1 A borrowing and debt repayment plan can be developed which fits an individual farm business. The budget can prevent excessive borrowing and shows how repaying debts as quickly as possible will save interest.

2 A cash flow budget may suggest ways to rearrange purchases and scheduled debt repayments to minimize borrowing. For example, capital expenditures and insurance premiums might be moved to months where a large cash inflow is expected.

3 A cash flow budget combines both business and personal financial affairs into one complete plan.

4 A bank or other lending agency is better able to offer financial advice and spot potential weaknesses or strengths in the business based on a completed cash flow budget.

5 By planning ahead and knowing cash will be available, managers can obtain discounts on input purchases by making a prompt cash payment.

6 A cash flow budget can also have a payoff in tax planning by pointing out the income tax effects of the timing of purchases, sales, and capital expenditures.

7 A cash flow budget can help spot an imbalance between short, intermediate, and long-term credit and suggest ways to improve the situation. For example, too much short-term debt relative to long-term debt can create cash flow problems.

Other uses and advantages of a cash flow budget could be listed depending on the individual financial situation. The alert manager will find many ways to improve the planning and financial management of a business through the use of a cash flow budget.

MONITORING ACTUAL CASH FLOWS

A cash flow budget can also be used as part of a system for monitoring and controlling the cash flows during the year. The control function of management was discussed earlier and one control method using a cash flow budget will be illustrated here. A form such as the one shown in Table 11-4 provides an organized way to compare budgeted cash flows with the actual amounts.

Budgeted total annual cash flow for each item can be entered in the first column. The total budgeted cash flow to date is entered in the second column and the actual

TABLE 11-4   A FORM FOR MONITORING CASH FLOWS

Name _____   Year _____

	Annual Total	Budget to Date	Actual to Date	
1 Beginning cash balance				
2 Operating receipts: Grain and feed				
3 Feeder livestock				
4 Livestock products				
5 Other				
6				
7 Capital receipts: Breeding livestock				
8 Machinery and equipment				
9 Other				
10 Nonfarm income: Wages and salary				
11 Investments				
12 Other				
13 Total cash inflow (add lines 1–12)				
14 Operating expenses: Seed				
15 Fertilizer and lime				
16 Chemicals				
17 Other crop expenses				
18 Gas, oil, lubricants				
19 Hired labor				
20 Machine hire				
21 Feed and grain				
22 Feeder livestock				
23 Livestock expenses				
24 Repairs—machinery				
25 Repairs—buildings and improvements				
26 Cash rent				
27 Supplies				
28 Property taxes				
29 Insurance				
30 Utilities				
31 Auto and pickup (farm share)				
32 Other farm expenses				
33				
34				
35 Total cash operating expenses				

cash flow to date in the third column. If the figures are updated monthly or at least quarterly, it is easy to make a quick comparison of budgeted and actual cash flows.

This comparison is a means of monitoring and controlling cash flows, particularly cash outflows, throughout the year. Outflows which are exceeding budgeted amounts are quickly identified, and action can be taken to find and correct the causes. Estimates for the rest of the year can also be revised. Actual results at the end of the year can be used to make adjustments in budgeted values for next year, which will improve the accuracy of future budgeting efforts.

CASH FLOW BUDGET AND INCOME STATEMENT

While a cash flow budget has many uses, it should *not* be used for some purposes. One common misuse is to make projections of net farm income based on net cash flows. Such a projection can be very misleading as there are important differences between a cash flow budget and an income statement.

The first general difference is the omission of any noncash items from a cash flow budget. Depreciation and inventory changes, for example, can have a major impact on net farm income but are not part of a cash flow budget. The second general difference is the need to enter all cash flows on a cash flow budget even if they are not a business revenue or expense. Examples would be new loans received and principal payments on debt, full cost or revenue from purchase or sale of capital assets, family living expenses, income taxes, and other nonfarm revenues and expenses. As these examples indicate, there may be a large difference between net cash flow and net farm income.

INVESTMENT ANALYSIS USING A CASH FLOW BUDGET

The discussion to this point has been on using a cash flow budget that covers the entire business for one year. There is another important use of a cash flow budget. Any major new capital investment such as the purchase of land, machinery, or buildings can have a large effect on cash flows, particularly if additional capital is borrowed to finance the purchase.

Borrowed capital requires principal and interest payments which represent new cash outflows. The question to answer before making the new investment is: Will the new investment generate enough additional cash income to meet its additional cash requirements? In other words, is the investment financially feasible, as opposed to economically profitable?

A hypothetical example illustrates this use of a cash flow budget. Suppose a farmer is considering the purchase of an irrigation system to irrigate 120 acres of some crop. The following information has been gathered to use in completing the cash flow analysis:

Cost of irrigation system	$60,000
Down payment in cash	18,000
Capital borrowed (@ 12% for 3 years)	42,000
Additional crop income from yield increase ($140 per acre)	16,800
Additional crop expenses from higher input levels ($30 per acre)	3,600
Irrigation expenses ($25 per acre)	3,000

Table 11-5 summarizes this information in a cash flow format. Since this is a long-lived capital investment, it is important to look at the cash flow over a number of years rather than month by month for one year as was done for the whole farm cash flow budget.

TABLE 11-5   CASH FLOW ANALYSIS OF HYPOTHETICAL IRRIGATION INVESTMENT

	Year				
	1	2	3	4	5
Cash inflow:					
Increase in crop income	$16,800	$16,800	$16,800	$16,800	$16,800
Cash outflow:					
Additional crop expenses	3,600	3,600	3,600	3,600	3,600
Irrigation expenses	3,000	3,000	3,000	3,000	3,000
Principal payments	14,000	14,000	14,000	0	0
Interest payments	5,040	3,360	1,680	0	0
Total cash outflow	25,640	23,960	22,280	6,600	6,600
Net cash flow	(−8,840)	(−7,160)	(−5,480)	10,200	10,200

The new loan requires a principal payment of $14,000 each year for the three years of the loan plus interest on the unpaid balance. This obligation generates a large cash outflow requirement the first three years causing a negative net cash flow for these years. Once the loan is paid off in the third year, there is a positive net cash flow in the following years. This result is quite common when a large part of the purchase price is borrowed and the loan must be paid off in a relatively short time.

This investment is obviously going to cause a cash flow problem the first three years. Does this mean the investment is a bad one? Not necessarily. The irrigation system should last for more than the five years shown in the table and will continue to generate a positive cash flow in later years. Over the total life of the irrigation system there would be a positive net cash flow, perhaps a substantial one. The problem is how to get by the first three years.

At this point the purchase of the irrigation system should be incorporated into a cash flow budget for the entire farm. This budget may show that other parts of the farm business are generating enough excess cash to meet the negative cash flow resulting from the purchase of the irrigation system. If not, one possibility would be to negotiate with the lender for a longer loan with smaller annual payments. This solution would help reduce the cash flow problem but would extend principal and interest payments over a longer time period.

This cash flow analysis did not include any effects the investment would have on income taxes. However, a new investment can have a large impact on income taxes to be paid which in turn affects the net cash flows. Accuracy is improved if after-tax cash flows are used in this type of analysis. (See Chapter 14 for a discussion of income taxes.)

Land purchases often generate negative cash flows for a number of years unless a substantial down payment is made. New livestock facilities may require significant cash for construction, breeding stock, and feed before any additional cash income is generated. These examples illustrate the importance of analyzing a new investment with a cash flow budget. It is always better to know about and solve a cash flow problem ahead of time than be faced with an unpleasant surprise. In some cases there may be no way to solve a projected negative cash flow associated with a new investment. In this situation, the investment would be financially infeasible because the cash flow requirements cannot be met.

SUMMARY

A cash flow budget is a summary of all cash inflows and outflows for a given time period, usually in the future. The emphasis is on *cash,* and all cash flows regardless of type, source, or use, including both farm and personal cash items, are included on a cash flow budget. *No* noncash entries are included. A cash flow budget also includes the timing of the cash inflows and outflows to show how well they match up. The final result provides an estimate of the borrowing needs of the business, its debt repayment capacity, and the timing of both.

A cash flow budget can also be used to do a financial feasibility analysis of a proposed investment. The investment may cause major changes in cash income and cash expenses, particularly if new borrowing is used to finance the purchase. A cash flow analysis concentrating on just the cash flows resulting from the purchase will indicate if the investment will generate the cash needed to meet the cash outflows it causes. If not, the purchase needs to be reconsidered *or* further analysis done to be sure the cash needed is available from other parts of the business.

QUESTIONS FOR REVIEW AND FURTHER THOUGHT

1 Why is machinery depreciation not included on a cash flow budget?
2 Identify four sources of cash inflows which would not be included on an income statement but which would be on a cash flow budget. Why are they on the cash flow budget?
3 Identify four types of cash outflows which would not be included on an income statement but which would be on a cash flow budget. Why are they on the cash flow budget?
4 Discuss the truth or falsity of the following statement: A cash flow budget is used primarily to show the profit from the business.
5 Discuss how you would use a cash flow budget when applying for a farm business loan.
6 Assume you would like to make an investment in agricultural land in your area. Determine local land prices, cash rental rates, and the cash expenses you would have as the land owner. Construct a 5-year cash flow budget for the investment assuming you borrow 60 percent of the purchase price with a 20-year loan at 10 percent interest. Would the investment be financially feasible without some additional source of cash inflow?
7 A feasible cash flow budget should project a positive cash balance for each month of the year, as well as for the entire year. When making adjustments to the budget to achieve this, should you begin with the annual cash flow or the monthly values? Why?

REFERENCES

Bucher, Robert F., and Duane Griffith: *Cash Flow Budgets,* Circular 1129 (rev.), Cooperative Extension Service, Montana State University, 1983.

Calkins, Peter H., and Dennis D. DiPietre: *Farm Business Management,* Macmillan Publishing Co., New York, 1983, chap. 6.

Edwards, William: *Twelve Steps to Cash Flow Budgeting,* FM-1792 (rev.), Cooperative Extension Service, Iowa State University, 1987.

Oltmans, Arnold W., Danny A. Klinefelter, and Thomas L. Frey: *Agricultural Financial Reporting and Analysis,* Century Communications Inc., Niles, IL, 1992.

Perry, Robert E.: *Cash Flow Planning with the Aid of Your Record Book and Budgeting,* Circular EC 71-850, Cooperative Extension Service, University of Nebraska, 1971.

Prevatt, J. W.: *Cash-Flow Analysis: A Farm Management Technique,* Circular 488, Cooperative Extension Service, University of Florida, 1981.

PART FOUR

IMPROVING
MANAGEMENT SKILLS

HIGHLIGHTS

This part will take the basic management skills presented in Parts Two and Three and extend and apply them to specific areas of the farm or ranch business. In addition, some of the simplifying assumptions such as perfect knowledge will be relaxed and decision making will be discussed in more of a real-world environment.

A beginning farmer or rancher must decide on what type of business organization to use, and this decision should be reviewed throughout the life of the business. Business organization refers to the legal and operational framework in which the management decisions are made and carried out. The basic choices are sole proprietorship, partnership, and corporation, which are discussed in Chapter 12. Over time, family size and involvement, goals, tax rules, and financial condition will change requiring a periodic review of the organizational structure of the business.

Chapter 13 recognizes that few decisions will be made with full and accurate data available. Uncertainty about prices, yields, and other production and financial conditions is typical of many management decisions. This chapter presents some methods to improve decision making under uncertainty and some techniques for reducing the risk inherent in agricultural production.

A profitable farm or ranch business will lose part of that profit to the payment of income taxes. This affects the cash flow of the business and slows the growth in net worth. Chapter 14 emphasizes how management decisions affect income taxes and why a manager should consider the tax effects on all decisions made throughout the year. It also discusses some basic tax management strategies that may be useful to meet a goal of maximizing *after-tax* profit.

Many resources used in agriculture are long-term investments requiring a large amount of capital. These purchase decisions can affect the financial condition of the business for many years. Chapter 15 explains various investment analysis techniques such as capital budgeting, net present value, and internal rate of return. The use of these tools will help farm managers make better decisions about long-term capital investments and financing methods.

Many of the analysis tools used in the control function of management were presented in Part Two. These are consolidated in Chapter 16 and combined with some additional analytical methods to show how they can be used to perform a whole farm business analysis. Chapter 16 illustrates a logical framework for calculating measures of solvency, liquidity, profitability, and efficiency, which can then be compared against certain goals and standards. Areas of strength and weakness can be identified and any specific problems corrected in order to improve overall business profitability and efficiency.

FORMS OF FARM
BUSINESS ORGANIZATION

CHAPTER OBJECTIVES

1 To identify the sole proprietorship, partnership, and corporation as the three primary forms of business organization

2 To discuss the organization and characteristics of each form of business organization

3 To analyze the advantages and disadvantages of each form of business organization

4 To show how income taxes are affected by the form of business organization

5 To summarize the factors to be considered when selecting a form of business organization

6 To compare alternative arrangements for transferring the income, ownership, and management of a farm business

Any business, including a farm or ranch business, may be legally organized in a number of ways. While it is generally possible to change during the life of the business, a little thought and legal advice at the beginning may save time, trouble, and expense at a later date.

The three basic forms of business organization are: (1) *sole proprietorship (or individual proprietorship)*, (2) *partnership,* and (3) *corporation.* Each one has different legal and organizational characteristics, as well as different income tax regulations. According to data from the Census of Agriculture, nearly 87 percent of U.S. farms and ranches are organized as sole proprietorships. Approximately 10 percent are partnerships, and 3 percent are corporations. The proper choice of organization may depend on the size of the business, the number of people involved in it, the career stage of the primary operators, and the owners' desire for passing on their assets to their heirs. A final choice of business organization should be made only after analyzing all possible long-run effects on the business and on the individuals involved.

LIFE CYCLE

Each farm business has a life cycle, with four stages: entry, growth, consolidation, and exit. These stages are depicted in Figure 12-1. The entry stage includes selecting farming as a career, acquiring and organizing the necessary resources, and establishing a solid financial base.

The growth stage involves expansion in the size of the business, typically through purchasing or leasing additional land or by increasing the scale of the livestock enterprises. This stage often utilizes additional debt to finance the expansion and requires good financial planning and management. It may include merging with another operator. Following the growth stage, the emphasis often turns toward consolidation of the operation. Debt reduction becomes a priority and increased efficiency is preferred to increased size.

As the farm owner nears retirement, attention turns toward reducing risk, liquidating the business or transferring property to the next generation. In this exit stage, business size may actually decline. The income tax consequences of liquidation or transfer must be considered along with the need for adequate retirement income. Another consideration is equitable treatment of children who may choose farming as a career as well as those who do not farm. Early planning and merging of the next generation into the business may allow it to continue in the growth or consolidation stage for some time without showing a decline in size.

The life cycle stage in which the farm is currently operating is an important consideration in selecting the form of business organization. Total capital invested, size of

FIGURE 12-1   Illustration of the life cycle of a farm business.

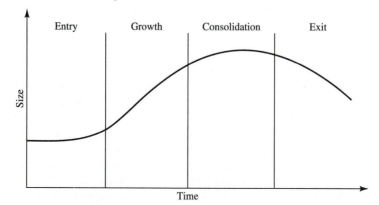

debt, income taxes, owner's goals, and other factors are likely to be different in each stage. Farmers and ranchers may find a different form of business organization is appropriate for two or more of the stages in the life cycle. However, careful and early thought must be given to how the change from one type of organization to another will be handled.

THE SOLE PROPRIETORSHIP

The sole proprietorship is the most common form of farm business organization. A sole proprietorship is easy to form and easy to operate, which accounts for much of its popularity.

Organization and Characteristics

In a sole proprietorship, the owner or sole proprietor owns and manages the business, assumes all the risks, and receives all the profits or losses. Its distinguishing characteristic is the single owner, who acquires and organizes the necessary resources, provides the management, and is solely responsible for the success or failure of the business as well as all business debts. Perhaps the sole proprietorship should be more accurately called a family proprietorship because on many farms and ranches husband, wife, and children may all be involved in ownership, labor, and management.

A sole proprietorship is established simply by starting the business as such. No special legal procedures, permits, or licenses are required. A sole proprietorship is not limited in size by either the amount of inputs which can be used or the amount of products produced. The business can be as large or small as the owner desires. There can be any number of employees, additional management may be hired, and property may even be co-owned with others. A sole proprietorship does not necessarily need to own any assets; one can exist even when all land and machinery are leased.

Advantages

The advantages of a sole proprietorship are its simplicity and the freedom the owner has in operating the business. No other operator or owner needs to be consulted when a management decision is made. The owner is free to organize and operate the business in any legal manner, and all business profits and losses belong to the owner who is the sole proprietor.

A sole proprietorship is also flexible. The owner can quickly make decisions regarding investments, purchases, sales, enterprise combinations, input levels, and so forth based solely on his or her best judgment. Assets can be quickly purchased or sold, money borrowed, or the business even liquidated if necessary, although the latter may require concurrence of a landlord or lender.

Disadvantages

The management freedom inherent in a sole proprietorship includes both responsibilities and disadvantages. Owners of sole proprietorships are personally liable for any legal difficulties and debts related to the business. Creditors have the legal right to attach not only the assets of the business but the personal assets of the owner in fulfillment of any unpaid financial obligations. This feature of a sole proprietorship can be an important disadvantage for a large, heavily financed business where the owner has substantial personal and nonfarm assets. Business failure can result in these assets being acquired by creditors to pay the debts of the sole proprietorship.

The size of a sole proprietorship is limited by the capital available to the single owner. If only a small amount is available, the business may be too small to realize any economies of size, making it difficult to compete with larger and more efficient farms. At the other extreme, the management abilities and time of the single owner may be insufficient for a large business, making it difficult to become an expert in any one area. Thus, a large sole proprietorship may need to hire additional management expertise.

Another disadvantage of a sole proprietorship is a lack of business continuity. It is difficult to bring children into the business on any basis other than as employees. Death of the owner also means the business may have to be liquidated or reorganized under new ownership. This can be time-consuming and costly, resulting in a smaller inheritance and less income for the heirs during the transition period.

Income Taxes

The owner of a business organized as a sole proprietorship pays income taxes on any business profit at the tax rates in effect for individual or joint returns. Business profits and capital gains are added to other taxable income earned to determine the individual's total taxable income.

Joint Operating Agreements

Sometimes two or more sole proprietors may carry on some joint farming activities while maintaining individual ownership of their own resources. Such an activity is often called a *joint operating agreement.* They tend to be informal, short-term arrangements.

In an operating agreement all parties generally pay all the costs related to ownership of their own assets, such as property taxes, insurance, maintenance, and interest on debts. Operating expenses, such as for seed, fertilizer, veterinarian fees, or utilities may be shared in a fixed proportion, often the same proportion as the value of fixed costs contributed by each party. In other cases, one party or another may pay for all of certain operating costs, such as fuel or feed, as a matter of convenience. In either case, the general principle for an operating agreement is to share income in the same proportion as resources are contributed.

An enterprise budget can be a useful tool for evaluating resource contributions. The example in Table 12-1 is based on the cow/calf enterprise budget found in Table 8-3. In this situation party A owns all the pasture, buildings, and fences for the cow herd as well as all the livestock. Party A will also pay the cost of maintaining the pastures and repairing fences and other livestock facilities.

TABLE 12-1 EXAMPLE BUDGET FOR A JOINT COW/CALF ENTERPRISE (ONE HEAD)

Item	Value	Party A	Party B
Variable costs:			
Hay	$ 90.00	$	$ 90.00
Grain and supplement	56.00		56.00
Salt and minerals	2.40		2.40
Pasture maintenance	22.50	22.50	
Veterinary and health exp.	10.00		10.00
Livestock facilities	8.00	8.00	
Machinery and equipment	5.00		5.00
Breeding expense	5.00	5.00	
Labor	30.00		30.00
Miscellaneous	10.00		10.00
Interest on variable costs	11.95	1.78	10.17
(10% for 6 months)			
Fixed costs:			
Interest on brdg. herd	75.00	75.00	
Livestock facilities			
Deprec. and interest	10.00	10.00	
Mach. and equipment			
Deprec. and interest	6.50	6.50	
Land charge	105.00	105.00	
Total cost	$447.35	$233.78	$213.57
Percent contribution	100%	52%	48%

Party B will provide all the feed and labor for the cow/calf enterprise and pay all other variable costs. The question, then, is how should the income from the cattle be divided? Table 12-1 shows that party A is contributing 52 percent of the total costs and party B is contributing 48 percent. Income can be divided in the same proportion or, for convenience, equally since the results are nearly a 50–50 division of costs.

An alternative would be to value the unpaid resources (land, livestock investment, and labor in this example) and share all cash expenses and income in the same proportion. This may not be convenient unless a joint operating bank account is established. However, the arrangement then begins to look like a partnership, which may not be intended nor desirable.

Two cautions should be observed when valuing resource contributions. The opportunity costs for all unpaid resources such as labor and capital assets must be included along with cash costs, and all costs must be calculated for the same time period, usually one year. It would not be correct, for example, to value land at its full, current market value when labor and operating costs are included for only one year. An annual interest charge or rental value for the land should be used.

PARTNERSHIPS

A partnership is an association of two or more persons who share the ownership of a business to be carried on for profit. Partnerships can be organized to last for a brief time or for a long duration.

Two types of partnerships are recognized in most states. The *general partnership* is the most common, but the *limited partnership* is used for some situations. Both types have the same characteristics with one exception. The limited partnership must have at least one general partner, but can have any number of limited partners. Limited partners cannot participate in the management of the partnership business and, in exchange, their financial liability for partnership debts and obligations is limited to their actual investment in the partnership. The liability of general partners, however, can extend even to their personal assets.

Limited partnerships are most often used for businesses such as land speculation and cattle feeding, where investors want to limit their financial liability and do not wish to be involved in management. Most farm and ranch operating partnerships are general partnerships. For this reason, the discussion in the remainder of this section will assume a general partnership unless otherwise indicated.

Organization and Characteristics

There are many possible patterns and variations in partnership arrangements. However, the partnership form of business organization has three basic characteristics:

1 A sharing of the business profits and losses
2 Shared control of property with possible shared ownership of some property
3 Shared management of the business

The exact sharing arrangement for each of these characteristics is flexible and should be outlined in the partnership agreement.

Oral partnership agreements are legal in most states but are not recommended. Important points can be overlooked, and fading memories over time on the details of an oral agreement can create friction between the partners at a later date. Problems can also be encountered when filing the partnership income tax return if arrangements are not well documented.

A written partnership agreement is of particular importance if the profits will not be divided equally. The courts and the Uniform Partnership Act, which is a set of regulations enacted by most states to govern the operation of partnerships, assume an equal partnership in everything including management decisions unless there is a written agreement specifying a different arrangement. For any situation not covered in the written agreement, the regulations in the state partnership act will apply.

The written agreement should cover at least the following points:

1 *Management*   Who is responsible for which management decisions and how will they be made?
2 *Property Ownership and Contribution*   This section should list the property each partner will contribute to the partnership and describe how it will be owned.
3 *Share of Profit and Losses*   The method for calculating profits and losses and the share going to each partner should be carefully described, particularly if there is an unequal division.

4 *Records*   Records are important for the division of profits and for maintaining an inventory of assets and their ownership. Who will keep what records should be part of the agreement.

5 *Taxation*   The agreement should contain a detailed account of the tax basis of property owned and controlled by the partnership and copies of the partnership information tax returns.

6 *Termination*   The agreement should contain the date the partnership will be terminated if one is known or can be determined.

7 *Dissolution*   The termination of the partnership on either a voluntary or involuntary basis requires a division of partnership assets. The method for making this division should be described to prevent disagreements and an unfair division.

Partners may contribute land, capital, labor, management, and other assets to the partnership. Profits are generally divided in proportion to the value of the assets, labor, and management contributed to the business; that is, if each of the two partners contributes one-half of the assets, labor, and management, they share the profits on a 50–50 basis. However, many other contributions are possible, and the profits may be shared on any basis agreeable to the partners regardless of their contributions.

Property may be owned by the partnership, or the partners may retain ownership of their individual property and rent it to the partnership. When the partnership itself owns property, any partner may sell or dispose of any asset without the consent and permission of the other partners. This aspect of a partnership suggests that retaining individual ownership may be desirable in some cases particularly if it does not affect use of the asset by the partnership.

A partnership can be terminated in a number of ways. The partnership agreement may specify a termination date. If not, a partnership will terminate upon the incapacitation or death of a partner, bankruptcy, or by mutual agreement between the partners. Termination upon the death of a partner can be prevented by placing provisions in the written agreement that allow the deceased partner's share to pass to the estate and hence to the legal heirs.

In addition to the three basic characteristics mentioned earlier, factors that indicate a particular business arrangement is legally a partnership include:

1 Joint ownership of assets
2 Operation under a firm name
3 A joint bank account
4 A single set of business records
5 Management participation by all parties
6 Sharing of profits and losses

Joint ownership of property does not by itself imply a partnership. However, a business arrangement with all or most of these characteristics may be a legal partnership even though a partnership was not intended. These characteristics should be kept in mind for landlord–tenant arrangements, for example. A partnership can be created by a leasing arrangement based on shares unless steps are taken to avoid most of the conditions mentioned above.

In the final analysis a partnership is a close relationship, both legal and business, between two or more persons. The partners need mutual understanding and trust if the business is to succeed. Personal characteristics and compatibility are very important, as personal conflicts over relatively small matters probably terminate more partnerships than any other reason. Each partner must believe the partnership will be more advantageous than operating alone. If the final agreement is not fair and equitable to each partner, there will be a temptation to remove the contributed capital, labor, and management to some other business and dissolve the partnership.

Advantages

The primary advantages of a partnership over a sole proprietorship involve capital and management. Pooling the capital of the partners allows a larger business, which can be more efficient than two or more smaller businesses. It may also increase the amount of credit available, allowing further increases in business size. Total management and labor are also increased by pooling the capabilities of all the partners. Management efforts can be divided, with each partner specializing in one area of the business. It is also easier for one operator to be absent from the business when a partner is available.

A partnership is easier and cheaper to form than a corporation. It may require more records than a sole proprietorship, but not as many as for a corporation. While each partner may lose some individual freedom in making management decisions, a carefully written agreement can maintain some of this freedom.

A partnership is a flexible form of business organization where many types of arrangements can be accommodated and included in a written agreement. It fits situations such as when parents desire to bring children and their spouses into the business. The children may contribute only labor to the partnership initially, but the partnership agreement can be modified over time to allow for their increasing contributions of management and capital.

Disadvantages

The unlimited liability of each general partner is an important disadvantage of a partnership. A partner cannot be held personally responsible for the personal debts of the other partners. However, each partner can be held personally and individually responsible for any lawsuits and financial obligations arising from the operation of the partnership. If the partnership does not have sufficient assets to cover its legal and financial obligations, creditors can bring suit against all partners individually to collect any money due them. In other words, a partner's personal assets can be claimed by a creditor to pay partnership debts.

This disadvantage takes on a special significance considering that any partner, acting individually, can act for the partnership in legal and financial transactions. For this reason if for no other, it is important to know and trust your partners and to have the management decision-making procedure included in the partnership agreement. Too many partners or an unstructured management system can easily create problems. The primary advantage of a limited partnership is that it limits the liability of the limited partners to only the capital they have actually contributed to the partnership.

Similar to a sole proprietorship, a partnership has the disadvantage of poor business continuity. It can be unexpectedly terminated by the death of one partner or a disagreement among the partners. The dissolution of any business is generally time-consuming and costly, particularly when it is caused by the death of a close friend and partner or by a disagreement resulting in bad feelings among the partners. The required sharing of management decision making and the loss of some personal freedom in a partnership are always potential sources of conflict between the partners.

Laws governing the formation and taxation of partnerships are less detailed than for corporations. Unfortunately, this allows many farm partnerships to operate with minimal records and documentation of how resources were contributed and income was divided. This can make a fair dissolution of the partnership very difficult.

Income Taxes

A partnership does not pay any income taxes directly. Instead it files an information income tax return reporting the income and expenses for the partnership. Income from the partnership is then reported by the individual partners on their tax returns, based on their respective shares of the partnership income as stated in the partnership agreement.

In a 50–50 partnership, for example, each partner would report one-half of the partnership income, capital gains, expenses, depreciation, losses, and so forth. These items would be included with any other income the partner had to determine the total income tax liability. Partnership income is, therefore, taxed at individual tax rates, with the exact rate depending on the amount of partnership income and other income earned by each partner/taxpayer.

Partnership Example

Since a partnership is usually formed for a relatively long period, contributions to the partnership can be valued based on the current market values of the assets contributed by each party. Table 12-2 shows an example for two people entering into a partnership for raising feeder pigs. The contribution of total assets is in the proportion of 60 to 40 percent. All operating expenses would be paid by the partnership, usually from a separate partnership account. If either or both partners contribute labor they should be paid a representative wage by the partnership.

In some partnerships major assets such as land and buildings are not considered to be partnership property. Instead, the partnership pays rent to the owners of these assets (who may also be the partners) for their use. In such cases the rented assets would not be included when calculating the relative ownership shares of the partnership.

Net income from the partnership is generally divided in the same proportion as ownership, although other splits can be agreed on. Net income may either be withdrawn or left in the partnership account to increase net worth. As long as all with-

TABLE 12-2 PARTNERSHIP EXAMPLE

Assets contributed	Value		
	Total	Party A	Party B
Operating capital	$ 10,000	$ 5,000	$ 5,000
Current feed inventory	15,600	9,500	6,100
Pigs on feed	16,000	10,000	6,000
Breeding livestock	2,900	1,900	1,000
Livestock equipment	11,500	10,200	1,300
Tractor and machinery	27,000	0	27,000
Buildings and land	33,000	33,000	0
Total	$116,000	$69,600	$46,400
Ownership share	100%	60%	40%

drawals (not including wages or rent) are made in the same proportion as the ownership of the partnership, the ownership shares will not change. However, if one partner leaves more profits in the business, such as by retaining more breeding stock, or contributes additional assets, the ownership percentages will change. Sometimes this is done as a way to move toward more equal ownership of the partnership.

When a partnership is liquidated the proceeds should be distributed in the same proportion as the ownership. For this reason it is important to keep detailed and accurate records of the property utilized by a partnership and how wages, rent, and net income are distributed to the partners.

CORPORATION

A corporation is a separate legal entity which must be formed and operated in accordance with the laws of the state in which it is organized. It is a legal "person" separate and apart from its owners, managers, and employees. This separation of the business entity from its owners distinguishes a corporation from the other forms of business organization. As a separate business entity a corporation can own property, borrow money, enter into contracts, sue, and be sued. It has most of the basic rights and duties of an individual.

Organization and Characteristics

The laws affecting the formation and operation of a corporation vary somewhat from state to state. In addition, several states have laws that prevent a corporation from engaging in farming or ranching, or place special restrictions on farm and ranch corporations. For this and other reasons competent legal advice should be sought before attempting to form a corporation.

While state laws do vary, there are some basic steps that will generally apply to forming a farm corporation. These are:

1 The incorporators file a preliminary application with the appropriate state official. This step may include reserving a name for the corporation.

2 The incorporators draft a preincorporation agreement outlining the major rights and duties of the parties after the corporation is formed.

3 The articles of incorporation are prepared and filed with the proper state office.

4 The incorporators turn property and/or cash over to the corporation in exchange for shares of stock representing their ownership share of the corporation.

5 The shareholders meet to organize the business and elect directors.

6 The directors meet to elect officers, adopt bylaws, and begin business in the name of the corporation.

These steps identify three groups of individuals involved in a corporation: shareholders, directors, and officers. The shareholders own the corporation. Stock certificates are issued to them for property or cash transferred to the corporation. As owners, the shareholders have the right to direct the affairs of the corporation, which is done through the elected directors and at annual meetings. Each shareholder has one vote for each share of voting stock owned. Therefore, any shareholder with 51 percent or more of the outstanding voting stock has majority control over the business affairs of the corporation. In some corporations contributions of capital are exchanged for bonds instead of stock. Bonds carry a fixed return, but no voting privileges.

The directors are elected by the shareholders at each annual meeting and hold office for the following year. They are responsible to the shareholders for the management of the business. The number of directors is normally fixed by the articles of incorporation. Meetings of the directors are held to conduct the business affairs of the corporation and to set broad management policy to be carried out by the officers.

The officers of a corporation are elected by the board of directors and may be removed by them. They are responsible for the day-to-day operation of the business within the guidelines established by the board. The officers' authority flows from the board of directors, to whom they are ultimately responsible. A corporation president may sign certain contracts, borrow money, and perform other duties without board approval but will normally need board approval before committing the corporation to large financial transactions or performing certain other acts.

The number of shareholders may be as few as one in some states, while three is the minimum number in several other states. In many small family farm corporations the shareholders, directors, and officers are all the same individuals. To an outsider, the business may appear to be operated like a sole proprietorship or partnership. Even the directors' meetings may be held informally around a kitchen table, but a set of minutes must be kept for each official meeting.

Two types of corporations are recognized for federal income tax purposes. The first is the regular corporation, which is also called a *C corporation.* A second type of corporation is the *S corporation,* sometimes referred to as a tax-option corporation. Both types have many of the same characteristics. However, there are certain restrictions on the formation of S corporations. Some of the more important restrictions include:

1 There must be no more than 35 shareholders (a husband and wife are considered one shareholder even though both own stock).

2 Shareholders are limited to individuals (foreign citizens are excluded), estates, and certain types of trusts. Other corporations cannot own stock in an S corporation.

3 There may be only one class of stock, but the voting rights may be different for some shareholders.

4 All shareholders must initially consent to operating as an S corporation.

There are also certain limitations on the types and amounts of income that a corporation can receive from specified sources and still maintain the S corporation status. The special taxation of an S corporation will be covered in a later section.

Advantages

The number of farm and ranch corporations, though still small, has been increasing. The majority are classified as family farm corporations with a relatively small number of shareholders all related by blood or marriage.

Corporations provide limited liability for all the shareholder/owners. They are legally responsible only to the extent of the capital they have invested in the corporation. Personal assets of the shareholders cannot be attached by creditors to meet the financial obligations of the corporation. This advantage may be negated if a corporation officer is required to personally cosign a note for corporation borrowing. In this case, the officer can be held personally responsible for the debt if the corporation cannot meet its responsibilities.

A corporation, like a partnership, provides a means for several individuals to pool their resources and management. The resulting business, with a larger size and the possibility of specialized management, can provide greater efficiency than two or more smaller businesses. Credit may also be easier to obtain because of the business continuity embodied in a corporation. The business is not terminated by the death of a shareholder, as the shares simply pass to the heirs and the business continues. However, a plan for management continuity should exist by having more than one person involved and capable of taking over complete management. Otherwise the death of a principal shareholder who may also be the president and sole manager will disrupt the business during the transition to new management.

A corporation provides a convenient way to divide and transfer business ownership. Shares of stock can be easily purchased, sold, or given as gifts without actually transferring title to specific parcels of land or other assets. Transferring shares does not disrupt or reduce the size of the business and is a convenient way for a retiring farmer to transfer part of the business to the next generation while maintaining an ongoing business.

There can be income tax advantages to incorporation, depending on the size of business, how it is organized, and the income level of the shareholders. A C corporation is a separate taxpaying entity, and these corporations are subject to different tax rates than individuals. Corporate tax rates have been changed from time to time and may change again. The personal and corporate tax rates in effect for 1993 are shown in Table 12-3. (These are subject to change annually.)

TABLE 12-3   PERSONAL AND CORPORATE INCOME TAX RATES FOR 1993 TAX YEAR

Personal (joint return)		Corporate	
Taxable income*	Marginal tax rate (%)	Taxable income	Marginal tax rate (%)
Up to $36,900	15	Up to $50,000	15
36,900 to 89,150	28	50,000 to 75,000	25
More than 88,150	31	75,000 to 100,000	34
		100,000 to 335,000	39
		More than 335,000	34†

*The personal tax brackets will be increased in future years to adjust for the effects of inflation.
†Corporations with a taxable income over $335,000 pay a flat, single rate of 34% on all taxable income.

One advantage of a C corporation is that the owners are usually employees as well. This allows the tax deductibility of certain fringe benefits provided to the shareholder/employees such as premiums for health, accident, and life insurance. It is also easier to allocate income among individuals by setting salaries, rents, and dividends.

Disadvantages

Corporations are more costly to form and maintain than sole proprietorships and partnerships. Certain legal fees are necessary when organizing a corporation, and legal advice will be needed on a continuing basis to ensure compliance with state regulations. An accountant may also be needed during the formation period and throughout the life of the corporation to handle financial records and tax-related matters. Most states require various fees when filing the articles of incorporation and some type of annual operating fee or tax on corporations, which are not assessed on the other forms of business organization.

Doing business as a corporation requires that shareholders and directors meetings are held, minutes are kept of directors meetings, and annual reports are filed with the state. However, if forming a corporation results in better business and financial records being kept, this might be viewed as an advantage rather than a disadvantage because better information for making management decisions will be available.

Another disadvantage of a C corporation is the potential double taxation of income. After the corporation pays tax on its taxable income, any after-tax income distributed to the shareholders as dividends is considered taxable income to the shareholders. It is taxed at their applicable individual rates. Many small farm corporations avoid some of this double taxation by distributing most of the corporation income to the shareholders as rent, wages, salaries, and bonuses. These items are tax-deductible expenses for the corporation but taxable income for the shareholders/employees. However, any wages and salaries paid must be reasonable, for bona fide work performed for the corporation and not just a means to avoid double taxation.

The S corporations do not pay income tax themselves but are taxed like a partnership. Hence, the name "tax-option" corporation. The corporation files an information tax return, but the shareholders report their share of the corporation income, expenses, and capital gains according to the proportion of the total outstanding stock they own. This income (or loss) is included with the shareholder's other income, and tax is paid on the total based on the applicable individual rates. The income tax treatment of an S corporation avoids the dividend double-taxation problem of a C corporation.

Only a partial treatment of the income tax regulations applicable to both types of corporations is included here. There are many rules pertaining to special situations and special types of income and expenses. All applicable tax regulations should be reviewed with a qualified tax consultant before the corporate form of business organization is selected.

Corporation Example

A corporation can be formed and operated in a manner similar to a partnership. Table 12-4 illustrates the same business venture as Table 12-2, only this time it is formed as a corporation. The same assets are contributed, but in this example the corporation also takes over a $16,000 debt which party A owed on the livestock buildings. It should be noted that partnerships can also receive contributions of debt from the partners as well as assets.

The shares of the corporation are divided in the same proportion as the original contributions of *net worth*. In this example party A has a smaller share of the business than in the partnership example, but the corporation will make the payments on the intermediate debt.

As with the partnership, the corporation can pay salaries for labor contributed by shareholders, as well as rent for the use of assets not owned by the corporation. Distributions of net income would be in the form of dividends.

TABLE 12-4 CORPORATION EXAMPLE

Assets contributed	Total	Party A	Party B
Operating capital	$ 10,000	$ 5,000	$ 5,000
Current feed inventory	15,600	9,500	6,100
Pigs on feed	16,000	10,000	6,000
Breeding livestock	2,900	1,900	1,000
Livestock equipment	11,500	10,200	1,300
Tractor and machinery	27,000	0	27,000
Buildings and land	33,000	33,000	0
Total assets	$116,000	$69,600	$46,400
Liabilities contributed			
Intermediate loan on buildings	$ 16,000	$16,000	0
Net worth contributed	$100,000	$53,600	$46,400
Number of shares @ $100	1,000	536	464

The column header "Value" spans Total, Party A, and Party B.

The number of shares to be issued is an arbitrary decision by the stockholders. Initial value of each share is found by dividing the beginning net worth of the corporation by the number of shares to be issued.

Table 12-5 summarizes the important features of each of the three forms of business organization. The advantages and disadvantages of each feature should be carefully evaluated before one is selected.

TRANSFERRING THE FARM BUSINESS

At the beginning of this chapter the life cycle of a farm business was illustrated (Figure 12-1). If the exit stage of one operator coincides with the entry stage of the next operator, the transfer process may be relatively simple. Livestock, equipment, and machinery may be sold to the new operator or dispersed by an auction. Land may be sold outright, sold on an installment contract, or rented. The size and structure of the business may change very little in the process.

In many family farming situations, however, the next generation is ready to enter the business while the current generation is still in the growth or consolidation stage. This brings up several important questions.

TABLE 12-5   COMPARISON OF FORMS OF FARM BUSINESS ORGANIZATION

Category	Sole proprietor	Partnership	Corporation
Ownership	Single individual	Two or more individuals	A separate legal entity that is owned by its shareholders
Life of business	Terminates on death	Agreed term or terminates at death of a partner	Forever unless fixed by agreement; in case of death, stock passes to heirs
Liability	Proprietor is liable	Each partner liable for all partnership obligations, even to personal assets	Shareholders not liable for corporate obligations; in some cases individual stockholders may be asked to cosign corporation notes
Source of capital	Personal investments, loans	Partnership contributions, loans	Shareholders' contributions, sale of stock, sale of bonds, and loans
Management decisions	Proprietor	Agreement of partners	Shareholders elect directors who manage business
Income taxes	Business income is combined with other income on individual tax return	Partnership files IRS information report; each partner's share of partnership income is added to her or his individual taxable income	Regular (C) corporation: Corporation files a tax return and pays income tax; salaries to shareholder employees are deductible; shareholders pay tax on dividends received Tax option (S) corporation: Shareholders report their share of income, operating loss, and long-term capital gain on individual returns; IRS information report filed by corporation

1 Is the business large enough to productively employ another person? If the labor supply will be decreased through retirement or reduction of hired labor, then a new operator may be efficiently utilized. If not, alternatives for expanding the business may have to be considered.

2 Is the business profitable enough to support another operator? Simply adding more labor does not necessarily produce more net income. If an additional source of income, such as social security payments, will become available to the older generation, then farm income can be more easily diverted to the new operator.

If adding another operator will require expanding the business, a detailed whole farm budget should be completed. Liquidity as well as profitability should be analyzed, especially if additional debt will be used to finance the expansion. The net cash flow that will be available in a typical year should be projected for each person as well as for the business as a whole to avoid unexpected financial stress.

3 Can management responsibilities be shared? If the people involved do not have compatible personalities and mutual goals, then even a profitable business may not provide a satisfying career. Farmers and ranchers who are accustomed to working alone and making decisions independently may find it very difficult to accommodate a partner.

Three key areas of a farm business must be transferred: income, ownership, and management. Typically they are not all transferred at the same time.

Income can be transferred simply by paying a wage which could include some type of bonus or profit sharing. As the younger partner acquires more assets, part of the income can be based on a return on investment as well.

Ownership can be transferred by allowing the younger generation to gradually acquire property, such as by saving breeding stock or investing in machinery. Longer life assets can be sold on contract or by outright sale. Property can also be gifted to the eventual heirs, although if annual giving exceeds certain limits, gift tax may have to be paid. Finally, careful estate planning will assure an orderly transfer of property upon the death of the owner.

Management may be the most difficult part of a farming operation to transfer. The new operator may be given responsibility for one particular enterprise or for a certain management area, such as feeding, breeding, or record keeping. Allowing younger operators to rent additional land or produce a group of livestock on their own while using the farm's machinery or facilities can also help them learn management skills without putting the entire operation at risk.

The specific arrangements chosen for transferring income, ownership, and management depend on the type of organization that is ultimately desired and how quickly the family wants to arrive there. A *testing stage* of one to five years is recommended, during which the entering operator is employed and receives a salary and bonus or incentive, and may own or rent some additional assets. Forms of business organization and acquisition of assets that would be difficult to liquidate should be avoided until both parties determine their ultimate goals, and whether or not they can work together. Following the testing stage at least three alternatives are available, as illustrated in Figure 12-2.

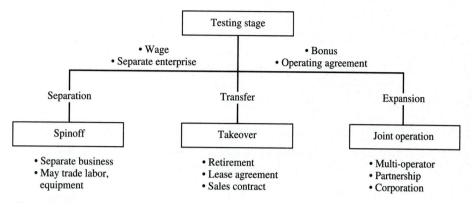

FIGURE 12-2 Alternatives for farm business transfer.

1 The *spin-off* option involves the separation of the operators into their own individual operations. A spin-off works best when both operators wish to be financially and operationally independent or expansion possibilities for the original business are limited. The separate operations may still trade labor and the use of equipment, or even own some assets together, in order to realize some economies of size.

2 The *takeover* option occurs when the older generation phases out of active labor and management, usually to retire or enter another occupation. Expansion of the business may not be necessary. Transfer can take place by renting or selling the farm or ranch to the younger generation. A detailed budget should be developed to make sure that enough income for living expenses will be available after lease or mortgage payments are made.

3 A *joint operation* may be developed when both generations wish to continue farming together. This arrangement often involves an expansion of the business in order to adequately employ and support everyone. The farm may be organized with a joint operating agreement, as a partnership, or as a corporation. These multioperator family farming operations include some of the most profitable and efficient farm businesses to be found, as long as effective personal working and management relationships can be developed.

Of course, the testing stage may also end in a decision by the younger generation to not enter farming as a career. This should not be considered a failure but as a realistic assessment of a person's values and goals.

SUMMARY

This chapter is an introduction to Part Four, which discusses acquiring and managing the resources necessary to implement a farm or ranch production plan. Before the resources to implement the plan are actually acquired, a form of business organization must be selected. The selection should be reviewed from time to time as the business grows and moves through the stages of the life cycle.

A farm or ranch business can be organized as a sole proprietorship, a partnership, or a corporation. There are advantages and disadvantages to each form of business organization, depending on the size of the business, stage of its life cycle, and the desires of the owners. The right business organization can make transferring farm income, ownership, and management to the next generation easier.

QUESTIONS FOR REVIEW AND FURTHER THOUGHT

1 Discuss the differences among the four stages in the life cycle of a farm business.
2 Explain why the sole proprietorship is the most common form of farm business organization.
3 How does a joint operating agreement differ from a partnership?
4 For the joint operating agreement example shown in Table 12-1, how would you divide income if party A and party B each owned one-half of the livestock?
5 Explain the importance of a written partnership agreement. What should be included in a partnership agreement?
6 Does a two-person partnership have to be 50–50? Can it be a 30–70 or a 70–30 partnership? How should the division of income be determined?
7 Explain the differences between a C and S corporation. When would each be advantageous?
8 Why might a partnership or corporation keep more and better records than a sole proprietorship?
9 Compare the personal and corporate tax rates shown in Table 12-3 in this chapter. Would a sole proprietor (personal) or a corporation pay more taxes on a taxable income of $25,000? On $50,000? On $150,000?
10 Would you expect most farm corporations to be above or below average in size? Why?
11 What form of business organization would you choose if you were just beginning a small farming operation on your own? What advantages would this form have for you? What disadvantages?
12 What form of business organization might be preferable if you had just graduated from college and were joining your parents in the operation and management of an existing farm? What advantages and disadvantages would there be to you? To your parents?

REFERENCES

Broadsworth, William R., and Ralph S. Winslade: *Farm Estate Planning,* Ontario Ministry of Agriculture and Food Publication 70, Guelph, Ontario, Canada, 1991.
Boehlje, Michael D., and Vernon R. Eidman: *Farm Management,* John Wiley & Sons, New York, 1984, chap. 9.
Census of Agriculture (1987), U.S. Department of Commerce, Washington, D.C., 1989.
Harl, Neil E.: *The Farm Corporation,* North Central Regional Extension Publication No. 11, rev., January, 1991.
Harl, Neil E.: *Farm Estate & Business Planning,* 11th ed., Century Communications Inc., Niles, IL, 1991.
Harl, Neil E.: *Organizing the Farm Business,* Iowa State University Extension Publication Pm-878, Ames, IA, 1991.

Thomas, Kenneth H., and Michael D. Boehlje: *Farm Business Arrangements: Which One for You?*, North Central Regional Extension Publication No. 50, University of Minnesota, rev. 1991.

Thomas, Kenneth, Robert Luening, and Ralph Hepp: *Planning Your General Farm Partnership Arrangement,* North Central Regional Extension Publication 224, University of Minnesota, rev. 1991.

13

MANAGING RISK AND UNCERTAINTY

CHAPTER OBJECTIVES

1 To discuss the importance of risk and uncertainty in decision making

2 To identify the sources of risk and uncertainty which affect farmers and ranchers

3 To illustrate some different methods for forming expectations about uncertain and risky events

4 To discuss the importance of variability and its relationship to risk and uncertainty

5 To demonstrate several methods that can be used to help make decisions under risk

6 To discuss strategies that can be used to reduce risk or control its effects

Decision making was discussed in Chapter 1 as the principal activity of management, and Chapters 5 through 11 introduced some principles and techniques useful in making management decisions. These chapters implicitly assumed that all the necessary information about input prices, output prices, yields, and other technical data was available, accurate, and known with certainty. This assumption of perfect knowledge simplifies the understanding of a new principle or concept, but it seldom holds true in the real world of agriculture.

We live in a world of uncertainty. There is an old saying that "nothing is certain except death and taxes." Managers find that their best decisions often turn out to be less than perfect because of changes which take place between the time the decision is made and the time the outcome of that decision is finalized. Many agricultural decisions have outcomes that take place months or years after the initial decision is made.

Crop farmers must make decisions about what crops to plant and what seeding rates, fertilizer levels, and other input levels to use early in the cropping season. The ultimate yields and prices obtained will not be known with certainty for several months, or even several years in the case of perennial crops. A rancher who has decided to expand a beef cow herd by raising replacement heifers must wait several years before the first income is received from the calves of the heifers kept for the herd expansion. Unfortunately farmers and ranchers can do little to speed up the biological processes in crop and livestock production.

When an outcome is more favorable than expected, a manager may regret not having implemented the decision more aggressively or on a larger scale. However, in this case the financial health of the operation is enhanced, not threatened. The real risk comes from unexpected outcomes with adverse results such as low prices, drought, or disease. Risk management is mostly concerned with reducing the possibility of unfavorable outcomes, or at least softening their effects.

SOURCES OF RISK AND UNCERTAINTY

There are many sources of risk and uncertainty in agriculture. What are the risks associated with selecting enterprises, determining the proper levels of feed and fertilizer to be used, and borrowing additional money? What makes the outcomes of these decisions difficult to predict? The more common sources of risk in agriculture can be conveniently summarized into three management areas: production, marketing, and financial.

Production and Technical Risk

Manufacturing firms know that the use of a certain collection of inputs will always result in a known quantity of output. This is not the case with most agricultural production processes. Crop and livestock performance depends on biological processes which are affected by weather, diseases, insects, weeds, feed conversion, and soil fertility. These factors cannot be predicted accurately.

Production functions for three possible weather conditions are shown in Figure 13-1. Inputs such as seed and fertilizer must be applied before the weather is known, and regardless of the input level selected, weather will affect the output level. This creates uncertainty about the output which will be received for each input level as well as uncertainty about what input level to use. This uncertainty is often further compounded by less than perfect knowledge about the true technical relationships in the production function.

Another source of production risk is new technology. There is always some risk involved when changing from an old and proven production technology to some new technology. Will the new technology perform as expected? Has it been thoroughly tested? Will it actually reduce costs and/or increase yields? These and other questions

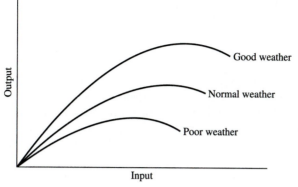

FIGURE 13-1 Production risk due to weather.

must be considered before adopting a new technology. However, not adopting a successful new technology means the operator may miss out on additional profits and become less competitive. The risk associated with adopting new technology is captured in an old saying: "Be not the first upon whom the new is tried nor the last to lay the old aside."

Marketing or Price Risk

Another major source of risk in agriculture is price variability. Commodity prices vary from year to year as well as day to day, due to reasons beyond the control of an individual farmer. The supply of a commodity is affected by production decisions made by many farmers and the resulting weather. Demand for a commodity is a result of consumer incomes, exchange rates, export policies, the strength of the general economy, and the supply of competing products. Some price movements follow seasonal or cyclical trends, which can be predicted, but even these movements exhibit a great deal of volatility.

Costs represent another source of price risk. Input prices tend to be less variable than output prices but still add to uncertainty. The cost of production per unit of output depends on both costs and yield and therefore can be highly variable.

Financial Risk

Financial risk is incurred when money is borrowed to finance the operation of the business. This risk is caused by uncertainty about future interest rates, a lender's willingness to continue lending at the levels needed now and in the future, changes in market values of loan collateral, and the ability of the business to generate the cash flows necessary for debt payments.

Production, marketing, and financial risk exist on most farms and ranches and are interrelated. The ability to repay debt depends on production levels and prices received for the production. Financing the production and storage of commodities depends on the ability to borrow the necessary capital. Therefore, all three types of risk need to be considered together, particularly when developing a whole farm plan.

RISK-BEARING ABILITY AND ATTITUDES

Farmers and ranchers vary greatly in their willingness to take risks and in their abilities to survive any unfavorable outcomes of risky actions. Therefore, the level of risk which a farm business should accept is very much an individual decision.

Ability to Bear Risk

Financial reserves play a big part in determining an operation's risk-bearing ability. Farms with a large *net worth* obviously can stand larger losses before they become insolvent. Highly *leveraged* farms, with a high debt relative to assets, can lose net worth quickly because their volume of production is high relative to their equity. These farms are also more vulnerable to financial risks such as rising interest rates.

Cash flow commitments also affect risk-bearing ability. Families with high fixed living expenses, educational expenses, or health care costs are less able to withstand a low income year and should not expose themselves to as much risk as other operations. Farms that have more of their assets in a liquid form such as a savings account, have secure off-farm employment, or can depend on relatives or friends to assist them in a financial emergency also have greater risk-bearing ability.

Willingness to Bear Risk

Some farmers and ranchers refuse to take risks even though they have no debt and a strong cash flow. They may have experienced financial setbacks in the past or be concerned about having enough income for retirement.

People can be classified as *risk lovers* or *risk avoiders*. Risk lovers prefer risky alternatives to safe ones even when the average outcomes are the same. They derive satisfaction from taking risks. Most farm and ranch operators are risk avoiders. They are willing to take some risks, but only if they can reasonably expect to increase their long-run returns by doing so. Some factors that influence to what degree farmers are willing to take risks include age, net worth, financial commitments, past financial experiences, the size of the gains or losses involved, family responsibilities, their familiarity with the risky proposition, emotional health, cultural values, and community attitudes.

EXPECTATIONS AND VARIABILITY

The existence of risk adds complexity to many problems and to the decision-making process. Yet decisions must be made, and made with consideration of risk and uncertainty. In an environment where risk exists, decisions are often made using some type of average or "expected" value for yields, costs, and prices. There is no assurance that this value will be the actual outcome each time, however. The actual outcome of the decision will only be known at some point in the future. To make decisions in a risky world, a manager needs to understand how to form expectations, how to use probabilities, and how to analyze the variability associated with the potential outcomes.

Probabilities are often needed or useful when forming an expectation. However, the true or actual probabilities are seldom known and *subjective probabilities* must be used. They are derived from whatever information is available plus the experience and judgment of the individual. The probability of rain in a weather forecast or the outcome of a sporting event are examples of subjective probabilities. Since each individual has had different experiences and may interpret available data differently, subjective probabilities will vary from individual to individual. This is one reason individuals make different decisions when faced with the same risky decision.

Forming Expectations

Several methods can be used to form expectations about future prices, yields, and other values which are not known with certainty. Once a value is chosen it can be used for planning and decision making as it becomes the "best estimate" of some unknown value which will be determined by future events.

Most Likely   One way to form an expectation is to choose the value that is most likely to occur. This procedure requires a knowledge of the probabilities associated with each possible outcome, either actual or subjective. The outcome with the highest probability is selected as the most likely to occur. An example is contained in Table 13-1 where four possible ranges of wheat yields are shown along with the estimated probabilities of the actual yield falling into each one. Using the "most likely" method to form an expectation, a yield of 21 to 27 bushels per acre would be selected. For budgeting purposes we can use the midpoint of the range, or 24 bushels. There is no assurance the actual yield will be between 21 and 27 bushels per acre in any given year, but if the probabilities are correct, it will occur 40 percent of the time over a long period.

TABLE 13-1   USING PROBABILITIES TO FORM EXPECTATIONS

Possible wheat yields	Midpoint	Probability	Probability × midpoint
9–15	12	0.1	1.2
15–21	18	0.3	5.4
21–27	24	0.4	9.6
27–33	30	0.2	6.0
Total		1.0	Expected value 22.2

Averages Two types of averages can be used to form an expected value. A *simple average,* or *mean,* can be calculated from a series of actual past prices or yields. The primary problem is selecting the length of the data series to use in calculating the simple average. Should the average be for the past 3, 5, or 10 years? As long as the fundamental conditions that affect the observed outcomes have not changed, as many years of data should be used as are available.

In some cases the fundamental conditions have changed. New technology may have increased potential crop yields and changes in demand may have affected prices. In these cases a method that gives recent values more importance than the older ones can be used to calculate a *weighted average.* A weighted average can also be used when true or subjective probabilities of the expected outcomes are available, but not all equal. A weighted average that uses probabilities as weights is called an *expected value.* The expected value is an estimate of what the simple average would be if the event were repeated many times. Of course, the accuracy of the expected value depends on the accuracy of the probabilities used.

Table 13-2 shows an example of using both simple and weighted average methods. Four years of price information is used, which gives a simple average of $2.96. As an alternative, the most recent value may be assigned a weight equal to the number of observations and the weight for each prior observation decreased by one. Each price is multiplied by its assigned weight, the results summed and then divided by the sum of the weights. The expected price would be $3.11 using the weighted average method. This method assumes that recent prices more accurately reflect current supply and demand conditions, while the simple average treats each year's results equally.

An example of calculating an expected value is shown in the right-hand column of Table 13-1. Subjective probabilities have been assigned to each yield interval. The midpoint of each possible outcome is multiplied by its associated probability, and the results are summed to find the expected value. Notice that the expected value of 22.2 bushels per acre is less than the most likely yield of 24 bushels because yields lower than 24 are more likely to occur than yields higher than 24.

TABLE 13-2 USING AVERAGES TO FORM AN EXPECTED VALUE

Year	Average annual price	Weight	Price × weight
4 years ago	$ 2.43	1	$ 2.43
3 years ago	3.02	2	6.04
2 years ago	2.94	3	8.82
Last year	3.46	4	13.84
Summation	$11.85	10	$31.13

Simple average: $\dfrac{\$11.85}{4} = \$2.96$ Weighted average: $\dfrac{\$31.13}{10} = \$3.11$

Variability

A manager who must select from two or more alternatives should consider another factor in addition to the expected values. The variability of the possible outcomes around the expected value is also important. For example, if two alternatives have the same expected value, most managers will choose the one whose potential outcomes have the least variability because there will be smaller deviations from its expected outcome with which to deal.

Range One simple measure of variability is the difference between the lowest and highest possible outcomes, or the *range.* Alternatives with a smaller range are preferred over those with a wider range if their expected values are the same. The range is not the best measure of variability, however, because it does not consider the probabilities associated with the highest and lowest values, nor the other outcomes within the range and their probabilities.

Standard Deviation One common statistical measure of variability is the standard deviation.[1] It can be estimated from a sample of actual outcomes for a particular event, such as historical price data for a certain week of the year. A larger standard deviation indicates a greater variability of possible outcomes and therefore a higher probability that the actual outcome will be quite different from the average or mean value.

Coefficient of Variation The standard deviation is difficult to interpret, however, when comparing two types of occurrences that have quite different means. The occurrence with the higher mean value often has a larger standard deviation. In this situation, it is more useful to look at the relative variability. The *coefficient of variation* measures variability relative to the mean and is found by dividing the standard deviation by the mean. Smaller coefficients of variation indicate the distribution has less variability in relation to its mean than other distributions.

$$\text{Coefficient of variation} = \frac{\text{standard deviation}}{\text{mean}}$$

Table 13-3 shows historical data for corn and soybean yields on an individual farm. If we assume that production potential has not changed over time, we can use the simple means as estimates of the expected yields for next year. Notice that corn had a

---

[1]Standard deviation is equal to the square root of variance. The equation for variance is

$$\text{Variance} = \frac{\sum_i \left(X_i - \overline{X} \right)^2}{n - 1}$$

where X_i is each of the observed values, $\overline{X}$ is the average of the observed values, and n is the number of observations.

TABLE 13-3 HISTORICAL CORN AND SOYBEAN YIELDS FOR AN INDIVIDUAL FARM

Year	Corn (bushels/acre)	Soybeans (bushels/acre)
1	125	35
2	145	45
3	141	38
4	88	28
5	105	33
6	129	44
7	118	40
8	75	21
9	132	37
10	127	48
Mean (expected value)	118.5	36.9
Standard deviation	21.5	7.8
Coefficient of variation	0.18	0.21

greater standard deviation than soybeans. However, comparing the coefficients of variation shows that soybean yields were actually more variable than corn yields when compared to their respective means. Thus, an operator who wished to reduce yield risk might prefer corn to soybeans.

Cumulative Distribution Function   Many risky events in agriculture have an almost unlimited number of possible outcomes, and the probability for each one becomes very small. A useful format for portraying a large number of possible outcomes is a *cumulative distribution function,* or CDF. The CDF is a graph of the values for all possible outcomes for an event and the probability that each outcome will be equal to or less than that value. The outcome with the smallest possible value has a cumulative probability of nearly zero, while the largest possible value has a cumulative probability of 100 percent.

The steps in creating a CDF are as follows:

1 List a set of possible values for the outcome of an event or strategy and estimate their probabilities. For example, the data from Table 13-3 can be used as possible values for corn and soybean yields. If it is assumed that each of the 10 historical observations has an equal chance of occurring again, each one represents 10 percent of the total possible outcomes.

2 List the values in order from lowest to highest, as shown in Table 13-4.

3 Assign a *cumulative* probability to the lowest value equal to one-half of the range it represents. Since each observation actually represents one segment or range out of the total distribution, it can be assumed that the observation falls in the middle of the range. For the example, the lowest yield observed represents the first 10 percent of the distribution, so it can be assigned a cumulative probability of 5 percent.

TABLE 13-4 CUMULATIVE PROBABILITY DISTRIBUTIONS FOR CORN AND SOYBEAN YIELDS

Cumulative probability (%)	Corn (bushels/acre)	Soybeans (bushels/acre)
5	75	21
15	88	28
25	105	33
35	118	35
45	125	37
55	127	38
65	129	40
75	132	44
85	141	45
95	145	48

4 Calculate the cumulative probabilities (probability of obtaining that value or a smaller one) for each of the other values by adding the probabilities of all the smaller values to one-half of that value's own probability. In the example the remaining observed yields would have cumulative probabilities of 15 percent, 25 percent, and so on.

5 Graph each pair of points and connect them, as shown in Figure 13-2.

The cumulative distribution function permits a view of all possible results for a certain event. The more vertical the graph, the less variability there is among the possible outcomes. Note that the upper portions of the graphs in Figure 13-2 are steeper than the lower portions, indicating the positive yield responses to good weather are not as significant as the negative responses to poor growing weather.

DECISION MAKING UNDER RISK

Making risky decisions requires careful consideration of the various strategies available and the possible outcomes of each. The process can be broken down into several steps:

1 Identify the possible sources of risk.

2 Identify the possible outcomes or events that can occur, such as variations in weather or prices.

3 Decide on the alternative strategies available.

4 Quantify the consequences or results of each possible outcome for each strategy.

5 Evaluate the trade-offs between risk and returns.

An example can be used to illustrate these steps. Assume a wheat farmer plants a given number of acres of wheat in the fall. Stocker steers can be purchased in the fall and grazed on the wheat over the winter and sold in the spring. For simplicity, we will assume all the steers must be purchased and sold at the same time. The farmer's problem is deciding how many steers to purchase without knowing what weather will

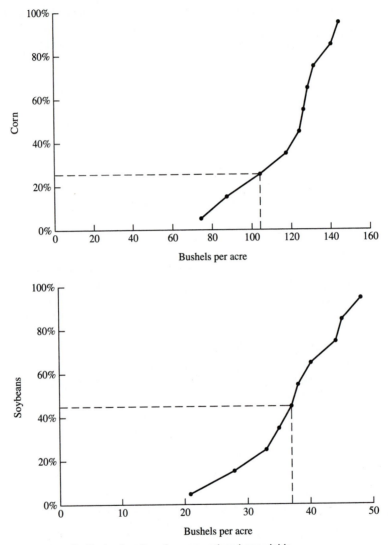

FIGURE 13-2 Cumulative distribution functions for corn and soybean yields.

occur and how much grazing will be available. If too few steers are purchased and the weather is *good*, there will be excess grazing available and an opportunity for additional profit will be lost. If too many steers are purchased and the weather is *poor*, there will not be enough grazing, feed may have to be purchased, and profit will be reduced or a loss incurred. A third possibility, of course, is for *average* weather to occur.

Further assume that the probabilities for good, average, or poor weather are estimated to be 20, 50, and 30 percent, respectively. The selection of probabilities is important and may be estimated by studying past weather events as well as near-term forecasts. The farmer is considering three strategies: purchase 300, 400, or 500 steers. These choices are the decision strategies. The same three weather outcomes are possible for each strategy, which creates nine potential combinations of results to consider.

Once the elements of the problem are defined, it is helpful to organize the information in some way. Two ways of doing this are with a decision tree or with a payoff matrix.

Decision Tree

A decision tree is a diagram that traces out all possible strategies, potential outcomes, and their consequences. Figure 13-3 is a decision tree for the example described above. Note that it shows three potential weather outcomes for each of three strategies, the probability for each outcome (which is the same regardless of the strategy chosen), and the estimated net returns for each of the nine potential consequences. For example, if 300 steers are purchased, the net return is $20,000 with good weather, $10,000 with average weather, and only $6,000 with poor weather.

The expected value for each strategy is the weighted average of the possible outcomes, as was demonstrated earlier in this chapter. Based on these values alone, it

FIGURE 13-3   Decision tree for stocker steer example.

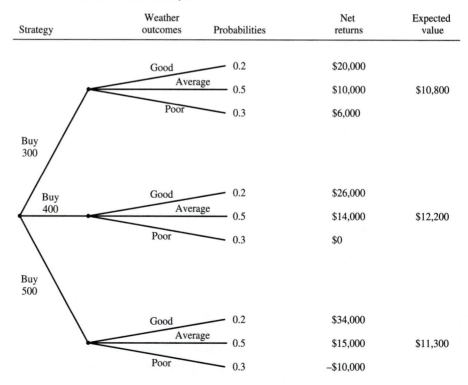

TABLE 13-5   PAYOFF MATRIX FOR STOCKER STEER PROBLEM

Weather outcomes	Probability	Purchase strategy		
		Buy 300	Buy 400	Buy 500
Good	0.2	$20,000	$26,000	$34,000
Average	0.5	10,000	14,000	15,000
Poor	0.3	6,000	0	–10,000
Expected value		10,800	12,200	11,300
Minimum value		6,000	0	–10,000
Maximum value		20,000	26,000	34,000
Range		14,000	26,000	44,000

might be expected that the farmer would select the "buy 400" strategy, as it has the highest expected value. However, this overlooks the possibility of only breaking even. Ways of making a decision taking this risk into account will be shown later.

Payoff Matrix

A payoff matrix contains the same information as a decision tree but is organized in the form of a contingency table. The top part of Table 13-5 shows the consequences of each strategy for each of the potential weather outcomes for the stocker steer example. The expected values as well as the minimum and maximum values are shown in the bottom part.

Decision Rules

A number of decision rules have been developed to help choose the appropriate strategy when faced with a problem involving risk. Using different rules may result in the selection of different strategies. The appropriate rule to use will depend on the decision maker's attitude toward risk, the financial condition of the business, cash flow requirements, and other factors. Since these factors will vary widely among decision makers, it is impossible to say that a certain rule is best for everyone.

Most Likely Outcome   This decision rule identifies the outcome that is most likely (has the highest probability) and chooses the strategy with the best consequence for that outcome. Table 13-5 indicates that average weather has the highest probability at 0.5 and the buy 500 strategy has the greatest net return for that outcome. This is a decision rule that does not consider the variability of the consequences nor the probabilities of the other possible outcomes, however.

Maximize Expected Value   This decision rule says to select the strategy with the highest expected value. The expected value represents the average result for a particular strategy *if* that strategy is used over a long period of time and the probabilities are correct.

Both Figure 13-3 and Table 13-5 show that the buy 400 strategy has the highest expected value, so it would be selected using this rule. This rule will result in the highest average net return over time, but it ignores the variability of the outcomes. In the example, poor weather has a probability of 0.3, or 3 years out of 10 on the average. However, there is no guarantee that poor weather will not occur 2 or 3 years in a row with the resulting $0 net return each year. The rule of choosing the maximum expected value ignores variability and should be used only by managers who have good risk-bearing ability and are neither risk lovers nor risk avoiders.

Risk and Returns Comparison   Managers who are risk averse and do not have complete risk-bearing ability must consider both the expected returns and the risk associated with various strategies. Any strategy that has both a lower expected return and higher risk than another strategy should be rejected. Such is the case with the buy 500 strategy in Table 13-5. It has a lower expected value than the buy 400 strategy, and also more risk, as measured by the range between the minimum and maximum results. The buy 300 strategy also has a lower expected return than the buy 400 strategy but is less risky. Therefore, managers who are very risk conscious might prefer this strategy.

One drawback to using measures such as the range, standard deviation, and coefficient of variation to measure risk is that they treat better than average results the same as worse than average results. However, only the unfavorable outcomes are of real concern to the decision maker. Thus, several decision rules have been developed that look only at the possible adverse results when assessing the riskiness of a decision.

Maxi-min   The *maxi-min* rule concentrates on the worst possible outcome for each strategy and ignores the other possible outcomes. It is as if the decision maker assumes that no matter which strategy is selected, the worst outcome will occur. Therefore, the strategy with the best of the worst possible results is selected, that is, the one with the maximum minimum value. Referring to Table 13-5, application of the maxi-min rule would result in selection of the buy 300 strategy, as its minimum consequence of a $6,000 profit is higher than the other minimums, which are a $0 profit for buying 400 steers and a loss of $10,000 for buying 500 steers.

The maxi-min rule assumes that above-average outcomes pose no problem, only the worst ones. However, it ignores the other outcomes and the probabilities associated with them. This rule would be appropriate for a business in such poor financial condition that it could not survive the consequences of one of the poorer outcomes.

Minimum Returns   Because of the financial condition of the business, the decision maker may want a reasonably high probability of some minimum net return to ensure its survival. In our example, if the decision maker wanted 100 percent assurance of a $5,000 net return or greater, only the buy 300 strategy would be acceptable. On the other hand, if the decision maker wanted a $10,000 minimum net return with a probability of at least 0.9, none of the strategies would meet this requirement. Each of them has a 0.3 probability of a net return lower than $10,000. This strategy is sometimes called the "safety first" strategy and requires a careful estimate of the minimum return that the business needs in order to survive. If more than one strategy has an acceptable chance of exceeding the minimum desired return, then the one with the highest expected value is preferred.

Break-even Probability   Knowing the probability that a certain strategy will result in a financial loss can also help the decision maker choose. Suppose, for example, that the farmer with the corn and soybean yield histories shown in Table 13-3 calculated that a yield of 106 bushels per acre for corn or 37 bushels per acre for soybeans was needed to just pay all production costs. From the CDF graphs in Figure 13-2, the probability of realizing less than the break-even level of production can be estimated by drawing a vertical line from the break-even yield to the CDF line, then a horizontal line to the cumulative probability scale. In this example, the probability of suffering an economic loss (producing a below break-even yield) is approximately 26 percent for corn and 45 percent for soybeans. Thus, soybeans carry more financial risk. However, the risk of a loss must also be weighed against the expected average return from each crop.

CONTROLLING RISK AND UNCERTAINTY

There are three general and perhaps related reasons why a manager would be interested in taking steps to reduce risk and uncertainty. The first is to reduce the variability of income over time. This allows more accurate planning for items such as debt payment, family living expenses, and business growth. Second, there may be a need to ensure some minimum income level to meet family living expenses and other fixed expenses. A third reason for minimizing risk is to enhance the survival of the business. Several consecutive years of low income may threaten business survival or result in bankruptcy. Some recent studies show many managers rate business survival as their most important goal. They are willing to accept a lower expected income if it reduces income variability and hence the risk of business failure.

There are four general classes of management techniques for controlling risk. Some reduce the amount of risk the manager faces while others help soften the impact of an undesirable result.

1 Some techniques reduce the variability of possible outcomes. The probability of a bad result is reduced, but the probability of a good result may also be reduced.

2 A minimum income or price level is set, usually for a fixed charge. Most insurance programs operate this way. The cost of the risk reduction is known, and the probability of a very good result is not affected.

3 Flexibility of decision making is maintained. Managers do not "lock in" decisions for long periods of time in case price or production prospects change.

4 The risk-bearing ability of the business is improved, so that an adverse result is less likely to affect the survival of the farming operation.

The various methods which can be used to reduce risk will be discussed under the three management areas identified earlier in this chapter: production, marketing, and financial.

Production Risk

Production risk is caused by variability of crop yields, livestock production rates, and other environmental occurrences. Several strategies can be used to reduce production risk.

Diversification   Many business firms produce more than one product to avoid having their income totally dependent on the production and price of one product. If profit from one product is poor, profit from producing and selling other products may prevent total profit from falling below acceptable levels. In agricultural production, diversifying by producing two or more commodities may reduce income variability if all prices and yields are not low or high at the same time.

Table 13-6 shows an example of how diversification can work to reduce income variability. Based on the average income for the past 5 years, specialization in crop B would provide the greatest average income, $20,000. However, this crop also has the greatest range between the highest and lowest results. Producing only crop A would reduce income variability but also reduce the average income. A third possibility is to diversify by utilizing half of the farm's resources in each crop. This results in an intermediate average income but a large reduction in the range and variability of income. Notice that whenever the income for crop A is below average the income for crop B is above average and vice versa. This negative correlation "smooths out" the annual income under diversification and reduces income variability.

To what extent will diversification reduce income variability in an actual farm situation? The answer depends on the price and yield correlations for the enterprises selected. If prices or yields for both of the enterprises tend to move up and down together, little is gained by diversifying. The less these values tend to move together or the

TABLE 13-6   THEORETICAL EXAMPLE OF DIVERSIFICATION

	Annual income ($)		
Year	Crop A	Crop B	Diversification (1/2 of each)
1	5,000	50,000	27,500
2	20,000	0	10,000
3	5,000	30,000	17,500
4	50,000	−5,000	22,500
5	10,000	25,000	17,500
Average	18,000	20,000	19,000
Range	45,000	55,000	17,500
Standard deviation	18,908	22,638	6,519
Coefficient of variation	1.05	1.13	0.34

more they move in opposite directions, the more income variability will be reduced by diversifying.

Weather is the primary factor influencing crop yields. Crops with the same growing season experience the same weather, and their yields tend to have a strong positive correlation. The yield correlation for crops that have different growing seasons and are susceptible to different insects and diseases will be somewhat lower. Production rates among different types of livestock are less closely correlated, and there is little correlation between crop yields and livestock performance.

Most studies on the price correlations for major agricultural commodities show that pairs of commodities with a strong yield correlation often have a positive price correlation as well, since year-to-year production changes have a major impact on prices. Some specialty crops such as fruits and vegetables, however, may show a weak or even negative price correlation with some of the major field crops.

Diversification plans can include nonfarm activities as well. Investing in stocks and bonds, carrying out a part-time business unrelated to agriculture, or holding a nonfarm job can all improve the stability of family income. Diversification may mean giving up the benefits of specializing in one enterprise in order to gain the benefits from less variability in net income.

Stable Enterprises   Some agricultural enterprises have a history of more stable income than others. Modern technology may be able to control the effects of weather on production or government programs and marketing orders may control prices or the quantity of a commodity that can be sold. For example, dairy enterprises tend to have relatively stable incomes as do farrow-to-finish hog enterprises. Irrigation will provide more stable crop yields than dryland farming in areas where rainfall is marginal or highly variable during the growing season. At the other extreme, enterprises such as livestock finishing, where both buying and selling prices can vary, and perishable products such as vegetables tend to have highly variable incomes. Production risk can therefore be reduced by selecting stable enterprises for the whole farm plan.

Insurance   There are several types of insurance that will help reduce production and financial risk. Formal insurance can be obtained from an insurance company to cover many types of risk which, should a bad outcome occur, could have a serious impact on business equity and survival. The alternative is for the business to be self-insured, that is, to maintain some type of readily available or liquid financial reserves in case a loss does occur. Without these financial reserves a crop failure, major windstorm, or fire might cause such a large financial setback that the business could not continue.

Of course, financial reserves held in a form that can be easily liquidated, such as a savings account, often earn a lower rate of return than the same capital would if it were invested in the farm business or some other long-term investment. This sacrifice in earnings is the opportunity cost of being self-insured and should be compared to the premium cost for an insurance policy that would provide the same coverage.

There are several common types of insurance:

1 *Life Insurance*   Life insurance is available to provide protection against losses that might result from the untimely death of the farm operator or a member of the family. The insurance proceeds can be used to meet family living expenses, pay off existing debts, pay inheritance taxes, and meet other expenses related to transferring management and ownership of the business. Care should be taken to select the most suitable type of life insurance for each individual's needs and interests.

2 *Property Insurance*   Property insurance protects against the loss of buildings, machinery, livestock, and stored grain from fire, lightning, windstorm, theft, and other perils. Property insurance is relatively inexpensive, while the loss from a serious fire or windstorm can be devastating. Therefore, most farmers and ranchers choose to carry at least a minimum level of property insurance on their more valuable assets.

3 *Liability Insurance*   Liability insurance protects against lawsuits by third parties for personal injury or property damage for which the insured or employees may be liable. Liability claims on a farm may occur when livestock wander onto a road and cause an accident or when someone is injured on the farm property. The risk of a liability claim may be small, but some of the damages awarded by the courts in recent years have been very large. Most people find liability insurance an inexpensive method to provide some peace of mind and protection against unexpected events.

4 *Crop Hail Insurance*   Crop hail insurance makes payments only for crop losses caused by hailstorms. A number of private companies provide this type of coverage, and the cost will depend on the amount of coverage desired and the past frequency and intensity of hailstorms in the area.

5 *Multiple Peril Crop Insurance*   Multiple peril (all risk) crop insurance is an insurance program backed by the U.S. Department of Agriculture with policies sold through private insurers. Several levels of yield protection are available based on 35, 50, 65, or 75 percent of the "proven yield" for the insured farm. This insurance provides protection against essentially everything except neglect, poor management practices, and theft. Examples of covered losses would be those due to drought, floods, hail, an early freeze, and insects.

Extra Production Capacity   When poor weather delays planting or harvesting of crops, many farmers depend on having excess machinery or labor capacity to help them catch up. They incur higher than necessary machinery ownership costs or wages some years as insurance against crop losses that could occur because of late planting or harvesting in other years. Some operators prefer to own new machinery to reduce the risk of breakdowns at crucial times or unexpected repair bills.

Share Leases   In many states, crop share or livestock share leases are common. The landowner usually pays part of the operating expenses and receives a portion of the crop or livestock produced instead of a cash rental payment. In this way the risk of poor production, low selling prices, or high input costs is shared between the tenant and the owner. The tenant also needs less operating capital under a share lease.

Custom Farming   Rather than risk uncertain prices and yields some operators prefer to custom farm cropland. They perform all machinery field operations for a landowner in exchange for a fixed payment. The landowner takes all the price and yield risk. Custom feeding is a similar arrangement. Livestock producers feed animals or poultry owned by investors in their own lot for a fixed price. Contracts for producing feeder livestock are also available. All custom contracts should be analyzed carefully to compare their potential risks and returns to those of being an independent producer.

Market Risk

Market risk exists because of the variability of commodity prices and because the manager does not know what future market prices will be when making the decision to produce a commodity. Several methods can be used to reduce price variability or to set a satisfactory price in advance of when crops or livestock are ready for sale.

Spreading Sales   Instead of selling all of a crop at one time, many farmers prefer to sell part of the crop at several times during the year. For example, 25 percent of the crop may be sold every 3 months or one-sixth every 2 months. Spreading sales avoids selling all of the crop at the lowest price of the year but also prevents selling all of it at the highest price. The result of spreading sales should be an average price received that is near the average annual price for the commodity.

Livestock sales can also be spread throughout the year. This can be accomplished by feeding several groups during the year or by calving or farrowing several times. Spreading sales of dairy and egg products occurs naturally, due to the continuous nature of their production.

Contract Sales   Producers of some specialty crops such as seed corn and vegetables often sign a contract with a buyer or processor before planting the crop. This contract will usually specify certain management practices to be followed as well as the price to be received for the crop and possibly the amount to be delivered. A contract of this type removes the price risk as the price is known at planting time.

It is also possible to obtain a forward price contract for many field crops and some types of livestock. Many grain and livestock buyers will contract to purchase a given amount of these commodities at a set price for delivery in a later month. These contracts are often available during the growing season as well as after the crop has been harvested and stored. Contract sales remove price uncertainty but do not allow selling at a higher price if prices rise later in the year.

Hedging   A market price can be established within a fairly narrow range by hedging on the commodity futures market. Hedging is possible before the crop is planted as well as during the growing season. Livestock can also be hedged at the time of purchase or any time during the feeding period.

Hedging involves selling a commodity futures contract instead of the actual commodity, usually because the manager is currently unable or unwilling to deliver the commodity. When it is time to deliver the commodity, the futures contract is repurchased and the commodity is sold on the cash market. Cash and futures prices usually move up or down together. Thus, any gain or loss that occurred because the cash market went up or down is offset by a corresponding loss or gain on the futures contract that was being held.

Before beginning a hedging program, a manager should carefully study the hedging process and the futures market and have a good understanding of *basis,* or the normal difference between the futures contract price and the local cash market price. The basis can become wider or narrower while the futures contract is being held, meaning that gains and losses in the cash and futures markets do not exactly offset each other. This variation is called *basis risk.*

Hedging can also be used to lock in the price of inputs to be purchased in the future, such as feedstuffs or feeder livestock. In this case a futures contract is *purchased* in advance, before it is feasible to take delivery of the input.

Commodity Options   Many marketers dislike the fact that although hedging protects them against falling prices, it also prevents them from benefiting from rising prices. They prefer to use commodity options that allow them to set a minimum price in exchange for paying a set fee or premium.

A manager who wants to protect against a price decline will generally buy the right to sell a futures contract at a specified price, called a *put.* If the market goes down (both cash and futures), the value of the put goes up and offsets the loss in value of the commodity the farmer is holding. If the market goes up, the value of the put goes down and may eventually reach zero. If the market rises even further, the value of the commodity being held continues to rise, but there is no further decline in the value of the put (below zero), and the farmer gains. When it is time to sell the commodity the put is also sold, if it still has some value. Thus, for the cost of buying the put (the premium), the farmer is protected against falling prices but can still benefit from rising prices. The commodity option provides a type of "price insurance."

Government Farm Programs   Government programs are always subject to change, but in the past they have provided ways to guarantee a minimum price for certain commodities. Participation may require setting aside or idling a percentage of the crop acres in exchange for some type of price guarantee. This guarantee or minimum price may be a direct payment to the producer to make up the difference between the actual market price and a predetermined "target" price. This is called a deficiency payment.

Some programs offer the producer a short-term loan equal to a fixed rate per bushel produced. The producer cannot sell the commodity until the loan is repaid. The loan can also be canceled by turning over the commodity to the government. Thus, the loan rate effectively becomes a guaranteed minimum price.

Flexibility   Some management strategies allow the operator to change a decision if price trends or weather conditions change. Planting annual crops instead of perennial or permanent crops is one example. Investing in buildings and equipment that can be used for more than one enterprise is another. Many grain producers build storage bins so that they can delay marketing until prices are more favorable. In the case of livestock, animals may be sold as feeder livestock or finished to slaughter weight. Renting certain assets such as land or machinery instead of buying them is another example of maintaining flexibility.

Financial Risk

Reducing financial risk requires strategies to maintain liquidity and solvency. Liquidity is needed to provide the cash to pay debt obligations and to meet unexpected financial needs in the short run. Solvency is related to long-run business survival, or having enough assets to adequately secure the debts of the business.

Fixed Interest Rates   Many lenders offer loans at either fixed or variable interest rates. The fixed interest rate may be higher when the loan is made but prevents the cost of the loan from increasing if interest rates trend upward.

Self-Liquidating Loans   Self-liquidating loans are loans that can be repaid from the sale of the loan collateral. Examples are loans for the purchase of feeder livestock and crop production loans. Personal loans and loans for land or machinery are examples of loans that are not self-liquidating. The advantage of self-liquidating loans is that the source of the cash to be used for repayment is known and relatively dependable.

Liquid Reserve   Holding a reserve of cash or other assets that are easily converted to cash will help the farm weather the adverse results of a risky strategy. However, there may be an opportunity cost for keeping funds in reserve rather than investing them in the business.

Credit Reserve   Many farmers do not borrow up to the limit imposed on them by their lender. This unused credit or credit reserve means additional loan funds can be obtained in the event of some unfavorable outcome. This technique does not directly reduce risk but does provide a safety margin. However, it also has an opportunity cost, equal to the additional profit this unused capital might have earned in the business.

Net Worth   In the final analysis, it is the net worth of the business that provides the solvency and much of the liquidity. Therefore, net worth should be steadily increased, particularly during the early years of the business.

SUMMARY

We live in a world of uncertainty. Rarely do we know the exact what, when, where, how, and how much of any decision and its possible outcomes. Decisions must still be made, however, using whatever information and techniques are available. No one will make a correct decision every time, but decision making under uncertainty can be improved by knowing how to identify possible strategies and events, estimate expected values, and analyze the variability of possible outcomes.

Decision trees and payoff matrices can be used to organize the outcomes of different strategies when faced with a risky decision. Several decision rules can be used to choose among risky alternatives. Some consider only expected returns, some take into account the variability of outcomes both above and below the mean, and some look only at adverse results.

Production, marketing, and financial risk can be reduced or controlled using a number of techniques that reduce the range of possible outcomes, guarantee a minimum result in exchange for a fixed cost, provide more flexibility for making decisions, or increase the risk-bearing ability of the business.

QUESTIONS FOR REVIEW AND FURTHER THOUGHT

1 List at least five sources of risk and uncertainty for farmers in your area. Classify them as production, price, or financial risk. Which are the most important? Why?

2 How might a young farmer with heavy debt view risk compared with an older, established farmer with little debt?

3 Might these same two farmers have different ideas about the amount of insurance they need? Why?

4 How do subjective probabilities differ from true probabilities? What sources of information are available to help form them?

5 Suppose you have developed the data in Table 13-2 for a crop you grow and want to use it to make production and marketing decisions. You are offered a contract which will guarantee you a price of $3.00. Would you accept it? Why or why not?

6 Select five prices which you feel might be the average price for choice fed steers next year, being sure to include both the lowest and highest price you would expect. Ask a classmate to do the same.

  a Who has the greatest range of expected prices?

  b Calculate the simple average for each set of prices. How do they compare?

  c Assign your own subjective probabilities to each price on your list, remembering that the sum of the probabilities must equal 1. Compare them.

  d Calculate the expected value for each set of prices and compare them.

  e Would you expect everyone in the class to have the same set of prices and subjective probabilities? Why?

7 Assume that average annual wheat prices in your region for the past 10 years have been as follows:

$3.45	$2.56
3.51	3.28
3.39	3.87
3.08	2.64
2.42	2.80

  a Calculate the simple average or mean. Calculate the coefficient of variation given that the standard deviation is 0.478.

  b Draw a cumulative distribution function for the price of wheat.

  c From the CDF estimate the probability that the annual average price of wheat will be $3.00 or less.

8 Identify the steps in making a risky decision.

9 Suppose that a crop consultant estimates that there is a 20 percent chance of a major insect infestation which could reduce your per acre gross margin by $60. If no insect

damage occurs, you expect to receive a gross margin of $120 per acre. Treatment for the insect is effective but costs $15 per acre. Show the possible results of treating or not treating in a decision tree and in a payoff matrix. What is the expected gross margin for each choice? What is the range between the high and low outcomes for each choice?

10 Would a manager who did not need to consider risk choose to treat or not treat for insects in Question 9? What about a manager who could not afford to earn a gross margin of less than $90 per acre?

11 Give three examples of risk management strategies that fall into each of the following general categories:
a Reduce the range of possible outcomes
b Guarantee a minimum result for a fixed cost
c Increase flexibility of decision making
d Improve risk-bearing ability

REFERENCES

Anderson, Jock R., John L. Dillon, and J. Brian Hardaker: *Agricultural Decision Analysis,* Iowa State University Press, Ames, IA, 1977.

Boehlje, Michael D., and Vernon R. Eidman: *Farm Management,* John Wiley & Sons, New York, 1984, chap. 11.

Calkins, Peter H., and Dennis D. DiPietre: *Farm Business Management,* Macmillan Publishing Co., New York, 1983, chaps. 8, 9.

Harsh, Stephen B., Larry J. Connor, and Gerald D. Schwab: *Managing the Farm Business,* Prentice-Hall, Englewood Cliffs, NJ, 1981, chap. 14.

Nelson, A. Gene, George L. Casler, and Odell L. Walker: *Making Farm Decisions in a Risky World: A Guidebook,* Oregon State University Extension Service, July 1978.

Osburn, Donald D., and Kenneth C. Schneeberger: *Modern Agricultural Management,* 2d ed., Reston Publishing Co., Reston, VA, 1983, chap. 20.

Patrick, George: *Managing Risk in Agriculture,* NCR Extension Publication 406, Purdue University, West Lafayette, IN, 1992.

14

MANAGING INCOME TAXES

CHAPTER OBJECTIVES

1 To show the importance of income tax management on farms and ranches

2 To identify the objective of tax management

3 To discuss the differences between the cash and accrual methods of computing taxable income and the advantages and disadvantages of each

4 To explain how the marginal tax system and social security taxes are applied to taxable income

5 To review a number of tax management strategies that can be used by farmers and ranchers

6 To illustrate how depreciation is computed for tax purposes and how it can be used in tax management

7 To note the difference between ordinary and capital gain income

Federal income taxes have been imposed since 1913 and affect all types of businesses including farms and ranches. Income taxes are an unavoidable result of operating a profitable business. However, management decisions made over time can have a large impact on the amount of income tax due. For this reason, a farm manager needs a basic understanding of the tax regulations in order to analyze possible tax consequences of management decisions. Since these decisions are made throughout the year, tax management is a year-round problem and not something to be done only when the tax return is completed.

Farm managers need not and cannot be expected to have a complete knowledge of the tax regulations in addition to all the other knowledge they need. However, a basic understanding and awareness of the tax topics to be discussed in this chapter enables a manager to recognize the possible tax consequences of some common business decisions. This basic information also has another function. It helps a manager identify those decisions which may have large and complex tax consequences. This should be the signal to obtain the advice of a tax accountant or attorney who is experienced in farm tax matters. Expert advice *before* the management decision is implemented can be well worth the cost and save time, trouble, and taxes at a later date.

The income tax regulations are numerous, complex, and subject to change as new tax legislation is passed by Congress. Only some of the more basic principles and regulations can be covered in this chapter. They will be discussed from the standpoint of a farm business operated as a sole proprietorship. Farm partnerships and corporations are subject to different regulations in some cases, and anyone contemplating using these forms of business organization should obtain competent tax advice before making the decision. The reader is also advised to check the *current* tax regulations for the latest rules. Tax rules and regulations can and do change frequently.

OBJECTIVES OF TAX MANAGEMENT

Any manager with a goal of profit maximization needs to modify that goal when considering income taxes. This goal should now become "the maximization of long-run, *after-tax* profit." Income tax payments represent a cash outflow for the business, leaving less cash available for other purposes. The cash remaining *after* paying income taxes is the only cash available for family living expenses, debt payments, and new business investment. Therefore, the goal should be the maximization of after-tax profit, since it is this amount which is finally available for use at the manager's or owner's discretion. Notice the goal is not to minimize income taxes. This goal could be met by arranging to have little or no taxable income.

A very *short-run* objective of tax management is to minimize the taxes due on a given year's income at the time the tax return is completed and filed. However, this income is the result of decisions made throughout the year and perhaps in previous years. It is too late to change these decisions. Therefore, good tax management requires a continuous evaluation of how each decision will affect income taxes not only in the current year but in future years. It should be a long-run rather than a short-run objective and certainly not something to be thought of only once a year at tax time. Therefore, tax management should be the managing of income, expenses, and the purchases and sales of capital assets in a manner which will maximize long-run after-tax profits.

Tax management is not tax evasion but is avoiding the payment of any taxes which are not legally due and postponing the payment of taxes whenever possible. Several tax management strategies are available which tend to postpone or delay tax payments but not necessarily reduce the amount paid over time. However, any tax payment which can be put off until later represents additional cash available for business use for one or more years.

Tax management, or minimizing taxes, is sometimes equated with identifying and using "tax loopholes." This is generally an unfair assessment. While some inequities undoubtedly exist in the tax regulations, what some people refer to as tax loopholes were often intentionally legislated by Congress for a specific purpose. This purpose may be to encourage investment and production in certain areas or to increase investment in general as a means of promoting an expanding economy and full employment. Taxpayers should not be reluctant to take nor feel guilty about taking full advantage of these regulations.

TAX ACCOUNTING METHODS

A unique feature of farm business income taxes is the choice of tax accounting methods allowed. Farmers and ranchers are permitted to report their taxable income using either the cash or accrual method of accounting. All other taxpayers engaged in the production and sale of goods where inventories may exist are required to use the accrual method. Farmers and ranchers may elect to use either one. The choice is made when the individual or business files the first farm tax return. Whichever method is used on this first tax return must be used in the following years.

It is possible to change tax accounting methods by obtaining permission from the Internal Revenue Service and paying a fee of $500. The request must be filed on a special form within 180 days from the beginning of the tax year in which the change is desired. It must also include a reason for the desired change along with other information. A request may be denied and, if granted, may add complexity to the accounting procedures during the change-over period. Therefore, it is advisable to study the advantages and disadvantages of each method carefully and make the correct choice when the first farm tax return is filed.

The Cash Method

Under the cash accounting method, income is taxable in the year it is received as cash or "constructively received." Income is constructively received when it is credited to your account or was available for use before the end of the taxable year. An example of the latter situation would be a check for the sale of grain which the elevator was holding for the customer to pick up. If the check was available on December 31 but was not picked up until several days later, it would still be income constructively received in December. The check was available in December, and the fact it was not picked up does not make it taxable income in the following year.

Expenses under cash accounting are tax deductible in the year they are actually paid regardless of when the item was purchased or used. An example would be a December feed purchase which was charged on account and not paid for until the following January. This would be a tax-deductible expense in the next year and not in the year the feed was actually purchased and used. An exception to this rule is the cost of items purchased for resale, which includes feeder livestock. These expenses can be deducted only in the taxable year in which they are sold. This means the expense of purchasing feeder cattle or other items purchased for resale needs to be held over into the next year if the purchase and sale do not occur in the same year.

Inventories are *not* used to determine taxable income with the cash method of tax accounting. Income is taxed when received as cash and not as it accumulates in an inventory of crops and livestock. The cash method has several advantages and disadvantages.

Advantages   A large majority of farmers and ranchers use the cash method for calculating taxable income. There are a number of advantages which make this the most popular method.

1 *Simplicity*   The use of cash accounting requires a minimum of records, primarily because there is no need to maintain inventory records for tax purposes.

2 *Flexibility*   Cash accounting provides maximum flexibility for tax planning at the end of the year. Taxable income for any year can be adjusted by the timing of sales and payment of expenses. A difference of a few days can put the income or expense into one of two years.

3 *Capital Gain from the Sale of Raised Breeding Stock*   More of the income from the sale of raised breeding stock will often qualify as long-term capital gain income under cash accounting. This income is not subject to social security taxes and may be taxed at a lower rate than ordinary income. This subject will be discussed in detail in a later section.

4 *Delaying Tax on Growing Inventory*   A growing business with an increasing inventory can postpone paying tax on the inventory until it is actually sold and converted into cash.

Disadvantages   There are several disadvantages to cash accounting which may make it less desirable in certain situations.

1 *Poor Measure of Income*   As discussed in Chapter 4, changes in inventory values are needed to measure net farm income accurately for an accounting period. Because inventory changes are not included when determining taxable income under cash accounting, the resulting value does not accurately reflect net farm income for the year.

2 *Potential for Income Variation*   Market conditions may make it desirable to store one year's crop for sale the following year and to sell that year's crop at harvest. The result is no crop income in one year and income from two crops in the following year. Because of the progressive nature of the income tax rates, more total tax may be paid than if the income had been evenly distributed between the years.

3 *Declining Inventory*   During years with declining inventory values, more tax will be paid because there is no inventory decrease to offset the cash sales from inventory.

The Accrual Method

Income under the accrual method of tax accounting is taxable in the year it is earned or produced. Accrual income includes any change in the value of crops and livestock in inventory at the end of the taxable year as well as any cash income received. Therefore, any inventory increase during the tax year is included in taxable income, and inventory decreases reduce taxable income. An account receivable such as income earned for custom work for a neighbor is "earned" income and therefore taxable income even though payment has not been received.

Expenses under the accrual method are deductible in the tax year in which they are "incurred" whether or not they are paid. In essence this means any accounts payable at the end of the tax year are deductible as expenses, since they represent expenses which have been incurred but not yet paid. Another difference from cash accounting is the cost of items purchased for resale which can be deducted in the year of purchase even though the items have not been resold. Their cost is offset by the increase in ending inventory value for the same items.

Advantages   Accrual tax accounting has several advantages over cash accounting which should be considered when choosing a tax accounting method.

1 *Better Measure of Income*   Taxable income under accrual accounting is calculated much like the net farm income calculated in Chapter 4. For this reason, it measures net farm income better than cash accounting and keeps income taxes paid up on a current basis.

2 *Reduces Income Fluctuations*   The inclusion of inventory changes prevents wide fluctuations in taxable income if the marketing pattern results in the production from two years being sold in one year.

3 *Declining Inventory*   During years when the inventory value is declining, as might happen when a farmer is slowly retiring from the business, less tax will be paid under accrual accounting because the decrease in inventory value offsets some of the cash receipts.

Disadvantages   The disadvantages of accrual accounting offset many of the advantages for many farm businesses. This makes cash accounting the more popular method.

1 *Increased Record Requirement*   Complete inventory records must be kept for proper determination of accrual income. They should also be kept for use in calculating net farm income, but the additional complexity of a tax inventory and valuation of that inventory may discourage farmers from using accrual accounting.

2 *Increasing Inventory*   When the inventory is increasing, accrual accounting results in tax being paid on the inventory increase before the actual sale of the products. As no cash has yet been received for these products, this can cause a cash flow problem.

3 *Loss of Capital Gain on Raised Breeding Stock*   Accrual accounting will result in losing much if not all of the capital gain income from the sale of raised breeding stock. This factor alone generally makes accrual accounting less desirable for taxpayers with a breeding herd who raise most of their herd replacements. (See later section on "Capital Gain and Livestock".)

4 *Less Flexibility*   Accrual tax accounting is less flexible than cash accounting. It is more difficult to adjust taxable income at year-end through the timing of sales and paying of expenses.

Tax Record Requirements

Regardless of the accounting method chosen, complete and accurate records are essential for good tax management and properly reporting taxable income. Both farming and the tax regulations have become too complex for records to be a collection of receipts and canceled checks tossed in a shoe box. Poor records often result in two related and undesirable outcomes: the inability to verify receipts and expenses in case of a tax audit and paying more taxes than may be legally due. In either case, good records do not cost; they pay in the form of lower income taxes.

Record books suitable for farm and ranch use are generally available through county or state agricultural extension service offices. Several states offer a computerized record-keeping service through their agricultural college or state extension service. These services may also be available from a local bank, other private businesses, or a farm cooperative.

Complete tax records include a listing of receipts and expenses for the year. A tax depreciation schedule is also needed for all depreciable property to determine annual tax depreciation and to calculate any gain, loss, or depreciation recapture when the item is sold. Permanent records should be kept on real estate and other capital items including purchase date and cost, depreciation taken, cost of any improvements, and selling price. These items are important to determine any gain or loss from sale or for gift and inheritance tax purposes.

A computerized record system can be very helpful in keeping tax records. Accuracy, speed, and convenience are some of the advantages. In addition to keeping a record of income and expenses, some computer accounting programs also have the ability to maintain a depreciation schedule and calculate each year's tax depreciation. Some can print out the year's income and expenses in the same format as a farm tax return and others can actually print a copy of the completed tax return.

FARM TAX RETURN

Taxable income and tax-deductible expenses for a farm or ranch business are reported on Schedule F (Form 1040). A copy of this form is shown in Table 14-1. The net farm profit or loss on line 37 is transferred to the basic Form 1040, where it is combined with any income from other sources to determine the taxpayer's total taxable income.

Schedule F uses a single section (Part II) to record expenses for both cash and accrual basis taxpayers. However, income is reported in Part I for cash basis taxpayers and Part III for accrual basis taxpayers. The inclusion of inventory value changes in taxable income when using accrual accounting can be seen by careful study of these parts of Schedule F.

THE TAX SYSTEM AND TAX RATES

The income tax system in the United States is based on a number of marginal tax rates (the additional tax on an additional dollar of income within a certain range). While the actual tax rates and the range of taxable income each rate applies to change from time to time, the rates have always been marginal rates. For 1993, the income tax rates for a married couple filing a joint tax return were as follows:

Taxable income[1]	Marginal tax rate
$0–$36,900	15%
$36,900–$89,150	28%
Over $89,150	31%

Taxable income includes farm income as well as income earned from other sources *minus* personal exemptions and deductions.

In addition to income taxes, self-employed individuals such as farmers and ranchers are subject to social security or self-employment tax. Only income from a farm or ranch or other business activity is subject to self-employment tax. Unlike income tax, *all* eligible income is subject to this tax. It is applied to income before personal exemptions or deductions.

Self-employment tax includes both social security and Medicare taxes. For 1993, these combined tax rates were:

Income subject to self-employment tax[2]	Rate
$0–$57,600	15.3%
$57,600–$135,000	2.9%
Over $135,000	0%

As discussed above and because capital gains are not included, the income subject to self-employment tax will be different from what is subject to income taxes. However, the combination of the two taxes creates a number of income brackets and a wide range of total combined tax rates.

SOME POSSIBLE TAX MANAGEMENT STRATEGIES

A number of tax management strategies exist because of the marginal nature of the income tax rates and other features of the tax system. Tax management strategies are basically one of two types: those which reduce taxes and those which may only postpone

---

[1]The dividing points between each marginal tax rate will be adjusted upward each year by an amount equal to the inflation rate.

[2]These values are set by congressional legislation and have historically increased nearly every year.

TABLE 14-1 SCHEDULE F (FORM 1040) FOR FILING A FARM TAX RETURN

SCHEDULE F (Form 1040) Department of the Treasury Internal Revenue Service (X)	**Profit or Loss From Farming** ▶ Attach to Form 1040, Form 1041, or Form 1065. ▶ See Instructions for Schedule F (Form 1040).	OMB No. 1545-0074 **1992** Attachment Sequence No. **14**

Name of proprietor

Social security number (SSN)

A Principal product. Describe in one or two words your principal crop or activity for the current tax year.

B Enter principal agricultural activity code (from page 2) ▶

D Employer ID number (Not SSN)

C Accounting method: **(1)** ☐ Cash   **(2)** ☐ Accrual

E Did you "materially participate" in the operation of this business during 1992? If "No," see page F-1 for limitations on losses.   ☐ Yes ☐ No

Part I Farm Income—Cash Method—Complete Parts I and II (Accrual method taxpayers complete Parts II and III, and line 11 of Part I.)
Do not include sales of livestock held for draft, breeding, sport, or dairy purposes; report these sales on Form 4797.

1	Sales of livestock and other items you bought for resale	**1**	
2	Cost or other basis of livestock and other items reported on line 1	**2**	
3	Subtract line 2 from line 1		**3**
4	Sales of livestock, produce, grains, and other products you raised		**4**
5a	Total cooperative distributions (Form(s) 1099-PATR) **5a**	**5b** Taxable amount	**5b**
6a	Agricultural program payments (see page F-2) **6a**	**6b** Taxable amount	**6b**
7	Commodity Credit Corporation (CCC) loans (see page F-2):		
a	CCC loans reported under election		**7a**
b	CCC loans forfeited or repaid with certificates **7b**	**7c** Taxable amount	**7c**
8	Crop insurance proceeds and certain disaster payments (see page F-2):		
a	Amount received in 1992 **8a**	**8b** Taxable amount	**8b**
c	If election to defer to 1993 is attached, check here ▶ ☐	**8d** Amount deferred from 1991	**8d**
			9
9	Custom hire (machine work) income		
10	Other income, including Federal and state gasoline or fuel tax credit or refund (see page F-3)		**10**
11	**Gross income.** Add amounts in the right column for lines 3 through 10. If accrual method taxpayer, enter the amount from page 2, line 51. ▶		**11**

Part II Farm Expenses—Cash and Accrual Method (Do not include personal or living expenses such as taxes, insurance, repairs, etc., on your home.)

12	Car and truck expenses (see page F-3—also attach **Form 4562**) . . **12**		25	Pension and profit-sharing plans **25**	
13	Chemicals **13**		26	Rent or lease (see page F-4):	
14	Conservation expenses. Attach **Form 8645**. **14**		a	Vehicles, machinery, and equipment. **26a**	
15	Custom hire (machine work). . **15**		b	Other (land, animals, etc.) . . **26b**	
16	Depreciation and section 179 expense deduction not claimed elsewhere (see page F-3) . . **16**		27	Repairs and maintenance . . **27**	
			28	Seeds and plants purchased . **28**	
			29	Storage and warehousing . . **29**	
17	Employee benefit programs other than on line 25 . . . . **17**		30	Supplies purchased **30**	
18	Feed purchased **18**		31	Taxes **31**	
19	Fertilizers and lime **19**		32	Utilities **32**	
20	Freight and trucking **20**		33	Veterinary, breeding, and medicine . **33**	
21	Gasoline, fuel, and oil . . . **21**		34	Other expenses (specify):	
22	Insurance (other than health) . **22**		a	.... **34a**	
23	Interest:		b	.... **34b**	
a	Mortgage (paid to banks, etc.) . **23a**		c	.... **34c**	
b	Other **23b**		d	.... **34d**	
24	Labor hired (less jobs credit) . **24**		e	.... **34e**	
			f	**34f**	

35	**Total expenses.** Add lines 12 through 34f . . . . ▶		**35**
36	**Net farm profit or (loss).** Subtract line 35 from line 11. If a profit, enter on Form 1040, line 19, and on Schedule SE, line 1. If a loss, you MUST go on to line 37 (fiduciaries and partnerships, see page F-5)		**36**
37	If you have a loss, you MUST check the box that describes your investment in this activity (see page F-5). If you checked 37a, enter the loss on Form 1040, line 19, and Schedule SE, line 1. If you checked 37b, you MUST attach **Form 6198**.	**37a** ☐ All investment is at risk. **37b** ☐ Some investment is not at risk.	

TABLE 14-1 *(continued)*

Schedule F (Form 1040) 1992 — Page **2**

Part III | **Farm Income—Accrual Method** (see page F-5)

Do not include sales of livestock held for draft, breeding, sport, or dairy purposes; report these sales on Form 4797 and do not include this livestock on line 46 below.

38	Sales of livestock, produce, grains, and other products during the year.	**38**	
39a	Total cooperative distributions (Form(s) 1099-PATR) **39a** _____	39b Taxable amount	**39b**
40a	Agricultural program payments **40a** _____	40b Taxable amount	**40b**
41	Commodity Credit Corporation (CCC) loans:		
a	CCC loans reported under election		**41a**
b	CCC loans forfeited or repaid with certificates **41b** _____	41c Taxable amount	**41c**
42	Crop insurance proceeds		**42**
43	Custom hire (machine work) income		**43**
44	Other income, including Federal and state gasoline or fuel tax credit or refund		**44**
45	Add amounts in the right column for lines 38 through 44 ▶		**45**
46	Inventory of livestock, produce, grains, and other products at beginning of the year.	**46**	
47	Cost of livestock, produce, grains, and other products purchased during the year.	**47**	
48	Add lines 46 and 47	**48**	
49	Inventory of livestock, produce, grains, and other products at end of year	**49**	
50	Cost of livestock, produce, grains, and other products sold. Subtract line 49 from line 48*		**50**
51	**Gross income.** Subtract line 50 from line 45. Enter the result here and on page 1, line 11 ▶		**51**

*If you use the unit-livestock-price method or the farm-price method of valuing inventory and the amount on line 49 is larger than the amount on line 48, subtract line 48 from line 49. Enter the result on line 50. Add lines 45 and 50. Enter the total on line 51.

Part IV | **Principal Agricultural Activity Codes**

*Caution: File **Schedule C** (Form 1040), Profit or Loss From Business, or **Schedule C-EZ** (Form 1040), Net Profit From Business, instead of Schedule F if:*

● *Your principal source of income is from providing agricultural services such as soil preparation, veterinary, farm labor, horticultural, or management for a fee or on a contract basis, or*

● *You are engaged in the business of breeding, raising, and caring for dogs, cats, or other pet animals.*

Select one of the following codes and write the 3-digit number on page 1, line B:

120 **Field crop,** including grains and nongrains such as cotton, peanuts, feed corn, wheat, tobacco, Irish potatoes, etc.

160 **Vegetables and melons,** garden-type vegetables and melons, such as sweet corn, tomatoes, squash, etc.

170 **Fruit and tree nuts,** including grapes, berries, olives, etc.

180 **Ornamental floriculture and nursery products**

185 **Food crops grown under cover,** including hydroponic crops

211 **Beefcattle feedlots**

212 **Beefcattle,** except feedlots

215 **Hogs, sheep, and goats**

240 **Dairy**

250 **Poultry and eggs,** including chickens, ducks, pigeons, quail, etc.

260 **General livestock,** not specializing in any one livestock category

270 **Animal specialty,** including bees, fur-bearing animals, horses, snakes, etc.

280 **Animal aquaculture,** including fish, shellfish, mollusks, frogs, etc., produced within confined space

290 **Forest products,** including forest nurseries and seed gathering, extraction of pine gum, and gathering of forest products

300 **Agricultural production,** not specified

the payment of taxes. However, any tax payment which can be delayed until next year makes that sum of money available for business use during the coming year. That amount is available for investment or to reduce the amount of borrowing needed.

Form of Business Organization

As was discussed in Chapter 12, the form of business organization will have an effect on the taxes paid over time. The different forms of business organization should be analyzed both when starting a farm or ranch business and whenever there is a major change in the business. While taxes may not be the only nor always the most important reason for changing the type of business organization, it is always a factor to consider in the decision.

Income Leveling

There are two reasons for trying to maintain a steady income level. First, years of higher than normal income may put the taxpayer in a higher marginal tax bracket. More tax will be paid over time than with a level income taxed at a lower rate. Second, in years of low income, some personal exemptions and deductions may be lost.

For the 1993 tax year, each taxpayer gets a personal exemption of $2,350 and the same amount for each dependent. There is also a standard deduction of $6,200 for a married couple filing a joint return.[3] Taxpayers who "itemize" their personal deductions may have an even larger deduction. Therefore, a married couple with two dependent children has exemptions and a minimum deduction of ($2,350 × 4) + $6,200 = $15,600. *Taxable income* is only the income above this amount. If the total family income is less than $15,600, no income taxes are due, but some (or all if income is $0 or less) of the allowable exemptions and deductions are lost. They cannot be carried over to use next year. Even though no income tax may be due, there may still be some self-employment tax.

Deferring or Postponing Taxes

Taxes can be deferred or postponed by delaying taxable income until a later year. This saves taxes in the current year and the money saved can be invested or used to reduce borrowing for the next year. Taxable income can be deferred by increasing expenses in the current year, deferring sales until the next year, or both. Tax regulations limit the amount of prepaid expenses which can be deducted but still allow considerable flexibility for many farmers and ranchers.

To level income or defer taxes requires flexibility in the timing of income and expenses. The taxpayer must be able to make the expense deductible and the income tax-

---

[3]Both the personal exemption and standard deduction are due to be adjusted upward each year by an amount equal to the rate of inflation.

able in the year of choice. Flexibility is greatest with the cash method of computing taxable income. With the accrual method, many of the timing decisions on purchases and sales are offset by a change in inventory values.

Taxpayers must be cautious and consider other than tax reasons when adjusting the timing of purchases and sales. It is very easy to make a poor economic decision in an effort to save or postpone a few dollars in taxes. Expected market trends are one factor to consider. For example, if commodity prices are expected to move lower into next year, a farmer may be ahead by selling now and paying the tax in the current year. Remember the objective of tax management is to "maximize long-run after-tax profit" and not to minimize income taxes paid in any given year.

Net Operating Loss (NOL)

The wide fluctuations in farm prices and yields can cause a net operating loss some years despite the best efforts to level out annual taxable income. Special provisions relating to net operating losses allow them to be used to reduce past and/or future taxable income. Any net operating loss from a farm business can first be used to off-set taxable income from other sources. If the loss is greater than the nonfarm income, any remaining loss can be carried back for 3 years and then forward for up to 15 years.

When carried back, the net operating loss is used to reduce taxable income for these years, and a tax refund is requested. When carried forward, it is used to reduce taxable income and therefore income taxes in future years. A taxpayer can elect to forgo the 3-year carry-back provision and apply the net operating loss only against future taxable income for up to 15 years. However, this election must be made at the time the tax return is filed, and it is irrevocable for that tax year. In making this election, care should be taken to receive maximum advantage from the loss by applying it to the years with above-average taxable income.

A net operating loss is basically an excess of allowable tax-deductible expenses over gross income. However, there are certain adjustments and special rules which apply when calculating a net operating loss. Expert tax advice may be necessary to properly calculate and use a net operating loss to its best advantage.

Tax-Free Exchanges

The sale of an asset such as land will generally incur some taxes whenever the sale price is above the asset's cost or tax basis (its value for tax purposes). If the land being sold is to be replaced by the purchase of another farm, a tax-free exchange can eliminate or at least reduce the tax caused by the sale. There are some specific regulations regarding tax-free exchanges so competent tax advice is necessary to be sure the exchange qualifies as tax-free.

A tax-free exchange involves trading property. The farm you wish to own is purchased by the taxpayer who wants to purchase your farm. The two individuals then exchange or trade farms so each ends up with the farm they wish to own. Completion of a tax-free exchange simply transfers the basis on the original farm to the new farm, and no tax is due if the sale price of both farms are equal. In practice, this rarely happens and some tax may be due. Again, good tax advice is necessary to compute any tax due on the exchange. Because of the strict requirements, tax-free exchanges do not fit every situation. However, they are something to consider whenever an existing asset is to be replaced by a new or different one.

DEPRECIATION

Depreciation plays an important role in tax management for two reasons. First, it is a noncash but tax-deductible expense, which means it reduces taxable income without being a cash outflow. Second, some flexibility is permitted in calculating tax depreciation, making it another tool which can be used for income leveling and postponing taxes. A working knowledge of tax depreciation, depreciation methods, and the rules for their use is necessary for good tax management.

The difference between tax depreciation and other calculations of depreciation should be kept in mind throughout the following discussion. Tax depreciation is depreciation expense computed for tax purposes. Depreciation for accounting and management purposes would normally use one of the depreciation methods discussed in Chapter 2. As will be shown, the usual tax depreciation method is different and may use different useful lives and salvage values than would be used for management accounting. Therefore, there may be a large difference in the annual depreciation for tax versus management uses.

Tax Basis

Every business asset has a *tax basis,* which is simply the asset's value for tax purposes at any point in time. At the time of purchase it is called a beginning tax basis, and when this value changes it becomes an adjusted tax basis. Any asset purchased directly, either new or used, has a beginning tax basis equal to its purchase price. When a used asset is traded in as partial payment on a new purchase, the beginning tax basis on the newly acquired asset is equal to the adjusted tax basis of the trade-in plus any cash difference or "boot" paid. For example, if a used tractor with an adjusted tax basis of $10,000 plus $40,000 cash is exchanged for a new tractor with a list price of $60,000, the beginning tax basis on the new tractor is $50,000 and not $60,000. Since the beginning basis represents the value of the asset for tax purposes, it is the starting point for calculating tax depreciation.

An asset's basis is adjusted downward each year by the amount of depreciation taken. The result is its *adjusted tax basis.* In the example above, if $7,500 of tax depreciation is taken the first year, the tractor would have an adjusted tax basis of $50,000 − $7,500 = $42,500 at the end of the year. The basis would be adjusted downward each following year by the amount of tax depreciation taken that year. Note that the resulting value may not be anywhere close to market value.

The basis on a nondepreciable asset such as land generally remains at its original cost for as long as it is owned. An exception would be some capital improvement to the land itself when the cost of the improvement may not be tax deductible in the current year nor depreciable. Terraces or earthen dams would be examples. In this case, the cost of the improvement is added to the original basis to find the new adjusted basis.

An accurate record of the basis of each asset is important when it is sold or traded. The current adjusted basis is used to determine the profit or loss on the sale or to determine the beginning basis for the newly acquired asset when a trade-in is part of the transaction. Cost and basis information on assets such as land and buildings, which have a long life and are sold or traded infrequently, may need to be kept for many years. This is another reason for a complete, accurate, and permanent record-keeping system.

MACRS Depreciation

For many years, the standard accounting depreciation methods were used for tax depreciation purposes. This was changed by the Economic Recovery Act of 1981, which made broad changes in tax depreciation by implementing the Accelerated Cost Recovery System (ACRS). New regulations modified this system for depreciable assets purchased after January 1, 1987. The result is the current tax depreciation system, called the Modified Accelerated Cost Recovery System or MACRS. There are several alternatives under MACRS, and the discussion here will concentrate on the regular MACRS, which is used more frequently than the others. Alternate MACRS will be discussed briefly in a later section.

MACRS contains a number of property classes which determine an asset's useful life. Each depreciable asset is assigned to a particular class depending on the type of asset. Farm and ranch assets generally fall into the 3-, 5-, 7-, 10-, 15-, or 20-year classes. Some examples of assets in each class are:

3-year: breeding hogs
5-year: cars, pickups, breeding cattle and sheep, dairy cattle, computers, trucks
7-year: most machinery and equipment, fences, grain bins, silos, furniture
10-year: single-purpose agricultural and horticultural structures, trees bearing fruit or nuts
15-year: paved lots and drainage tile
20-year: general-purpose buildings such as machine sheds and hay barns

These are only some examples. The tax regulations should be checked for the proper classes for other specific assets.

Another characteristic of MACRS is an assumed salvage value of zero for all assets. Therefore, taxpayers do not have to select a useful life and salvage value when using MACRS. This method also includes an automatic one-half year's depreciation in the year of purchase regardless of the purchase date. This is called the half-year convention or half-year rule. It benefits a taxpayer who purchases an asset late in the year but is a disadvantage whenever an asset is purchased early in the year. This half-year rule applies *unless* more than 40 percent of the assets purchased in any year are placed in service in the last quarter of the year. In this case, the midquarter rule applies, which requires the depreciation on each newly purchased asset to begin in the middle of the quarter in which it was actually purchased.

TABLE 14-2 REGULAR MACRS RECOVERY RATES FOR FARM AND RANCH ASSETS IN THE 3-, 5-, AND 7-YEAR CLASSES

Recovery percentages

Recovery year	3-year class	5-year class	7-year class
1	25.00	15.00	10.714
2	37.50	25.50	19.133
3	25.00	17.85	15.033
4	12.50	16.66	12.249
5		16.66	12.249
6		8.33	12.249
7			12.249
8			6.124

MACRS depreciation is computed using fixed percentage recovery rates, which are shown in Table 14-2 for the 3-, 5-, and 7-year classes. For agricultural assets, these percentages are based on the use of 150 percent declining balance for each successive year until the percentage using straight-line depreciation results in a higher percentage. The rate for year 1 for the 5-year class was determined in the following manner. Remember that the straight-line percentage rate for a 5-year useful life is 20 percent:

$$20\% \times 1.5 \times \tfrac{1}{2}\,\text{year} = 15\%$$

The same procedure will result in 25 percent recovery for year 1 of the 3-year class.

Regular MACRS depreciation for each year of an asset's assigned life is found by multiplying the asset's beginning tax basis by the appropriate percentage. For example, if a new pickup is purchased (5-year class) and has a beginning tax basis of $18,000, the depreciation would be computed as follows:

Year 1: $18,000 \times 15.00\% = \$2,700$
Year 2: $18,000 \times 25.50\% = \$4,590$
Year 3: $18,000 \times 17.85\% = \$3,213$
Year 4: $18,000 \times 16.66\% = \$2,999$
Year 5: $18,000 \times 16.66\% = \$2,999$
Year 6: $18,000 \times \ \ 8.33\% = \$1,499$

Depreciation for assets in other classes would be computed in a similar manner using the appropriate percentages for the class.

Alternative Depreciation Methods

Some flexibility is allowed in computing tax depreciation since there are alternatives to the regular MACRS method discussed above. These alternatives will all result in slower depreciation than regular MACRS, which is one reason they are not widely used. However, taxpayers who wish to delay some depreciation until later years can choose from one of the alternate methods:

- Regular MACRS class life or recovery period with straight-line depreciation
- Alternate MACRS recovery periods (usually longer than regular MACRS) and 150 percent declining balance method
- Alternate MACRS recovery periods with straight-line depreciation

The latter would be the "slowest" of the four possible methods.

If an alternate method is used for one asset, all assets in that class purchased in the same year must be depreciated using the alternate method. Also, once an alternate method has been selected, it cannot be changed in a later year.

Expensing

Section 179 of the tax regulations provides for a special deduction called "expensing." While not specifically called depreciation, for practical purposes it is treated in the same manner. Expensing is an *optional* deduction which can be elected only in the year an asset is purchased and which reduces taxable income if elected. The maximum Section 179 expensing which can be taken any year is $10,000.

Property eligible for expensing is defined by the tax regulations somewhat differently than property eligible for depreciation. However, for practical purposes, the result is that most 3-, 5-, 7-, and 10-year class property is eligible. The amount eligible for expensing is equal to the cost or beginning basis on an outright purchase, but only the cash difference or "boot" is eligible when there is a trade.

Since expensing is treated as depreciation, it reduces an asset's tax basis. The beginning basis must be reduced by the amount of expensing taken before computing any remaining regular depreciation. From this procedure it can be seen that an election to take expensing does not increase the total amount of depreciation taken over the asset's life. It only puts more depreciation in the first year (expensing *plus* regular depreciation for year 1), which leaves less annual depreciation for later years.

Using the previous example, if expensing was taken on the pickup, its basis would be reduced to $18,000 - $10,000 = $8,000. Regular depreciation would then be computed as follows:

Year 1: $8,000 \times 15.00\% = $1,200
Year 2: $8,000 \times 25.50\% = $2,040

Depreciation, Income Leveling, and Tax Deferral

The choices taxpayers are allowed in computing tax depreciation permit depreciation to be used as a means to level out taxable income or defer taxes. If eligible assets are purchased in a year when taxable income is high, expensing can be elected and the regular MACRS method used for depreciation to get maximum current tax deductions. In years when taxable income is below average, expensing would not be taken (remember it is optional), and one of the alternative depreciation methods can be used. However, a taxpayer should always remember that due to the time value of money, a dollar of current tax savings is worth more than a dollar of tax savings in the future.

Depreciation Recapture

The combination of expensing, fast depreciation under the regular MACRS method, and the assumed zero salvage value means an asset's adjusted tax basis will often be less than its market value. Whenever a depreciable asset is sold for more than its adjusted basis, the difference is called depreciation recapture. This amount represents excess depreciation taken since the asset did not lose market value as fast as it was being depreciated for tax purposes. Depreciation recapture is included as taxable income for the year of sale but it is not subject to social security taxes.

When a used asset is traded in for a new asset of the same type, there will be no depreciation recapture. Remember that the beginning basis on the new asset is the adjusted basis of the old asset plus the cash difference, or "boot," paid. If the old asset has a market value above its adjusted basis, the boot will be smaller and the beginning basis of the new asset smaller than it would have been with a cash purchase. In essence, the smaller beginning basis replaces depreciation recapture when there is an eligible trade.

CAPITAL GAINS

The tax regulations recognize two types of income, ordinary income and capital gain. Ordinary income includes wages and salaries, interest, dividends, cash rents, and revenue from the sale of crops and feeder livestock. *Capital gain* can result from the sale or exchange of certain types of qualified assets. In simplified terms, capital gain is the gain or profit made by selling an asset at a price which is higher than the original purchase price. Technically, it is the difference between the selling price and the *higher* of the asset's cost or adjusted basis. For depreciable assets, that means depreciation recapture applies first, and capital gain is only the amount the sales price exceeds the original cost or beginning basis. It is also possible to have a capital loss if the selling price is less than the adjusted basis.

Two types of assets can qualify for capital gain treatment. *Capital assets* include primarily nonbusiness investments such as stocks and bonds. Of more importance to most farmers and ranchers are *Section 1231 assets,* which are defined as property used in a trade or business. In a farm or ranch business, Section 1231 assets would include land, buildings, fences, machinery, equipment, breeding livestock, and other depreciable assets used in the business.

For income to qualify as long-term capital gain, the asset must have been owned for a minimum period of time or else the gain will be short-term capital gain. The holding periods have changed from time to time but are now one year for most assets. One exception is certain types of livestock, which will be discussed later.

As one example of long-term capital gain income, assume farmland was purchased for $100,000 and sold 3 years later for $150,000. A $50,000 long-term capital gain would result from this transaction. In a similar manner, there can be a long-term capital gain (or loss) from the sale or exchange of other qualified assets provided they have been owned long enough to qualify.

Taxation of Long-term Capital Gains

The distinction between ordinary income and long-term capital gain income is important because of the different way the two types of income are taxed. These differences have been more important in the past than now but are always subject to change. The current tax advantages for long-term capital gain income are:

1 Capital gain income (both short and long term) is not subject to self-employment or social security tax.

2 Long-term capital gain income has a maximum rate of 28 percent as compared with 31 percent for ordinary income. This advantage only applies to taxpayers in the 31 percent tax bracket as long-term capital gain income is taxed at the same rate as ordinary income in the lower brackets.

Long-term capital gain income has a history of favorable tax treatment. It has ranged from only 40 percent of it being taxed to, for a short time, 100 percent of it taxed at ordinary rates. The current tax treatment is some of the least favorable that has been applied to long-term capital gain income. This is an area which is always subject to change by future legislation.

Except for not being subject to social security tax, short-term capital gain income has not received any special tax treatment. This makes knowledge of and a record of the required holding period an important factor. Any existing advantages for long-term capital gain income will be lost by selling even a day or two short of the required holding period. Also, if there are short-term as well as long-term gains and losses in a given tax year, special rules apply for offsetting the various gains with the losses. These rules should be carefully studied to obtain maximum advantage from the losses.

Capital Gain and Livestock

Farmers and ranchers will generally have more frequent opportunities to receive capital gain income from livestock sales than from selling other assets. Livestock are classified as Section 1231 assets, and their sale or exchange may result in a long-term capital gain or loss provided they were held for draft, dairy, breeding, or sporting purposes. In addition, cattle and horses must have been owned 24 months and hogs and sheep for 12 months. Eligibility depends on both requirements, the purpose and the holding period.

Raised Livestock The above tax provision is of special importance to cash basis farmers and ranchers who raise their own replacements for a breeding or dairy herd. Raised replacements do not have a tax basis established by a purchase price nor do they have a basis established by an inventory value as would occur with accrual tax accounting. Therefore, cash basis taxpayers have a zero basis on raised replacement animals. If they are used for draft, dairy, breeding, or sporting purposes and are held for the required length of time, the entire income from their sale is long-term capital gain (sale price minus the zero basis).

This provision of the tax law makes cash basis accounting very desirable for taxpayers with beef, swine, sheep, or dairy breeding herds who raise their own replacements. It will be an even more important consideration if more favorable tax treatment is given to long-term capital gain income some time in the future. An accrual basis taxpayer loses most of this capital gain on raised animals because they have a tax basis derived from their inventory value.

Purchased Livestock   It is also possible to receive capital gain income from the sale of purchased breeding livestock but only if the selling price is above the original purchase price. For example, assume a beef cow was purchased for $500 and depreciated to an adjusted basis of $0 over the 5-year class life. If this cow is then sold for $600, the sale would result in $500 of depreciation recapture and only $100 of long-term capital gain.

The capital gain cash basis taxpayers can receive from selling raised breeding and dairy animals does not necessarily mean replacements should be raised rather than purchased. There are also some tax advantages from purchasing replacements. Purchased replacements can be depreciated including the use of the optional expensing. A replacement method should be selected after carefully considering the costs and production factors as well as the income tax effects. This is another example of the importance income taxes and the various tax regulations play in a manager's decision-making environment. To maximize long-run after-tax profit, a manager is forced to consider income taxes along with the costs and technical production factors of the various alternatives being considered.

SUMMARY

A business can spend or invest only that portion of its profit remaining after income taxes are paid. Therefore, the usual goal of profit maximization should become maximization of long-run after-tax profit. For a farm or ranch business, the tax management necessary to achieve this goal begins with selecting either the cash or accrual tax accounting method. It should continue on a year-round basis because of the many production and investment decisions which have tax consequences.

A number of tax management strategies are possible. They include ways to take full advantage of existing tax regulations and to postpone or defer taxes. Some examples include income leveling, tax deferral, proper use of a net operating loss, and tax-free exchanges. Appropriate use of depreciation and the related expensing is another common tax management strategy.

Long-term capital gain income is not subject to social security taxes and is taxed at a lower top rate than the top rate for ordinary income. This type of income may result from the sale or exchange of a capital asset or a Section 1231 asset. A common source of capital gain income for a cash basis taxpayer is income from the sale of raised breeding or dairy animals. This and other sources of capital gain income may require careful planning to qualify the sale or exchange for the reduced tax rates applicable to long-term capital gain income.

QUESTIONS FOR REVIEW AND FURTHER THOUGHT

1 How does a farmer choose a tax accounting method? Can it be changed? How?
2 Which accounting method would you recommend for each of the following? Why?
   a A crop farmer whose marketing policies cause wide variations in cash receipts and inventory values from year to year.
   b A rancher with a beef breeding herd who raises all the necessary replacement heifers.
3 What advantage is there to deferring income taxes to next year? If they must be paid, why not pay them now?
4 Assume Fred Farmer purchases a new planter for $24,500 on March 1. Compute the expensing and regular depreciation for the year of purchase using the regular MACRS method and the planter's adjusted tax basis at the end of the year.
5 Repeat the calculations in Question 4 without expensing.
6 In a year when farm prices and yields are very good, machinery and equipment dealers often experience large sales toward the end of the year. How would you explain this increase in sales?
7 Explain how a cash basis farmer can raise or lower taxable income by purchase and selling decisions made in December. Can an accrual basis farmer do the same thing? Why?
8 Assume a rancher purchased a young herd bull for $3,000. Three years later when the bull had an adjusted tax basis of $1,785, he was sold for $5,000. How much depreciation recapture and/or capital gain income will result from this sale?
9 Suppose a farmer made a mistake and counted $4,000 of long-term capital gain income as ordinary income on the farm tax return. How much additional income and social security tax would this mistake cost if the farmer had taxable income of $28,000? Taxable income of $95,000? (Use the tax rates in the "Tax System and Tax Rate" section.)

REFERENCES

Farmer's Tax Guide, 1992 Edition, Publication 225, Department of the Treasury, Internal Revenue Service.

Harris, Philip E., and Myron P. Kelsey: *Tax Planning When Buying or Selling a Farm,* North Central Regional Publication 43, University of Wisconsin, rev. 1986.

Harris, Philip E., and W. A. Tinsley: *Income Tax Management for Farmers,* Southern Farm Management Publication 17, rev. 1990.

O'Byrne, John C., and Charles Davenport: *Doane's Tax Guide for Farmers,* Doane Information Services, St. Louis, MO, 1988.

15

INVESTMENT ANALYSIS

CHAPTER OBJECTIVES

    1 To understand the time value of money and its use in decision making and investment analysis

    2 To illustrate the process of compounding or determining the future value of a sum of money or series of payments

    3 To demonstrate the process of discounting or determining the present value of a future sum or series of payments

    4 To discuss the payback period, simple rate of return, net present value, and internal rate of return as different methods to analyze investments

    5 To learn how to apply these methods to different types of investment alternatives

    6 To compare the advantages and disadvantages of each method for analyzing investment opportunities

Capital available on a farm or ranch can be used in two general types of investments. The first is annual operating inputs such as feed, seed, fertilizer, chemicals, and fuel, and the second is capital assets such as land, machinery, buildings, and orchards. Different methods are used to analyze each type of investment because of differences in the timing of the expenses and associated returns. Both expenses and returns from investing in annual operating inputs occur within one production cycle of a year or less. Investing in a capital asset typically means a large expense (the initial purchase) occurs in one time period with additional operating expenses, and the resulting returns spread over a number of future time periods. Partial budgeting can be used to analyze these investments by looking at changes in costs and returns for an average year. However, other investment analysis methods are often more appropriate.

The marginal principles used to find the profit-maximizing level or allocation of operating inputs often ignore time. Expenses and returns are assumed to fall within the same year or production cycle so there is a short time span. However, time can be incorporated into the analysis by including the opportunity cost of capital as part of the cost of the input. Enterprise and partial budgets may recognize time by including opportunity costs on annual operating inputs, but the amounts are typically small due to the short investment period.

While time may be of minor importance when analyzing investments in annual operating inputs, it becomes of major importance for capital assets. The latter often represent large sums of money where the expenses and returns are in different time periods and spread over many years. Alternative and competing investments may have different annual returns received over different lengths of time. A proper analysis of these capital investments requires the recognition of the time value of money and its use in different analytical methods.

There are other reasons for carefully and properly analyzing potential capital investments *before* they are made. Decisions about operating inputs can be changed annually. However, capital investments are, by definition, long-lasting assets; thus investment decisions are made less frequently and often involve large sums of money. It is often difficult or costly to change, alter, or reverse a capital investment decision once the asset is purchased or constructed. Therefore, sufficient time and proper analytical techniques should be used when making these decisions.

TIME VALUE OF MONEY

We all prefer to have a dollar today than a dollar at some time in the future. People instinctively recognize that a dollar today is worth more than a dollar at some future date. Why? First, a dollar received today can be invested to earn interest and will therefore increase to a dollar *plus* interest by the future date. In other words, the interest represents the opportunity cost of receiving the dollar in the future rather than now. This is the investment explanation for the time value of money.

The second explanation is from the viewpoint of consumption. If the dollar was to be spent for consumer goods such as a new TV, stereo, automobile, or furniture, we would prefer to have the dollar now so we could enjoy the new item now rather than later. A third reason for preferring the dollar now rather than later is the risk that some unforeseen circumstance will prevent us from collecting it in the future.

This chapter will focus on the investment reason for the time value of money. It is important in business investment decisions and for a manager who must make investments on a farm or ranch.

Terms, Definitions, and Abbreviations

A number of terms and abbreviations will be used in the discussion and equations throughout this section. They are as follows with the abbreviations shown in parentheses:

Present Value (PV)   The number of dollars available or invested at the current time *or* the current value of some amount(s) to be received at some time in the future.

Future Value (FV)   The amount of money to be received at some time in the future *or* the amount a present value will become at some future date when invested at a given interest rate.

Payment (PMT)   The number of dollars to be paid or received at the end of each of a number of time periods.

Interest Rate (i)   Also called the *discount rate* in some applications, it is the interest rate used to find present and future values. It will often be the opportunity cost of capital.

Time Periods (n)   The number of time periods to be used for computing present and future values. Time periods are often a year in length but can be shorter such as a month. The annual interest rate *i* must be adjusted to correspond to the length of the time periods, that is, a monthly interest rate must be used if the time periods are months.

Annuity   A term used to describe a series of equal periodic payments (PMT). The payments may be either receipts or expenditures.

Finding Future Values

The future value of money refers to the value of an investment at a specific date in the future. This concept is based on the investment earning interest during each time period, which is then reinvested at the end of each period so it will also earn interest in later time periods. Therefore, the future value will include the original investment, the interest it has earned, plus interest on the accumulated interest. The procedure for determining future values when accumulated interest also earns interest is called *compounding*. It can be applied to a one-time lump sum investment (a PV) or to an investment which takes place through a series of payments (PMT) over time. Each will be analyzed separately.

Future Value of a Present Value   Figure 15-1*a* graphically illustrates this situation. Starting with a given amount of money today, a PV, what will it be worth at

FIGURE 15-1   Illustration of the concept of a future value for a present value and for an annuity.

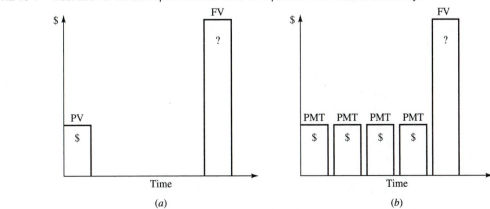

(a)   (b)

some date in the future? The answer depends on three things: the PV, the interest rate it will earn, and the length of time it will be invested.

Assume you have just invested $100 in a savings account which earns 8 percent interest compounded annually, and you would like to know the future value of this investment after 3 years. The following table illustrates how the balance in the account changes from year to year.

Year	Value at beginning of year ($)	Interest rate (%)	Interest earned ($)	Value at end of year ($)
1	100.00	8	8.00	108.00
2	108.00	8	8.64	116.64
3	116.64	8	9.33	125.97

Notice the assumption that all interest earned is allowed to accumulate in the account so it also earns interest for the remaining years. In this example, a present value of $100 has a future value of $125.97 when invested at 8 percent interest for 3 years. Interest is compounded when accumulated interest also earns interest in the following time periods.

Allowing the accumulated interest to earn interest is an integral part of the compounding procedure to find a future value. If the interest had been withdrawn, only $100 would have earned interest each year. A total of $24.00 in interest would have been withdrawn compared to the $25.97 of interest earned in this example.

This procedure for finding a future value can become very tedious if the investment is for a long period of time. Fortunately the calculations can be simplified by using the mathematical equation

$$FV = PV(1 + i)^n$$

where the abbreviations are as defined earlier. Applied to the example, the calculations would be

$$FV = \$100(1 + 0.08)^3$$
$$= \$100(1.2597)$$
$$= \$125.97$$

giving the same future value as before.

This equation can be quickly and easily applied using a calculator which will raise numbers to a power. Without such an aid, the equation can be difficult to use, particularly when n is large. To simplify the calculation of future values, tables have been constructed giving values of $(1 + i)^n$ for different combinations of i and n. Appendix Table 2 is a table for finding the future value of a present value. Any future value can be found by multiplying the present value by the factor from the table corresponding to the appropriate interest rate and length of the investment. The value for 8 percent and 3 years is 1.2597, which when multiplied by the $100 in the example, gives the same future value of $125.97 as before.

An interesting and useful rule of thumb is the *Rule of 72*. The approximate time it takes for an investment to double can be found by dividing 72 by the interest rate. For example, an investment at 8 percent interest would double in approximately 9 years. (Notice the table value is 1.999.) At 12 percent interest a present value would double in approximately 6 years (table value of 1.9738).

The concept of future value can be useful in a number of ways. For example, what is the future value of $1,000 deposited in a savings account at 8 percent interest for 10 years? The table value is 2.1589, which gives a future value of $2,158.90. Another use is to estimate future land values. If land values are expected to increase at an annual compound rate of 6 percent, what will land with a present value of $1,500 per acre be worth 5 years from now? The table value is 1.3382, so the estimated future value is $1,500 × 1.3382, or $2,007.30 per acre.

Future Value of an Annuity   Figure 15-1b illustrates this concept. What is the future value of a number of payments (PMT) made at the end of each year for a given number of years? Each payment will earn interest from the time it is invested until the last payment is made. Suppose $1,000 is deposited at the end of each year in a savings account which pays 8 percent interest. What is the future value of this investment at the end of 3 years? It can be calculated in the following manner:

$$
\begin{aligned}
\text{First} \quad \$1,000 &= 1,000(1 + 0.08)^2 = 1,166.40 \\
\text{Second} \quad \$1,000 &= 1,000(1 + 0.08)^1 = 1,080.00 \\
\text{Third} \quad \$1,000 &= 1,000(1 + 0.08)^0 = \underline{1,000.00} \\
\text{Future value} \quad & \qquad\qquad\qquad\qquad\;\; \$3,246.40
\end{aligned}
$$

Since the money is deposited at the end of each year, the first $1,000 earns interest for only 2 years, the second $1,000 earns interest for 1 year, and the third $1,000 earns no interest as the future value is being determined at the end of 3 years. A total of $3,000.00 is invested and a total of $246.40 of interest is earned.

The future value of an annuity can be found using the above procedure, but an annuity with many payments requires many computations. An easier method is to use the equation

$$
\text{FV} = \text{PMT} \times \frac{(1 + i)^n - 1}{i}
$$

where PMT is the amount invested at the end of each time period. It is even easier to use table values such as those in Appendix Table 3, which cover a range of interest rates and years. Continuing with the same example, the table value for 8 percent interest and 3 years is 3.2464. Multiplying this factor by the annual payment or annuity of $1,000 confirms the previous future value of $3,246.40.

Present Value

The concept of present value refers to the current value of a sum of money to be received or paid in the future. Present values are found using a process called *discounting,* meaning the future value is discounted back to the present to find its current or present value. This discounting is done because a sum to be received in the future is worth less than the same amount available today. A present value can be interpreted as the sum of money which would have to be invested now at the given interest rate to equal the future value on the same date. When used to find present values, the interest rate is often referred to as the *discount rate.*

Compounding and discounting are opposite or inverse procedures, as shown in Figure 15-2. A present value is compounded to find its future value and a future value is discounted to find its present value. These inverse relationships will become more apparent in the following discussion.

Present Value of a Future Value   Figure 15-3*a* illustrates this concept. The future value is known and the problem is to find the present value of that amount. Notice that Figure 15-3*a* is exactly like Figure 15-1*a* except for the placement of the dollar sign and the question mark. This also illustrates the inverse relationship between compounding and discounting.

The present value of a future value depends on the interest rate and the length of time before the payment will be received. Higher interest rates and longer time periods will reduce the present value and vice versa. The equation to find the present value of a single payment to be received in the future is

$$PV = \frac{FV}{(1+i)^n} \quad \text{or} \quad FV \frac{1}{(1+i)^n}$$

where the abbreviations are as defined earlier.

FIGURE 15-2   Relationship between compounding and discounting ($1 at 8 percent interest).

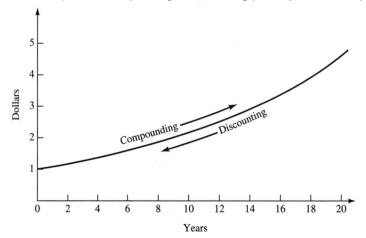

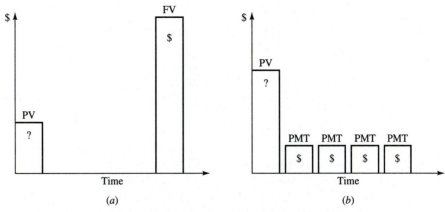

FIGURE 15-3   Illustration of the concept of a present value for a future value and for an annuity.

This equation can be used to find the present value of $1,000 to be received in 5 years using an interest rate of 8 percent. The calculations would be

$$PV = 1,000 \, \frac{1}{(1 + 0.08)^5}$$
$$= 1,000(0.6806)$$
$$= \$680.60$$

A payment of $1,000 to be received in 5 years has a present value of $680.60 at 8 percent interest. Stated differently, $680.60 invested for 5 years at 8 percent compound interest would have a future value of $1,000. This again shows the relationship between compounding and discounting. A more practical explanation is that an investor should not pay more than $680.60 for an investment which will return $1,000 in 5 years if there are other alternatives which will pay 8 percent interest or more. Investment analysis makes heavy use of present values as will be shown later in this chapter.

Tables are also available to assist in calculating present values as shown in Appendix Table 4. The factor for the appropriate interest rate and number of years is multiplied by the future value to obtain the present value. For example, the present value of $1,000 to be received in 5 years at 8 percent interest is equal to $1,000 multiplied by the table value of 0.6806 or $680.60.

Present Value of an Annuity   Figure 15-3b illustrates the common problem of determining the present value of an annuity or a number of payments to be received over time. Suppose an annuity of $1,000 will be received at the end of each year for 3 years and the interest rate is 8 percent. The present value of this income stream or annuity can be found as shown in the table below using values from Appendix Table 4.

Year	Amount ($)	Present value factor	Present value ($)
1	1,000	0.926	926
2	1,000	0.857	857
3	1,000	0.794	794
		Total	2,577

A present value of $2,577 represents the maximum an investor should pay for an investment which will return $1,000 at the end of each year for 3 years if an 8 percent return is desired. Higher interest rates will reduce the present value and vice versa.

Computing the present value of an annuity as shown in the table above becomes very tedious if there are a number of time periods. The present value can be found directly from the equation

$$PV = PMT \ \frac{1 - (1 + i)^{-n}}{i}$$

As before, it is much easier to use table values such as those in Appendix Table 5. The value corresponding to the proper interest rate and number of years is simply multiplied by the annual payment to find the present value of the annuity. The table value for 8 percent and 3 years is 2.5771, which is multiplied by $1,000 to find the present value of $2,577.10.

INVESTMENT ANALYSIS

Investment, as the term will be used in this section, deals with other than short-term or annual investments. It refers to the addition of intermediate and long-term assets to the business. Because these assets last a number of years, or forever in the case of land, these investment decisions will have long-lasting consequences and often involve large sums of money. Such investments should be thoroughly analyzed before the investment decision is made.

Investment analysis, or *capital budgeting* as it is sometimes called, is the process of determining the profitability of an investment or comparing the profitability of two or more alternative investments. A thorough analysis of an investment requires four pieces of information: (1) the annual net cash revenues from the investment, (2) its initial cost, (3) the terminal or salvage value of the investment, and (4) the interest or discount rate to be used.

Net cash revenues or cash flows must be estimated for each year in the life of the investment. Cash receipts minus cash expenses equals the additional net cash revenue generated by the proposed investment. Depreciation is not included as an annual expense as it is a noncash expense and is already accounted for by the difference between cost and terminal value. Any interest and principal payments on a loan needed to finance the purchase of the investment are also omitted from the calculation of net cash revenue. Investment analysis methods are used to determine the profitability of an investment without considering the method or amount of financing needed to purchase it.

The cost of the investment should be the actual total expenditure for its purchase and not the list price or just the down payment if it is being financed. Of the four types of information required for investment analysis, this will generally be the easiest to obtain. The terminal value will often need to be estimated and may be set equal to the salvage value for a depreciable asset. For a nondepreciable asset such as land, the terminal value should be the estimated market value of the asset at the time the investment is terminated.

The discount rate is often one of the more difficult pieces of information to estimate. It is the opportunity cost of capital representing the minimum rate of return required to justify the investment. If the proposed investment will not earn this minimum, the capital should be invested in an alternative investment. If money will be borrowed to finance the investment, the discount rate can be set equal to the cost of borrowed capital. A weighted average of the interest rate on borrowed capital and the opportunity cost of equity capital can also be used. Since risk must also be considered, the discount rate should be equal to the rate of return expected from an alternative investment of equal risk. Adjustment of the discount rate for risk and other factors will be discussed later in this chapter.

An example of the information needed for investment analysis is contained in Table 15-1. For simplicity, the terminal values were assumed to be zero. Whenever terminal values exist, they should be added to the net cash revenue for the last year, as they represent an additional cash receipt. The information for the two potential investments in Table 15-1 will be applied to illustrate four methods used for analyzing and comparing investments. They are (1) the payback period, (2) the simple rate of return, (3) net present value, and (4) the internal rate of return.

TABLE 15-1 NET CASH REVENUES FOR TWO $10,000 INVESTMENTS (NO TERMINAL VALUE)

Year	Net cash revenues ($) Investment A	Investment B
1	3,000	1,000
2	3,000	2,000
3	3,000	3,000
4	3,000	4,000
5	3,000	6,000
Total	15,000	16,000
Average return	3,000	3,200
Less annual depreciation	2,000	2,000
Net revenue	1,000	1,200

Payback Period

The payback period is the number of years it would take for an investment to return its original cost through the annual net cash revenues it generates. If the net cash revenues are constant each year, the payback period can be calculated from the equation

$$P = \frac{I}{E}$$

where P is the payback period in years, I is the amount of the investment, and E is the expected annual net revenue. For example, investment A in Table 15-1 would have a payback period of $10,000 divided by the net cash revenue of $3,000 per year or 3.33 years. When the annual net cash revenues are not equal, they should be summed year by year to find the year where the total is equal to the amount of the investment. For investment B in Table 15-1, the payback period would be 4 years because the accumulated net cash revenues reach $10,000 in the fourth year. Investment A is preferred over investment B as it has the shorter payback period.

The payback method can be used to rank investments according to payback period as was done above. Limited capital can be invested first in the highest ranked investment and then on down the list until the investment capital is exhausted. Another application is to establish a maximum payback period and reject all investments with a longer payback. For example, a manager may select a 4-year payback and invest only in alternatives with a payback of 4 years or less.

The payback method is easy to use and quickly identifies the investments with the most immediate cash returns. However, it also has several serious disadvantages. This method ignores any cash flows occurring after the end of the payback period as well as the timing of cash flows during the payback period. Selecting investment A using this method ignores the higher returns from investment B in years 4 and 5 as well as its greater total return. The payback method does not really measure profitability but is more a measure of how quickly the investment will contribute to the liquidity of the business. For these reasons, it can easily lead to poor investment decisions and is not the best method of investment analysis.

Simple Rate of Return

The simple rate of return expresses the average annual net revenue as a percentage of the investment. Net revenue is found by subtracting the average annual depreciation from the average annual net cash revenue. The simple rate of return is calculated from the equation

$$\text{Rate of return} = \frac{\text{average annual net revenue}}{\text{cost of the investment}} \times 100$$

Applying the equation to the example in Table 15-1 gives the following results:

Investment A: $$\frac{\$1,000}{\$10,000} \times 100 = 10\%$$

Investment B: $$\frac{\$1,200}{\$10,000} \times 100 = 12\%$$

This method would rank investment B higher than A, which is a different result than from the payback method. The simple rate of return method is better than the payback because it considers an investment's earnings over its entire life. However, it uses *average* annual earnings, which fails to consider the size and timing of annual earnings and can therefore cause errors in selecting investments. This is particularly true when there are increasing or decreasing net revenues. For example, investment A would have the same 10 percent rate of return if it had no net cash revenue the first 4 years and $15,000 in year 5, as the average return is still $3,000 per year. However, a consideration of the time value of money would show these to be greatly different investments. Because of this shortcoming, the simple rate of return is not generally recommended for analyzing investments.

Net Present Value

A serious shortcoming of both the payback and simple rate of return methods is their failure to account for the time value of money. For this reason, the net present value method is a preferred method because it does consider the time value of money as well as the stream of cash flows over the entire life of the investment. It is also called the *discounted cash flow method.*

Net present value of an investment is the sum of the present values for each year's net cash flow (or net cash revenue) less the initial cost of the investment. The equation for finding the present value of an investment is

$$\text{NPV} = \frac{P_1}{(1+i)^1} + \frac{P_2}{(1+i)^2} + \cdots + \frac{P_n}{(1+i)^n} - C$$

where NPV is net present value, P_n is the net cash flow in year n, i is the discount rate, and C is the initial cost of the investment. An example of calculating net present values using an 8 percent discount rate is shown is Table 15-2. The present value factors are table values from Appendix Table 4.

Investments with a positive net present value would be accepted using this procedure, those with a negative net present value would be rejected, and a zero value would make the investor indifferent. The rationale behind accepting investments with a positive net present value can be explained two ways. First, it means the actual rate of return on the investment is greater than the discount rate used in the calculations. In other words, the return is greater than the opportunity cost of capital if that was used as the discount rate. A second explanation is that the investor can afford to pay more for the investment and still achieve a rate of return equal to the discount rate used in

TABLE 15-2   NET PRESENT VALUE CALCULATIONS FOR TWO INVESTMENTS OF $10,000
(8% Discount Rate and No Terminal Values)

Year	Investment A			Investment B		
	Net cash flow ($) ×	Present value factor =	Present value ($)	Net cash flow ($) ×	Present value factor =	Present value ($)
1	3,000	0.926	2,778	1,000	0.926	926
2	3,000	0.857	2,571	2,000	0.857	1,714
3	3,000	0.794	2,382	3,000	0.794	2,382
4	3,000	0.735	2,205	4,000	0.735	2,940
5	3,000	0.681	2,043	6,000	0.681	4,086
		Total	11,979		Total	12,048
		Less cost	10,000		Less cost	10,000
		Net present value	1,979		Net present value	2,048

calculating the net present value. In Table 15-2, an investor could pay up to $11,979 for investment A and $12,048 for investment B and still receive an 8 percent return or more on invested capital. This method assumes the annual cash flows can be reinvested to earn a rate of return equal to the discount rate being used.

Both investments in Table 15-2 show a positive net present value using an 8 percent discount rate. In any determination of a present value, the selection of the discount rate affects the result. The net present values in Table 15-2 would be lower if a higher discount rate was used and vice versa. At some higher discount rate the net present values would be zero, and at an even higher rate they would become negative. Therefore, care must be taken to select the appropriate discount rate. It has a major influence on the results.

Internal Rate of Return

The time value of money is also reflected in another method of analyzing investments, the *internal rate of return,* or IRR. It provides some information not available directly from the net present value method. Both investments A and B have a positive net present value using the 8 percent discount rate. But what is the actual rate of return on these investments? It is the discount rate which makes the net present value equal to zero and can be found with the net present value method only by trial and error.

The actual rate of return on an investment with proper accounting for the time value of money is the internal rate of return, which is also called the *marginal efficiency of capital* or *yield on the investment.* The equation for finding IRR is

$$NPV = \frac{P_1}{(1+i)^1} + \frac{P_2}{(1+i)^2} + \cdots + \frac{P_n}{(1+i)^n} - C$$

where NPV is set equal to zero and the equation is solved for i, the discount rate.[1] A moment's reflection will show this to be a difficult equation to solve.

The IRR equation can be solved by a number of computer software programs or one of the more sophisticated financial calculators. In the absence of either of these, the IRR must be found through trial and error and some approximation. The procedure is shown in Table 15-3 for investment A. The relatively high net present value with the 8 percent discount rate indicates the return may be considerably higher than 8 percent, so 14 percent is arbitrarily chosen for the first estimation of IRR. The calculations show a positive net present value of $296, indicating the IRR is still higher. A 16 percent discount rate is used next, resulting in a net present value of *minus* $178. Whenever the net present value is negative, the actual IRR is less than the discount rate used so the actual IRR must be between 14 and 16 percent. The determination of IRR could continue using a 15 or 15.5 percent discount rate, or linear interpolation would estimate it at approximately 15.2 percent.[2]

The internal rate of return method can be used several ways in investment analysis. Any investment with an IRR greater than the opportunity cost of capital would be a profitable investment. However, some investors select an arbitrary cutoff value and invest only in those projects with a higher IRR. Unlike the net present value method, the IRR can be used to rank investments which have different initial costs and lives.

In addition to the rather difficult calculations involved, there is another potential limitation on the use of IRR. It implicitly assumes the annual net returns or cash flows from the investment can be reinvested to earn a return equal to the IRR. If the IRR is fairly high, this may not be possible causing the IRR method to overestimate the actual rate of return.

---

[1]Since NPV is set equal to zero, this equation is often rearranged with C to the left of the equal sign to make the net present value of the net revenue flows equal to the cost of the investment.

[2]The actual IRR is 15.24 percent for investment A and 13.76 percent for investment B, making A preferred over B using the internal rate of return method.

TABLE 15-3 ESTIMATION OF INTERNAL RATE OF RETURN FOR INVESTMENT A

Year	Net cash revenues ($)	14%		16%	
		Factor	Present value ($)	Factor	Present value ($)
1	3,000	0.877	2,631	0.862	2,586
2	3,000	0.769	2,307	0.743	2,229
3	3,000	0.675	2,025	0.641	1,923
4	3,000	0.592	1,776	0.552	1,656
5	3,000	0.519	1,557	0.476	1,428
		Totals	10,296		9,822
		Less cost	10,000		10,000
		Net present value	296		−178

Financial Feasibility

The investment analysis methods discussed so far are methods to analyze *economic profitability*. They are meant to answer the question, "Is the investment profitable?" However, investments identified as profitable may have years of negative cash flows depending on the pattern of net cash revenues over time and the method and amount of financing used to make the investment. Thus, when financing is used, an equally important question may be: "Is the investment *financially feasible?*" In other words will the investment generate sufficient cash flows at the right time to meet the required cash outflows? This potential problem was discussed in Chapter 11 but is worthy of further discussion here. Determination of financial feasibility should be the final step in any investment analysis.

A potential problem is illustrated in Table 15-4 for the example investments A and B used throughout this chapter. The data assume that each investment is totally financed with a $10,000 loan at 8 percent interest repaid over 5 years with equal annual principal payments. Both interest and principal are included in the debt payment column as they are both important cash outflows affecting financial feasibility.

Investment A shows a positive cash flow for each year as the net cash revenue is greater than debt payments. However, investment B has lower net cash revenues the first 2 years, which cause negative cash flows. Both investments had positive net present values using an 8 percent discount rate, and investment B actually had a slightly higher value than A. However, it is not unusual to find profitable investments which have negative cash flows in the early years if the net cash revenues are low and the investment requires a large amount of borrowed capital. The problem can be further compounded if the loan must be paid off in a relatively short period of time.

If a project such as investment B is undertaken, something must be done to make up for the negative cash flows. There are several possibilities which can be used individually or in combination. First, some equity capital can be used for part or all of the investment to reduce the size of the loan and annual debt payments. Second, the payment schedule for the loan may be adjusted to make debt payments more nearly equal to the net cash revenues. Smaller payments with a balloon payment at the end or interest only payments for the first few years would be possibilities.

TABLE 15-4   CASH FLOW ANALYSIS OF INVESTMENTS A AND B

	Investment A ($)			Investment B ($)		
Year	Net cash revenue	Debt payment*	Difference	Net cash revenue	Debt payment*	Difference
1	3,000	2,800	200	1,000	2,800	−1,800
2	3,000	2,640	360	2,000	2,640	− 640
3	3,000	2,480	520	3,000	2,480	520
4	3,000	2,320	680	4,000	2,320	1,680
5	3,000	2,160	840	6,000	2,160	3,840

*Assumes a $10,000 loan at 8% interest with equal principal payments over 5 years.

If neither of these methods is feasible, the cash deficits will have to be made up by excess cash available from other parts of the business. This will require a cash flow analysis of the entire business to see if cash will be available in sufficient amounts and at the proper times to meet the deficits from the investment.

OTHER FACTORS IN INVESTMENT ANALYSIS

The discussion has so far presented only the basic procedures and methods used in investment analysis. Several additional factors must be included in a thorough analysis of any investment or when comparing alternative investments. These include income taxes, inflation, and risk.

Income Taxes

The examples used to illustrate the methods of investment analysis did not consider the effects of income taxes on the net cash revenues. Taxes were omitted to simplify the introduction and discussion of investment analysis. However, investments are better analyzed using *after-tax* net cash revenues. Income taxes can substantially change the net cash revenues depending on the investor's marginal tax bracket and the type of investment. Because different investments may affect income taxes differently, they can only be compared on an after-tax basis.

Income taxes will reduce the net cash revenues with the actual amount depending on the marginal tax bracket of the taxpayer. For example, a taxpayer in the 28 percent marginal tax bracket will have 28 percent less net cash revenues on an after-tax basis. However, there may also be some tax savings. Depreciation was not included in calculating net cash revenue because it is a noncash expense. However, depreciation is a tax-deductible expense which reduces taxable income and therefore income taxes. The tax savings resulting from any depreciation associated with the new investment should be added to the net cash revenues. Certain investments will further reduce income taxes if they are eligible for any special tax credits or deductions that may be in effect at the time of purchase.

Whenever after-tax net cash revenues are used, it is important that an after-tax discount rate also be used. This places the entire analysis on an after-tax basis. The after-tax discount rate can be found from the equation

$$R = I(1 - t)$$

where R is the after-tax discount rate, I is the before-tax discount rate, and t is the marginal tax rate. For example, what is the after-tax discount rate if the before-tax discount rate is 12 percent and the taxpayer is in the 28 percent marginal tax bracket? The answer is 12 percent $\times$ (1 − 28 percent) = 8.64 percent. Notice that the *after-tax* cost of borrowed capital can be found from the same equation, with R and I changed to after- and before-tax interest rates. Interest is a tax-deductible expense, and every dollar of interest expense would reduce taxes by 28 cents for a taxpayer in the 28 percent marginal tax bracket. A taxpayer paying 12 percent interest on borrowed capital would have an 8.64 percent after-tax cost of capital as computed from the equation above.

Inflation

Inflation is a general increase in price levels over time, and it affects three factors in investment analysis: net cash revenues, terminal value, and the discount rate. Net cash revenues will change over time due to changes in the input and output prices, which may increase at different rates. Terminal values will also be higher than would be expected in the absence of inflation.

When the net cash revenues and terminal value are adjusted for inflation, the discount rate should also reflect the expected inflation. If the opportunity cost of an alternative investment is used for the discount rate and that opportunity cost already includes an expectation of inflation, the adjustment has implicitly been made. In other cases the discount rate will need adjustment. An interest or discount rate can be thought of as consisting of at least two parts: (1) a real interest rate, or the interest rate which would exist in the absence of inflation, either actual or expected, and (2) an adjustment for inflation or an inflation premium. During inflationary periods, a dollar received in the future will purchase fewer goods and services than it will at the present time. The inflation premium compensates for this reduced purchasing power through a higher interest or discount rate. An inflation-adjusted discount rate could then be estimated in the following manner:

Real interest rate	4%
Expected inflation rate	6%
Adjusted discount rate	10%

There may also be other factors to consider in estimating the discount rate as we will see in the next section.

In summary, there are two basic procedures to use in capital budgeting. The first is to estimate net cash revenues for each year and the terminal value using current prices, and then discount them using a *real* discount rate. This eliminates inflation from all elements of the equation. The second is to estimate net cash revenues and terminal value based on the expected inflation rate, and then use an *inflation-adjusted* discount rate. This method incorporates inflation into all parts of the equation. If the same inflation rate is used throughout, the net present value will be the same under either method. However, if some revenues or expenses are assumed to inflate at different rates than others, the inflation-adjusted NPV will be different.

Risk

Risk in investment analysis exists because the estimated net cash revenues and the terminal value depend on future prices and costs, which can be highly variable and difficult to predict. Unexpected changes in these values can quickly make a potentially profitable investment unprofitable. The longer the life of the investment, the more difficult it is to estimate future prices and costs.

One method of incorporating risk into the analysis is to add a risk premium to the discount rate. Investments with greater risk would have a higher risk premium. This is based on the concept that the greater the risk associated with a particular investment, the higher the expected return must be before an investor would be willing to accept the risk. In other words, there is a positive correlation between the risk involved and the return an investor would demand before accepting that risk.

The real rate of return discussed in the last section can be better defined as the rate of return which would exist for a *risk-free* investment in the absence of inflation. Insured savings accounts and U.S. government securities are typically considered to be nearly risk-free investments. Continuing with the example from the last section, we can add risk to our discount rate as follows:

Risk-free real rate of return	4%
Expected inflation rate	6%
Risk premium	5%
Adjusted discount rate	15%

Investments with a greater amount of risk would be assigned a higher risk premium and vice versa. This procedure is consistent with an earlier statement that the discount rate should be the rate of return expected from an alternative investment with *equal risk.* However, the risk premium is a subjective estimate and different individuals may have different estimates for the same investment.

Adding a risk premium to the discount rate does not eliminate risk. It is simply a way to build in a margin for error. If in our example the investment is acceptable using the higher discount rate of 15 percent, it simply means the net cash revenues over time could be somewhat lower than projected and still provide a 10 percent return, which would be acceptable for a risk-free investment. The risk premium simply adds in a margin for error.

Sensitivity Analysis

Given the uncertainty that may exist about the future prices and costs used to estimate net cash revenues and the terminal value, it is often useful to look at what would happen to net present value if the prices and costs were different. Sensitivity analysis is a process of asking a number of "What if" questions. What if the net cash revenues were higher or lower? What if the timing of net cash revenues was different? What if the discount rate was higher or lower? Sensitivity analysis involves changing one or more values in the net present value equation and recalculating the NPV. Recalculating the NPV for a number of different values gives the investor an idea of how "sensitive" the NPV, and therefore the investment decision, is to changes in the values being used. A sensitivity analysis will often give the investor better insight into the profitability of the investment, the effects of price changes, and therefore the amount of risk involved.

Recalculating the NPV can be tedious and time-consuming, particularly for investments with a long life and variable net cash revenues. These repetitive calculations are ideally suited for computer applications. With specialized financial software or spreadsheet programs, a computer can quickly and easily compute the NPV for many different combinations of net cash revenues, terminal values, and discount rates.

SUMMARY

Money has a time value because of the interest or rate of return it can earn over time. The future value of a present value is greater because of the interest the present value can earn over time. Future values are found through a process called compounding. The present value of a future value is smaller than the future value because money invested today at compound interest will grow into the larger future value. Discounting is the process of finding present values for amounts to be received in the future.

Investments can be analyzed by one or more of four methods: payback, simple rate of return, net present value, and internal rate of return. The first two are easy to use but have the disadvantage of not incorporating the time value of money into the analysis. This may cause errors in selecting or ranking alternative investments, and these methods are not recommended. The net present value method is widely recommended, as it properly accounts for the time value of money. Any investment with a positive net present value is profitable, as the rate of return is higher than the discount or interest rate being used. The internal rate of return method also considers the time value of money but is more difficult to use than net present value. It requires estimation of the actual rate of return, which can be difficult without a computer or financial calculator.

All four methods of investment analysis require estimation of net cash revenues over the life of the investment as well as terminal values. The net present value method also requires selecting a discount rate. All values should be on an after-tax basis in a practical application of these methods. The discount rate may also need to be adjusted for risk and inflation. A final step in analyzing any investment should be a financial feasibility analysis, particularly when a large amount of borrowed capital is used to finance the purchase.

QUESTIONS FOR REVIEW AND FURTHER THOUGHT

1 Put the concepts of future value and present value into your own words. How would you explain these concepts to someone hearing about them for the first time?

2 Explain the relationship between compounding and discounting.

3 Assume someone wishes to have $10,000 ten years from now as a college education fund for a child.

   a How much money would have to be invested now at 6 percent compound interest? At 8 percent?

   b How much would have to be invested annually at 6 percent compound interest? At 8 percent?

4 If land is currently worth $1,500 per acre and is expected to increase in value at a rate of 5 percent annually, what will it be worth in 5 years? In 10 years? In 20 years?

5 If you require an 8 percent rate of return, how much could you afford to pay for an acre of land which has annual net cash revenues of $60 and an expected selling price of $2,500 per acre in 10 years?

6 Suppose someone offers you the choice of receiving $1,000 now or $2,000 in 10 years and the opportunity cost of capital is 10 percent. Which option would you choose? What if the opportunity cost of your capital was only 6 percent?

7 Calculate the net present values for the investments in Table 15-2 using a 10 percent discount rate. Which investment would be preferred now? How would you explain the change?

8 Assume you have only $20,000 to invest and must choose between the two investments below. Analyze each using all four methods discussed in this chapter and an 8 percent opportunity cost for capital. Which investment would you select? Why?

	Investment A ($)	Investment B ($)
Initial cost	20,000	20,000
Net cash revenues:		
Year 1	6,000	5,000
Year 2	6,000	5,000
Year 3	6,000	5,000
Year 4	6,000	5,000
Year 5	6,000	5,000
Terminal value	0	8,000

9 Discuss economic profitability and financial feasibility. How are they different? Why should both be considered when analyzing a potential investment?

10 Why would capital budgeting be useful in analyzing an investment of establishing an orchard where the trees would not become productive until 6 years after planting?

11 What advantages would present value techniques have over partial budgeting for analyzing the orchard investment in Question 10?

REFERENCES

Barry, Peter J., John A. Hopkin, and C. B. Baker: *Financial Management in Agriculture,* 4th ed., The Interstate Printers & Publishers, Danville, IL, 1988, chaps. 9, 10, 11.

Bierman Jr., Harold, and Seymour Smidt: *Financial Management for Decision Making,* Macmillan Publishing Co., New York, 1986, chaps. 2, 6.

Boehlje, Michael D., and Vernon R. Eidman: *Farm Management,* John Wiley & Sons, New York, 1984, chap. 8.

Casler, G. L., B. Anderson, and R. D. Aplin: *Capital Investment Analysis Using Discounted Cash Flows,* 3d ed., Grid, Inc., Columbus, OH, 1984.

Lee, Warren F., Michael D. Boehlje, Aaron G. Nelson, and William G. Murray: *Agricultural Finance,* 8th ed., The Iowa State University Press, Ames, IA, 1988.

16

FARM BUSINESS ANALYSIS

CHAPTER OBJECTIVES

1 To introduce farm business analysis as part of the control function of management

2 To suggest several standards of comparison to use in the analysis of the farm business

3 To review measures which can be used to analyze the financial condition and profitability of the business

4 To identify some measures of business size

5 To illustrate the concept of economic efficiency and show how it is affected by physical efficiency, product prices, and input prices

6 To show how enterprise analysis can be used to identify strengths and weaknesses in a business

7 To outline a procedure for troubleshooting or locating problem areas in the farm or ranch business

Farm managers have always been interested in the reasons why some farms have higher net incomes than others of the same type and size. The observation and study of these differences and their causes began in the early 1900s and marked the beginning of farm management and farm business analysis. Differences in management are a common explanation for the different net farm incomes, but this explanation is not very complete. Unless the specific differences in management can be identified, there can be no precise recommendations for improving net income on the farms with "poor management."

TABLE 16-1   RETURNS TO MANAGEMENT PER FARM BY LEVEL OF GROSS SALES

Gross sales ($)	Returns to management per farm ($)		
	Low third	Average	High third
10,000 to 40,000	−22,005	−8,194	6,661
40,000 to 100,000	−25,087	−4,693	13,344
100,000 to 250,000	−23,744	1,260	27,894
Over 250,000	−63,647	2,701	70,200

Source: Iowa Farm Business Association, 1991.

The differences in profitability of similar farms can be summarized in several ways. Table 16-1 shows some observed differences in the returns to management per farm on a sample of central Iowa grain and livestock farms. Similar differences can be found in other states for other farm types. Notice that high gross sales does not ensure profitability. The average return to management per farm was −$63,647 for the low-third farms in the highest sales group while the high-third farms in the lowest sales group had an average return of $6,661. Several questions beg for an answer. What accounts for the differences? What can be done to improve returns on the less profitable farms? Differences in the operator's goals or the quality of resources available may be partial answers. However, complete answers can be found only by performing a complete farm business analysis.

This discussion emphasizes the need to perform the control function of management for the entire farm business as well as for individual enterprises. Three steps in the control function are (1) establishing standards for comparing results, (2) measuring the actual performance of the business, and (3) taking corrective action to improve the performance once problem areas have been identified. Improvement in performance is the payoff for the time spent on the control function of management.

TYPES OF ANALYSIS

A farm business analysis can be divided into five areas of investigation as follows:

1 *Financial*   Financial analysis concentrates on the capital position of the business including solvency, liquidity, and changes in net worth.

2 *Profitability*   Profitability is analyzed by comparing income and expenses. High net farm income is assumed to be an important, though not always the only, goal of the farm manager.

3 *Farm Size*   Not having adequate resources is often a cause for low profitability, although growing too rapidly or exceeding the scale that the operator can manage effectively can also reduce profits.

4 *Efficiency*   Low profitability can often be traced to poor efficiency in one or more areas of the business. Both economic and physical efficiency measures should be examined.

5 *Enterprises*   It is often important to estimate the profitability of individual enterprises. Low profits in some enterprises may offset high profits in others.

STANDARDS OF COMPARISON

The remainder of this chapter will examine some measures and ratios which can be used to conduct a complete farm analysis. Once a measure is calculated, the problem becomes one of evaluating the result. Is the value good, bad, or average? Compared with what? Can it be improved? These questions emphasize the need to have some standards against which the measures and ratios can be compared. There are three basic standards to use in analyzing farm business results: *budgets, comparative farms,* and *historical trends.*

Budgets   In budget analysis, the measures or ratios are compared against budgeted goals or objectives identified during the planning process. When results in some area consistently fall short of the budgeted objectives, the area needs additional managerial effort.

Comparative Analysis   Unfortunately, poor weather or low prices may prevent a farm from reaching its goals some years, even when no serious management problems exist. Therefore, a second source of useful standards for comparison is actual records of other area farms of similar size and type for the same year. These may not represent ideal standards but will indicate if the farm being analyzed was above or below average given the weather and price conditions that existed.

One of the advantages of participating in an organized record-keeping service such as those available through lending agencies, private firms, farm cooperatives, or the agricultural extension service is the annual record summary provided for all farms using the service. These summaries contain averages and ranges of values for different analysis measures for all farms as well as groups of farms sorted by size and type. Such results provide an excellent set of comparative standards as long as the farms being compared are of similar size and type, and in the same general geographical area.

Trend Analysis   When using trend analysis, the manager records the various measures and ratios for the same farm for a number of years and observes any trends in the results. The manager looks for improvement or deterioration in each measure over time. Those areas showing a decline deserve further analysis to determine the causes. This method does not compare results against any standards, only against historical values. The objective becomes one of showing improvement over the results of the past. However, trend analysis must take into account year-to-year changes in weather and prices.

Any of the analytical measures discussed in the remainder of this chapter can be used with budget, comparative, and trend analysis. These measures are shown in tables that contain blanks for entering the predetermined standards as well as space for recording actual results for a number of years.[1]

FINANCIAL MEASURES

The financial portion of a complete farm analysis is designed to measure the solvency and liquidity of the business, and identify weaknesses in the structure or mix of the various types of assets and liabilities. The balance sheet and the income statement are the primary sources of data for calculating the measures related to the financial position of the business. These financial measures and ratios were presented in Chapters 3 and 4 but will be reviewed briefly to show how they fit into a complete farm analysis.

It is convenient to record the established goals or standards along with actual values on a form such as that shown in Table 16-2. This permits a quick and easy comparison of current values against both the standard and historical values and identification of any trends which may have developed.

Solvency

Solvency refers to the value of assets owned by the business compared to the amount of liabilities owed.

Debt/Asset Ratio (FFSTF)   The debt/asset ratio is a common measure of business solvency and is calculated by dividing total farm liabilities by total farm assets using current market values for each:

[1]The analysis measures discussed in this chapter include those recommended by the Farm Financial Standards Task Force as well as a number of others. Those recommended by the task force are identified by the abbreviation (FFSTF) shown after their name.

TABLE 16-2   ANALYZING THE FINANCIAL CONDITION OF THE BUSINESS

Item	Budgeted goal or standard	Year or date		
1. Debt/asset ratio	————	————	————	————
2. Change in net worth	————	————	————	————
3. Current ratio	————	————	————	————
4. Working capital	————	————	————	————
5. Term debt and capital lease coverage ratio	————	————	————	————
6. Capital replacement and term debt repayment margin	————	————	————	————

$$\text{Debt/asset ratio} = \frac{\text{total farm liabilities}}{\text{total farm assets}}$$

Smaller values are usually preferred to larger ones, as they indicate a better chance of maintaining the solvency of the business should it ever be faced with a period of adverse economic conditions. However, low debt/asset ratios may also indicate that a manager is reluctant to use debt capital to take advantage of profitable investment opportunities. In this case, the low ratio would be at the expense of a potentially higher net farm income. A more complete discussion of the use of debt and equity capital is found in Chapter 17.

Two other common measures of solvency are the equity/asset ratio (FFSTF) and the debt/equity ratio (FFSTF). These were discussed in Chapter 3. They are simply transformations of the debt/asset ratio and provide the same information.

Change in Net Worth   An important measure of financial progress for a business is a net worth which is increasing over time. This is an indication of business growth, additional capital investment, and a greater borrowing capacity. As discussed in Chapter 3, net worth should be measured both with and without the benefit of inflation and its effect on the value of assets such as land. This requires a balance sheet prepared on both a cost and a market basis, so that any increase in net worth can be separated into the result of the production activities during the accounting period and the effect of changes in asset values.

Liquidity

The ability of a business to meet cash flow obligations as they come due is called *liquidity*. Maintaining liquidity is important to keep the financial transactions of the business running smoothly.

The current ratio (FFSTF) is a quick indicator of a farm's liquidity. Current assets will be sold or turned into salable products in the near future and will generate cash to pay debt obligations that come due. The current ratio is calculated by dividing current farm assets by current farm liabilities:

$$\text{Current ratio} = \frac{\text{current farm assets}}{\text{current farm liabilities}}$$

Working capital (FFSTF), or the difference between current assets and current liabilities, is another measure of liquidity. It represents the excess dollars available from current assets after current liabilities have been paid.

These liquidity measures are easy to calculate but fail to take into account upcoming operating costs, overhead expenses, capital replacement needs, and nonbusiness expenditures. They also do not include sales from products yet to be produced and do not consider the *timing* of cash receipts and expenditures throughout the year.

An operation that concentrates its production during one or two periods of the year, such as a cash grain farm, needs to have a high current ratio at the beginning of the year because no new production will be available for sale until near the end of the year. On the other hand, dairy producers or other livestock operations with continuous sales throughout the year can operate safely with lower current ratios and working capital margins.

Chapter 11 discussed how to construct a cash flow budget and how to use it to manage liquidity. Although developing a cash flow budget takes more time than computing a current ratio, it takes into account *all* sources and uses of cash as well as the timing of their occurrence throughout the year. Every farm or ranch that has any doubts about being able to meet cash flow needs should routinely construct cash flow budgets and monitor them against actual cash flows throughout the year. This allows early identification of liquidity problems.

Tests for Liquidity Problems

When overall cash flow becomes tight, there are several simple tests that can be carried out to try to identify the source of the problem.

1 Analyze the debt structure by calculating the percent of total farm liabilities classified as current and noncurrent. Do the same for total farm assets. If the debt structure is "top heavy," that is, a larger proportion of the liabilities are current than are assets, it may be advisable to convert some of the current debt to a longer term liability. This will reduce the annual principal payments.

2 Compare the changes in the current assets over time, particularly crop and livestock inventories. This can be done by looking at the annual balance sheets. Building up inventories can cause temporary cash flow shortfalls. On the other hand, if cash flow has been met by reducing current inventories over time, then the liquidity problem has only been postponed.

3 Compare purchases of capital assets to sales and depreciation. Continually increasing the inventory of capital assets may not leave enough cash to meet other obligations. On the other hand, liquidation of capital assets may be necessary to meet cash flow needs at times, but could affect profitability in the future.

4 Compare the amount of operating debt carried over from one year to the next for a period of several years. Borrowing more than is repaid each year will eventually reduce borrowing capacity, increase interest costs, and make future repayment even more difficult.

5 Compare the cash flow needed for land (principal and interest payments, property taxes, and cash rent) per acre to a typical cash rental rate for similar land. Excess land debt or high cash rents can cause severe liquidity problems.

6 Compare withdrawals for family living and taxes to the opportunity cost of unpaid labor. The business may be trying to support more people than are fully employed or at a higher level than possible from farm income.

Any of these symptoms can cause poor liquidity, even for a profitable farm, and can eventually lead to severe financial problems.

Measures of Repayment Capacity

Two measures are recommended to evaluate the debt repayment capacity of the business. The first is the *term debt and capital lease coverage ratio* (FFSTF) where term debt refers to noncurrent debt, which typically has scheduled, amortized payments. It is the ratio of the cash available for term debt payments, both principal and interest, divided by the total of the term debt payments and capital lease payments due. The cash available is computed by:

> net farm income from operations
> *plus* total nonfarm income
> *plus* depreciation expense
> *plus* interest paid on term debt
> *plus* interest on capital leases
> *minus* withdrawals for family living and taxes

A ratio greater than 1.0 means that there is sufficient cash flow to meet the term debt payments.

The second measure is called the *capital replacement and term debt repayment margin* (FFSTF) and is simply the *difference* between the cash available and the total term debt payments due, including principal payments on unpaid past operating debts and personal loans. These two measures are similar in concept to the current ratio and working capital measures but take into account *all* cash inflows and outflows.

MEASURES OF PROFITABILITY

A business which is both solvent and liquid as shown by a financial analysis will not necessarily be profitable. Profitability analysis starts with the income statement, as was explained in Chapter 4 where several measures of profitability were also discussed. They are shown in Table 16-3, which can be used as a form for recording and comparing them over time.

TABLE 16-3   MEASURES OF FARM PROFITABILITY

Item	Budgeted goal or standard	Year or date		
		———	———	———
1. Net farm income ($)	———	———	———	———
2. Return to labor and management ($)	———	———	———	———
3. Return to management ($)	———	———	———	———
4. Rate of return on farm assets (ROA %)	———	———	———	———
5. Rate of return on farm equity (ROE %)	———	———	———	———
6. Operating profit margin ratio	———	———	———	———

Net Farm Income (FFSTF)   Net farm income is a measure of return to the opera-tor's equity capital, unpaid labor, and management contributed to the farm business. One method for establishing a goal for net farm income is to estimate the income the owner's labor, management, and capital could earn in nonfarm uses. In other words, the total opportunity cost for these factors of production becomes the goal or standard for net farm income. In any year the income statement shows a lower net farm income, one or more of these factors did not earn its opportunity cost. Another goal would be to reach a specific income level at some time in the future, such as at the end of 5 years. Progress toward this goal can be measured over time.

Although net farm income is influenced heavily by the profitability of the farm, it also depends on what proportion of the total resources utilized are contributed by the op-erator. Replacing borrowed capital with equity capital, rented land with owned land with little or no debt, or hired labor with operator labor will all improve net farm in-come. Figure 16-1 shows how the gross farm revenue "pie" is divided among the parties who supply resources to the farm business. The size of the farmer's slice of the pie de-pends on how many resources he or she contributes as well as the size of the pie. Farms with high net incomes are often those that have accumulated a large net worth over time or which depend heavily on unpaid operator and family labor. To more accurately mea-sure profitability the opportunity costs for these resources must be considered.

Return to Labor and Management   Some farm businesses have more assets or borrow more money than others. As shown in Chapter 4, a fair comparison requires

FIGURE 16-1   How is the total revenue pie divided?

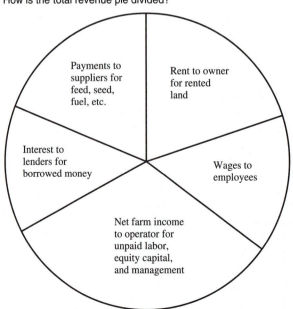

that the interest expense incurred by the business be added back to net farm income from operations to get adjusted net farm income. Next, an opportunity cost for total farm assets is subtracted. What remains is the return to the unpaid labor and management utilized in the business. One goal or standard for return to labor and management is the opportunity cost of the owner's labor and management in another use.

Return to Management   The return to management is that portion of adjusted net farm income remaining after the opportunity costs of both labor and capital have been subtracted. It represents the residual return to the owner for the management input, and it will be highly variable as prices and yields vary from year to year. Negative returns to management are not unusual, but the goal should be a positive return over the long run.

Rate of Return on Farm Assets (ROA) (FFSTF)   The rate of return on farm assets was discussed in Chapter 4. This value can be compared to rates of return on other long-term investments although differences in risk and potential gains in value must also be considered. It also allows comparison among farms of different sizes and types. Historical average rates of return to farm assets using annual cash income and current market values have generally been less than the opportunity cost of capital. However, when increases in asset values are included as part of the income, returns have been comparable to other investments.

Rate of Return on Farm Equity (ROE) (FFSTF)   This measure is perhaps more indicative of financial progress than ROA as it measures the percent return to owner's net worth or equity. If the farm business has no debts, the rate of return on farm equity will be the same as the rate of return on farm assets since they are one and the same. When debt capital is used interest must be paid. The amount available as a return on equity then becomes only that part of net farm income remaining after interest is paid. Opportunity costs for operator labor and management are also subtracted before computing the rate of return to equity. (See Chapter 4 for details on how to compute both ROA and ROE.)

Operating Profit Margin Ratio (FFSTF)   As explained in Chapter 4, the operating profit margin ratio (OPMR) measures the proportion of gross revenue that is left after paying expenses. It is found by dividing the dollar return to farm assets (adjusted net farm income minus the opportunity cost of labor and management) divided by the gross revenue of the farm.

Farms with large investments in fixed assets such as land and buildings and fewer operating expenses, such as for cash rent, will generally show higher operating profit margin ratios. Conversely, farms with more rented assets will often have a higher rate of return on farm assets (ROA) but a lower OPMR. Rented and owned assets are substitutes for each other. As more of one is used, the rate of return to the other will generally increase. Thus, these two measures of profitability must be considered together. An operation may be high in one ratio and low in the other simply because of the mix of owned and rented assets. If both measures are below the average or the goal, then problems of profitability are evident.

General Comments   There are two cautions regarding the procedure used to calculate returns to labor, management, assets, and equity. The first is the somewhat arbitrary nature of estimating the opportunity costs used in the calculations. There are no rigid rules to use, and individuals may have different opinions about the appropriate values to use. Obviously the calculated return to any factor will change if different opportunity costs are used for the others. The second caution is that these values are *average* returns to the factors and not the *marginal* returns. For example, the rate of return to assets is the average rate of return on all capital invested in the business and not the marginal return on the last dollar invested. The marginal rates of return would be much more useful for planning purposes.

Nevertheless, if the opportunity costs are carefully estimated and used consistently from year to year, and the average nature of the result is kept in mind, these measures provide a satisfactory means of historical and between-farm comparisons of profitability. They are, however, not as useful for evaluating marginal changes in the organization and operation of the farm business.

The measures of profitability included in Table 16-3 can be used to identify the existence of a problem. If profitability falls short of the goal or standard, however, these measures do not identify the exact source of the problem. Further analysis is needed to clarify the cause of the problem and suggest the corrective action needed. The next three sections proceed with a more in-depth analysis of the farm business to pinpoint more specific problem areas.

MEASURES OF SIZE

An income or profitability problem can exist in any single year due to low prices or to low crop yields. If the problem persists in years with above-average prices and yields, it may be caused by insufficient farm size. Historically, net farm income has been highly correlated with farm size, and a low income may be due entirely to operating a farm which is too small. Several common measures of farm size are shown in Table 16-4. Some measure the quantity or value of production while others measure the volume of resources used.

Volume of Sales   For specialized farms the number of units of output sold annually is a convenient measure of size. Examples are 3,000 head of farrow-to-finish hogs or 10,000 boxes of apples. On a diversified farm it may be necessary to record the units sold from several different enterprises to more accurately reflect farm size. Where a crop share or livestock share lease is being used, both the owner's and tenant's shares of production should be included when making comparisons with other farms.

Total or Gross Revenue   The volume of production from a farm producing several different products can be measured by total revenue. Table 4-1 shows the calculations where total revenue is defined as total sales and other cash income, adjusted for changes in inventories, accounts receivable, and any gains/losses on the sale of breeding livestock.

TABLE 16-4   MEASURES OF FARM SIZE

Item	Budgeted goal or standard	Year or date		
1. Volume of sales	———	———	———	———
2. Gross revenue	———	———	———	———
3. Value of farm production	———	———	———	———
4. Total farm assets	———	———	———	———
5. Total acres controlled	———	———	———	———
6. Livestock numbers				
———————	———	———	———	———
———————	———	———	———	———
———————	———	———	———	———
7. Total labor used (years)	———	———	———	———

Value of Farm Production   Some livestock operations buy feeder livestock and feed while others produce all their own feed and animals. In order to make a fair comparison of size of business, the value of feed and livestock purchases is often subtracted from gross revenue to arrive at the *value of farm production.* In this measure, only the value added to purchased feed and livestock is credited to the value of production. It is calculated as follows:

Gross revenue
minus livestock purchases
minus feed purchases
equals value of farm production

Value of farm production for a given farm may vary substantially from year to year, depending on weather and product prices. However, it is a convenient way to compare the size of different types of farms.

Total Farm Assets   Another measure of size is the total capital invested in land, buildings, machinery, livestock, crops, and other assets. The market value of total farm assets from the balance sheet is the easiest place to obtain this value. With all values in dollar terms, this measure allows an easy comparison of farm size across different farm types. However, it does not take into account rented assets.

Total Acres Controlled   The number of acres controlled includes both acres owned and acres rented. It is useful for similar types of farms but is not particularly good for comparing different types of farms. A California farm with 200 acres of irrigated vegetables is obviously not similar in size to an operation with 200 acres of dryland wheat in western Kansas or a 200-acre cattle feedlot in Texas. Number of acres is best used for comparing the size of crop farms of the same general type in an area with similar soil resources.

Livestock Numbers   It is common to speak of a 400-cow ranch or a 100-cow dairy farm or a 200-sow hog farm. These measures of size are useful, but only for comparing size among farms with the same class of livestock.

Total Labor Used   Labor is a resource common to all farms. Terms such as a one-person or two-person farm are frequently used to describe farm size. A full-time worker equivalent can be calculated by summing the months of labor provided by operator, family, and hired labor and dividing the result by 12. For example, if the operator works 12 months, family members provide 4 months of labor, and 5 months of labor is hired, the full-time worker equivalent is 21 months divided by 12, or 1.75 years of labor. This measure of size is affected by the amount of labor-saving technology used and should therefore be interpreted carefully when comparing farm sizes.

EFFICIENCY

Some of the measures of farm size presented here are measures of production, such as sales, gross revenue, and value of farm production. Others such as land area, total assets, labor employed, and number of breeding livestock owned measure the quantity of resources used. Will the farms or ranches with the most resources also have the most production? Generally speaking, yes, but there will always be exceptions. Some managers are able to generate more production or use fewer resources than their neighbors because they use their resources more efficiently. A general definition for physical efficiency is the quantity of production achieved per unit of resource employed.

$$\text{Efficiency} = \frac{\text{Production (output)}}{\text{Resources used}}$$

If a comparison with other farms or with a budget goal shows that an operation has an adequate volume of resources but is not reaching its production goals, then some resources are not being used efficiently. Because a farm or ranch business may use many types of resources, there are many ways to measure efficiency. Only the more useful and common ones will be discussed here. Efficiency measures are of two types, economic and physical.

Measures of Economic Efficiency

Economic efficiency measures differ from physical measures primarily because they are computed either as dollar values or as some rate or percentage relating to capital use. Measures of economic efficiency must be interpreted carefully. They measure average values, not the marginal values or the effect small changes would have on overall profit. The measures to be discussed are shown in Table 16-5.

Asset Turnover Ratio (FFSTF)   This ratio measures how efficiently capital invested in farm assets is being used. It is found by dividing gross revenue by total farm

TABLE 16-5 MEASURES OF ECONOMIC EFFICIENCY

Item	Budgeted goal or standard	Year or date		
		———	———	———
Capital efficiency:				
1. Asset turnover ratio	———	———	———	———
2. Operating expense ratio	———	———	———	———
3. Net farm income from operations ratio	———	———	———	———
Livestock efficiency:				
4. Returns per $100 of feed fed	———	———	———	———
5. Feed cost per 100 lb of gain (or milk)	———	———	———	———
Land efficiency:				
6. Crop value per tillable acre	———	———	———	———
Machinery efficiency:				
7. Machinery investment per tillable acre	———	———	———	———
8. Machinery cost per tillable acre	———	———	———	———

assets valued at market. For example, an asset turnover ratio equal to 0.25, or 25 percent, indicates the gross revenue was equal to 25 percent of the total capital invested in the business. At this rate, it would take 4 years to produce agricultural products with a value equal to the total assets.

The asset turnover ratio will vary by farm type. Dairy, hog, and poultry farms will generally have higher rates, beef cow farms tend to be lower, and crop farms usually have intermediate values. Farms or ranches that own most of their resources will generally have lower asset turnover ratios than those that rent land and other assets. Therefore, the asset turnover ratio should be compared only among farms of the same general type.

Operating Expense Ratio (FFSTF) Four operational ratios are recommended to show what percent of gross revenue went for operating expenses, depreciation, interest, and net income. The operating expense ratio is computed by dividing total operating expenses (minus depreciation) by gross revenue. Farms with a high proportion of rented land and machinery or hired labor will tend to have higher operating expense ratios.

Depreciation Expense Ratio (FFSTF) This ratio is computed by dividing total depreciation expense by gross revenue. Farms with a relatively large investment in newer machinery, equipment, and buildings will have higher depreciation expense ratios. A high ratio may indicate underutilized assets.

Interest Expense Ratio (FFSTF)   Total farm interest expense (adjusted for interest owed at the beginning and end of the year) is divided by gross revenue to find this ratio. Ratios higher than average or higher than desired may indicate too much dependence on borrowed capital or high interest rates on existing debt.

Net Farm Income from Operations Ratio (FFSTF)   Dividing net farm income from operations by gross revenue measures what percent of gross revenue was left after paying all expenses (but before any opportunity costs).

Note that these four operational ratios will sum to 1.00, or 100 percent. They can also be calculated using the value of farm production as a base instead of gross revenue. In that case the cost of purchased feed and livestock should not be included in operating expenses. The asset turnover ratio can also be calculated using the value of farm production. This is more accurate when comparing operations that purchase large quantities of feed and livestock to those that do not.

Livestock Production per $100 of Feed Fed   This is a common measure of economic efficiency in livestock production. It is calculated by dividing the value of livestock production during a period by the total value of all the feed fed and multiplying the result by 100. The value of livestock production is equal to:

Total cash income from livestock
plus or minus livestock inventory changes
minus value of livestock purchased
equals value of livestock production

Livestock production per $100 of feed fed is computed from the following:

$$\frac{\text{Value of livestock production}}{\text{Value of feed fed}} \times 100$$

A ratio equal to $100 indicates the livestock just paid for the feed consumed but no other expenses. The return per $100 of feed fed also depends on the type of livestock as shown in the following table:

Livestock enterprise	Production per $100 of feed fed (10-year average, 1982–1991)
Beef cow	$157
Dairy	183
Feeding cattle	148
Hogs (farrow to finish)	177

Source: Iowa Farm Costs and Returns, Iowa State University Extension Publication FM-1789, 1982–1991.

Higher values for some livestock enterprises do not necessarily mean these are more profitable. Enterprises with higher building and labor costs or other nonfeed costs such as dairy and feeder pig production need higher returns in order to be as profitable as enterprises with low nonfeed costs, such as finishing feeder animals. Because of these differences, the production per $100 of feed fed should be calculated for each individual livestock enterprise and compared only with values for the same enterprise.

Feed Cost per 100 Pounds of Gain The feed cost per 100 pounds of gain (or per 100 pounds of milk for a dairy enterprise) is found by dividing the total feed cost for each enterprise by the total pounds of production, and multiplying the result by 100. The total pounds of production for the year should be equal to the pounds sold and consumed minus the pounds purchased, and plus or minus any inventory changes. This measure is affected by both feed and livestock prices, and values should be compared only among the same type of livestock.

Crop Value per Tillable Acre This value measures the intensity of crop production and whether or not high-value crops are included in the crop plan. It is calculated by dividing the total value of all crops produced during the year by the number of tillable acres. It does not take into account production costs and therefore does not measure profit.

Machinery Investment per Tillable Acre Dividing the current value of all crop machinery and equipment by the number of tillable acres provides a measure of machinery efficiency. This is another measure which should be compared only with values for farms of the same type and size. Any additional machinery needed for livestock should be excluded. Larger farms tend to have lower investments per acre because of economies of size.

This is another measure which can be either too high or too low. A high value indicates there may be an excessive machinery investment while very low values may indicate the machinery is too old and unreliable or too small for the acres being farmed. There is an important interaction between machinery investment, labor requirements, timeliness of field operations, and the amount of custom work hired. These factors are examined more thoroughly in Chapter 20.

Machinery Cost per Tillable Acre This measure differs from the one above in that the total of all annual costs related to machinery is divided by the number of tillable acres. Total annual machinery costs include all fixed and variable costs plus the cost of custom work hired and machinery lease and rental payments.

Measures of Physical Efficiency

Poor economic efficiency can result from several types of problems. Physical efficiency, such as the rate at which seed, fertilizer, and water are converted into crops or the rate at which feed is turned into livestock products, may be too low. Below-average selling prices or above-average input prices can also cause low economic efficiency.

Physical efficiency measures include bushels harvested per acre, pigs weaned per litter, and pounds of milk sold per cow. A sample of physical efficiency measures is shown in Table 16-6. As with the measures of economic efficiency, they are average values and must be interpreted as such. An average value should not be used in place of a marginal value when determining the profit-maximizing input or output level. Sometimes high values are a result of using too much input and producing at a level which is beyond the profit-maximizing quantity.

Selling Prices

The average price received for a product may be lower than desired for several reasons. In some years supply and demand conditions cause prices in general to be low. However, if the average price received is lower than for other similar farms in the same year, the operator may want to look for different marketing outlets or spend some time improving marketing skills.

Purchase Prices

Sometimes physical efficiency is good and selling prices are acceptable, but profits are low because resources were acquired at high prices. Examples are buying land at a high price, paying high cash rents, paying high prices for seed and chemical products, buying overly expensive machinery, paying higher than necessary wage rates, borrowing money at high interest rates, and so forth. Input costs can be reduced by comparing prices at different suppliers, buying used or smaller equipment, buying large quantities when possible, and substituting raw materials for more refined or more convenient products.

Physical efficiency, selling price, and purchase price all combine to determine economic efficiency as shown in the following equations:

TABLE 16-6   EXAMPLES OF PHYSICAL EFFICIENCY MEASURES

Item	Budgeted goal or standard	Year or date		
Crops:				
1. Crop yield per acre	————	————	————	————
Livestock:				
2. Average daily gain	————	————	————	————
3. Feed per pound of gain	————	————	————	————
4. Calving percent	————	————	————	————
5. Pigs weaned per sow per year	————	————	————	————
6. Milk production per cow	————	————	————	————
7. Eggs per hen	————	————	————	————

$$\text{Economic efficiency} = \frac{\text{value of product produced}}{\text{cost of resources used}}$$

OR

$$\frac{\text{quantity of product} \times \text{selling price}}{\text{quantity of resource} \times \text{purchase price}}$$

Verifying Inventories

In order to compute many physical efficiency measures, an accurate count of the quantity of product produced is needed. This is usually measured in bushels, tons, pounds, or head. A check on the accuracy of such numbers can be made using the general rule that *sources* must equal *uses*.

Table 16-7 shows the relevant quantities for a crop enterprise. Most of the quantities and values can be measured directly or derived from sales receipts or purchase records. However, if all the quantities or values but one are known, the unknown one can be found by calculating the difference between the sources values and the uses values. This is sometimes done to estimate the quantity or value of feed fed to livestock or crop produced. However, greater accuracy will be achieved if all these quantities are measured directly and the equality relationship is used to verify that the sources equals the uses.

Livestock inventories can be verified with a similar form by adding an additional column for weight. For livestock, physical quantities are often measured in pounds or hundredweight as well as number of head. Calves, pigs, or lambs produced on the farm are assumed to have a beginning weight and value of zero. Some animals may also enter or leave the inventory when they are reclassified, such as females selected for breeding stock replacement.

TABLE 16-7 VERIFYING CROP INVENTORIES

Sources	Quantity (bu)	Value ($)
Beginning inventory	————	————
Purchased	————	————
Produced	————	————
Total	————	————

Uses	Quantity (bu)	Value ($)
Ending inventory	————	————
Sold	————	————
Used for seed	————	————
Used for feed	————	————
Spoilage, other losses	————	————
Total	————	————

ENTERPRISE ANALYSIS

A profitability analysis done for a whole farm may indicate a problem, but the source of the problem is often difficult to find if there are many enterprises. Several enterprises may be highly profitable while others are unprofitable or only marginally profitable. Enterprise analysis can identify the less profitable enterprises so that some type of corrective action can be taken.

An enterprise analysis consists of allocating all income and expenses among the individual enterprises being carried out. Hand account books usually use a separate column for each enterprise, while computer systems assign enterprise code numbers to enter with each transaction. The end result is similar to an income statement for each enterprise showing its gross revenue, expenses, and net income. The data gathered during an enterprise analysis are extremely useful in developing enterprise budgets for future years. They can also be used to calculate the total cost of production, which is helpful for making marketing decisions. Some analyses compute a cash flow break-even cost, which is useful for financial management, as well as an economic break-even cost.

Crop Enterprise Analysis

An example of an enterprise analysis for soybeans is contained in Table 16-8. Values are shown both for the whole farm and per acre. The first step is to calculate the total income, including both sales and any change in inventory value for soybeans. For crops which are also used for livestock feed, the value of any amounts fed must also be included in total income. Finally, any crop insurance payments or government program benefits received should also be added.

Costs are estimated next. The total costs of items such as seed, fertilizer, and chemicals are relatively easy to find in farm records or to estimate based on the quantities actually used. Costs such as fuel, machinery repairs, depreciation, interest, and labor are more difficult to allocate fairly among enterprises, unless records are kept on the hours of machinery and labor use by enterprise. If not, they might be allocated equally over all planted acres. Overhead and miscellaneous costs such as property and liability insurance, consultant fees, office expenses, and general building upkeep are difficult to assign to individual enterprises. They are often allocated in the same proportion as the contribution of each enterprise to gross revenue or all other expenses.

The example in Table 16-8 indicates the soybean enterprise had a total profit above all costs of $5,573, or $21.94 per acre. These values can be compared against similar values for other crop enterprises to determine which are contributing the most to overall farm profit. If any crop shows a consistent loss for several years, action should be taken either to improve its profitability or to shift resources to the production of some other more profitable crop.

Livestock Enterprise Analysis

An enterprise analysis of livestock enterprises is conducted in the same manner as for a crop enterprise. However, several special problems should be noted.

TABLE 16-8   EXAMPLE OF ENTERPRISE ANALYSIS FOR SOYBEANS

	Farm total	Per acre
Soybean production (254 acres)	11,430	45
Income:	$62,751	$247.05
Variable costs:		
Seed	$3,533	$13.91
Fertilizer	1,900	7.48
Pesticides	5,199	20.47
Fuel, lubrication	1,293	5.09
Machinery repairs	2,350	9.25
Custom hire, rental	1,354	5.33
Insurance, miscellaneous	1,786	7.03
Labor	5,113	20.13
Interest on variable costs	577	2.27
Fixed costs:		
Machinery ownership	8,026	31.60
Land charge	26,048	102.55
Total costs	$57,178	$225.11
Profit	$5,573	$21.94
Average income per bushel		$5.49
Average cost per bushel		5.00
Average profit per bushel		0.49

Source: Crop Enterprise Record Summary, Central Iowa, 1991, Iowa State University Extension publication AG-125.

Inventory changes are likely to be more important in determining income for a livestock enterprise and should be carefully estimated to avoid biasing the results. Current market prices at the beginning and end of the analysis period can be used to value market livestock, but per head values for breeding livestock should be held constant. Another problem is how to handle farm-raised feed fed to livestock. The amounts of grain, hay, and silage fed must be known or estimated and then valued according to current market prices or opportunity costs. Pasture can be valued using either the opportunity cost on the land itself or the income which could have been received from renting the pasture to someone else. The value of farm-raised feed is entered as a cost to the livestock enterprise that consumed it and as income to the crop enterprise that produced it.

Other internal transactions that occur in enterprise accounting include transferring manure value from a livestock to a crop enterprise and transferring weaned livestock from a breeding to a feeding enterprise. Table 16-9 shows an example of a net income statement using enterprise accounting, including several internal transactions. Net overhead (miscellaneous income minus unallocated expenses) was assigned to enter-

TABLE 16-9   INCOME STATEMENT EXAMPLE WITH ENTERPRISE ACCOUNTING

	Whole farm	Crops	Cattle	Machinery	Overhead
Income					
Cash crop sales	$22,644	$22,644			
Cash livestock sales	72,271		72,271		
Government payments	2,100	2,100			
Miscellaneous income	3,369				3,369
Home consumption	427		427		
Livestock inventory change	(2,870)		(2,870)		
Crop inventory change	13,835	13,835			
Total revenue	$111,776	$38,579	$69,828	$0	$3,369
Expenses					
Crop inputs	16,971	16,971			
Machine hire	4,693			4,693	
Fuel, lubrication	4,356			4,356	
Machinery repairs	3,780			3,780	
Building repairs	3,224	1,156	2,068		
Purchased feed	6,031		6,031		
Insurance, property taxes	3,462				3,462
Utilities	2,056	456	1,600		
Interest paid	19,433	15,000	3,000		1,433
Livestock health, supplies	1,228		1,228		
Miscellaneous	4,021				4,021
Depreciation	19,058	1,688	3,351	12,944	1,075
Total expenses	$88,313	$35,271	$17,278	$25,773	$9,991
Net farm income	$23,463	$3,308	$52,550	($25,773)	($6,622)
Internal transactions					
Raised crops fed	0	39,500	(39,500)		
Manure credit	0	(4,500)	4,500		
Machine work (allocated by hours)	0	(23,773)	(2,000)	25,773	
Allocation of net overhead (proportional to gross revenue)	0	(2,357)	(4,265)	0	6,622
Net farm income by enterprise	$23,463	$12,178	$11,285	$0	$0

prises in the same proportion as gross revenue. Machinery costs were recorded separately, then divided between crops and livestock according to estimated hours of use for each one.

USING COMPUTERS FOR FARM BUSINESS ANALYSIS

Calculating the measures and ratios used for farm business analysis can be done with a hand calculator. However, it can be done more easily and quickly on a computer. It is particularly convenient if the farm business analysis is combined with the software that is already being used to do the farm accounting. The required values can, in many cases, be drawn directly from the accounting data already stored in the computer. A program's ability to do at least a minimum level of whole farm analysis is a factor to consider when purchasing farm accounting software.

A number of farm accounting programs also have the ability to perform enterprise analysis. Receipts and variable expenses are identified by enterprise as they are entered. Some programs also have a procedure for allocating fixed costs among enterprises. The computer can then quickly sort through all receipts and expenses, collect and organize those that belong to a particular enterprise, and present the results in total dollars, cost per acre, or cost per unit of output. To do business and enterprise analysis correctly the program must track physical quantities of inputs and products as well as monetary values, and allow for transactions between enterprises, such as for farm-raised feed.

DIAGNOSING A FARM INCOME PROBLEM

This chapter has concentrated on the procedures and measures used to conduct an analysis of the whole farm business. A complete analysis is time-consuming, but some steps can be eliminated by using a systematic procedure to eliminate some possibilities while working toward identifying the source of a problem. Figure 16-2 illustrates a procedure which can reduce the number of measures to be calculated.

Profitability is generally the first area of concern. Low net farm income or returns to management may have several causes. The farm may not be large enough to generate the level of production needed for an adequate income. Measures such as gross revenue, value of farm production, or volume of crops and livestock sold should be compared to other farms.

If the level of production is too low, there may not be enough resources employed. Look for ways to farm more acres, increase the labor supply, expand livestock, or obtain more capital. If farm size cannot be increased, then fixed costs such as machinery and building depreciation, interest, and general farm overhead costs should be carefully evaluated. Steps should be taken to reduce those which will have the least effect on the level of production. Off-farm employment may also have to be considered as a way to improve family income.

If adequate resources are available but production levels are low, the resources are not being used efficiently. Computing several economic efficiency measures can

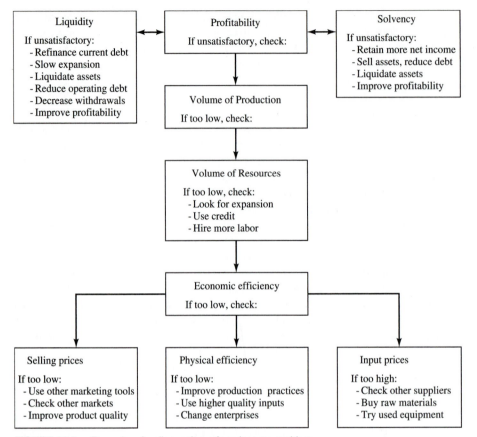

FIGURE 16-2   Procedure for diagnosing a farm income problem.

identify areas of inefficient resource use, such as capital, labor, or land. Poor economic efficiency may be due to low physical efficiency, low selling prices, or high input costs.

Low physical efficiency measures indicate a need to improve crop and livestock management practices or to obtain higher quality inputs. If physical efficiency measures for the current enterprises cannot be improved, it may mean the wrong enterprises are in the current farm plan. Enterprise analysis and enterprise budgeting can be used to identify more profitable enterprises and develop a new whole farm plan.

If the physical efficiency measures are satisfactory, another explanation for low profitability may be below-average selling prices, caused by low price cycles, poor marketing practices, or inferior product quality. Alternative marketing outlets and marketing tools should be considered.

Finally, low economic efficiency can also be caused by paying higher than necessary prices for land, feed, livestock, crop inputs, and even credit. Shopping around for different suppliers, leasing equipment or land, or buying more raw materials in place of finished inputs can help reduce input costs.

This procedure should help managers isolate and identify the causes of a profitability problem quickly and systematically. However, even profitable farms need to be concerned about liquidity and solvency. If cash flow always seems to be tight even when net farm income is satisfactory, the operation may need to refinance some current debt, slow down expansion, sell off some assets, try to reduce nonfarm withdrawals, or reduce operating debt. If solvency is not as strong as the operator would like it to be, more of the net farm income may need to be retained in the business each year, or some assets sold to reduce debt levels.

The three areas of farm business management, profitability, solvency, and liquidity, are all closely related. Outstanding farm managers are those who do at least an above-average job in all key areas of the business.

SUMMARY

A whole farm business analysis is much like a complete medical examination. It should be conducted periodically to see if symptoms exist which indicate the business is not functioning as it should. Standards for comparison can be budgeted goals, results from other farms, or past results from the same farm. Financial analysis uses balance sheet values to evaluate business solvency and liquidity, while profitability can be measured by net farm income and returns to labor, management, total assets, and equity.

If an income or profitability problem is found, the first area to look at is farm size, to see if there are enough resources for an adequate profit, and how many of them are being contributed by the operator. Next, various measures relating to efficient use of machinery, labor, capital, and other inputs can be computed. Economic efficiency depends on physical efficiency, the selling price of products, and the purchase price of inputs. Finally, enterprise analysis can be used to identify the less profitable enterprises so they can be improved or eliminated.

The various measures calculated for a whole farm analysis are part of the control function of management. As such, they should be used to identify and isolate a problem before it has a serious impact on the business.

QUESTIONS FOR REVIEW AND FURTHER THOUGHT

1 What are the steps in the control function of farm management?
2 What types of standards or values can be used to compare and evaluate efficiency and profitability measures for an individual farm? Discuss some advantages and disadvantages of each.
3 Can a farm have satisfactory measures of financial solvency but still show a poor profit? Why?
4 What are the differences between "value of farm production," "net farm income," and "return to management"?
5 Use the following data to calculate the profitability and efficiency measures listed below:

Net farm income from operations	$28,000
Tillable acres	620
Total asset value: beginning	$300,000
ending	330,000
Value of unpaid labor	$25,000
Gross revenue	$140,000
Value of farm production	$90,000
Interest expense	$18,000
Annual machinery expenses	$26,000
Farm net worth: beginning	$240,000
ending	252,000

a Rate of return on assets ⎯⎯
b Rate of return on equity ⎯⎯
c Asset turnover ratio ⎯⎯
d Machinery cost per tillable acre ⎯⎯
e Net farm income from operations ratio ⎯⎯

6 What three general factors determine economic efficiency?

7 Tell whether each of the following business analysis measures relates to volume of business, profitability, economic efficiency, or physical efficiency.

a Pounds of cotton harvested per acre
b Return to management
c Livestock return per $100 of feed fed
d Gross revenue
e Total farm assets
f Asset turnover ratio

8 Would a crop farm that cash rents 1,200 crop acres and owns 240 acres be more likely to have a better than average asset turnover ratio or operating expense ratio? Why?

9 Select an enterprise with which you are familiar and list as many appropriate physical efficiency measures as you can. Which measures do you think have the greatest impact on the profitability of that enterprise?

10 Given the inventory, purchase, and sales data below for a beef feedlot:

	Head	Weight (lb)	Value ($)
Beginning inventory	850	765,000	$ 612,000
Ending inventory	1,115	936,600	730,548
Purchases	1,642	1,018,040	865,334
Sales	1,340	1,586,000	1,064,630
Death loss	⎯⎯	0	0
Production increase	0	⎯⎯	⎯⎯

a What was the apparent death loss, in head?
b How much beef was produced, in pounds?
c What was the value of this production increase?

11 What is the purpose of enterprise analysis? What are "internal transactions"? How can overhead expenses be allocated among enterprises?

REFERENCES

Bennett, Myron: *Financial Analysis of the Farm Business,* University of Missouri Extension Agricultural Guide, Columbia, MO, 1983.

Boehlje, Michael D., and Vernon R. Eidman: *Farm Management,* John Wiley & Sons, New York, 1984, chaps. 16, 17, 18.

James, Sydney C., and Everett Stoneberg: *Farm Accounting and Business Analysis,* 3rd ed., Iowa State University Press, Ames, IA, 1986, chaps. 8, 9.

Libbin, James D., and Lowell B. Catlett: *Farm and Ranch Financial Records,* MacMillan Publishing Co., New York, 1987, chaps. 22, 23.

Osburn, Donald D., and Kenneth C. Schneeberger: *Modern Agricultural Management,* 2d ed., Reston Publishing Co., Reston, VA, 1983, chaps. 7, 8.

Recommendations of the Farm Financial Standards Task Force, American Bankers Association, Washington, D.C., 1991.

PART FIVE

ACQUIRING RESOURCES FOR MANAGEMENT

Although farm management skills are the central theme of this book, few people make a living in agriculture by applying management skills alone. There must be certain physical, financial, and human resources available before agricultural production can take place. The quantity of these resources used, how they are used, and how they are obtained often makes the difference between operating a business at a profit or a loss.

How to acquire and use resources are decisions a manager must make. Resources can be contributed by the operator and family or they can be obtained through borrowing, renting, or hiring. Determining the proper mix of owned and nonowned resources to use is a management decision. Net farm income is the return to resources contributed by the operator so one key to improving it is to increase the collection of resources owned by the operator over time. This requires a profitable business, gifts, or inheritances.

Chapter 17 discusses capital as an agricultural resource. Capital itself does not produce agricultural products, but it can be used to purchase or rent resources that do. Sources of capital include the operator's net worth, borrowing, or equity contributed by partners or investors. The use of credit to acquire capital is common in agriculture, but it requires careful planning and control to use it profitably and to prevent repayment problems.

In dollar terms, land is the most valuable resource used in agricultural production. Acquiring and using land is the subject of Chapter 18. Ownership versus leasing is discussed along with the various types of leases and their advantages and disadvantages. Soil conservation, sustainable agriculture, and how a manager should consider these topics in decision making are also discussed.

Agricultural labor is a resource which has evolved from requiring hard physical labor to performing highly skilled tasks using sophisticated equipment and technology. Even though the amount of labor used in agricultural production has been declining, it is still an important and necessary resource. Chapter 19 explains the concepts of planning and managing both hired labor and the operator's own time.

Mechanization has changed the occupation of farming more than any other innovation during the last century. It has caused a rapid increase in productivity per person, which has resulted in less labor needed in agricultural production. Machinery represents a large capital investment on many farms, and Chapter 20 explores alternatives for acquiring the use of machinery services. Methods for computing and controlling machinery costs and for improving machinery efficiency are also discussed.

17

CAPITAL AND THE USE OF CREDIT

1 To explain the importance of capital in agriculture

2 To illustrate the use of economic principles to determine the optimal use and allocation of capital

3 To compare different sources of capital and credit in agriculture

4 To describe different types of loans according to their length of repayment, use, and collateral

5 To show how to calculate principal and interest payments for various repayment plans

6 To outline considerations which are important for establishing and developing credit

7 To explain factors that affect the liquidity and solvency of a farm business

Many people think of capital as cash, balances in checking and savings accounts, and other types of liquid funds. This is a narrow definition of capital. In the broader definition, capital also includes money invested in livestock, machinery, buildings, land, and any other assets that can be bought and sold.

It is easy to see that capital is extremely important in agricultural production. Agriculture has one of the larger capital investment per worker of any major industry in the United States at over $400,000 per worker. This helps make U.S. farm operators very productive. Figure 17-1 shows the changes that have been taking place in the cap-

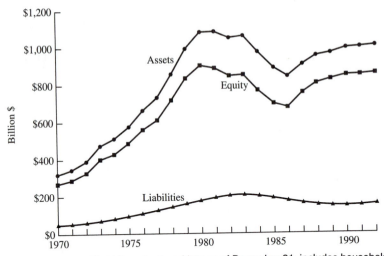

FIGURE 17-1 Capital investment in U.S. agriculture (data as of December 31; includes households) (*Source:* U.S. Department of Agriculture).

ital investment in agriculture. During the 1970s rapid increases in land prices accounted for much of the increase in asset values. As land prices declined in many parts of the United States in the early 1980s, total asset value declined slightly.

The capital investment *per farm* has risen even more rapidly than total investment in agriculture, since the number of farms and ranches in the United States has been declining. Many full-time, commercial farms and ranches have capital investments of over $1,000,000. Land accounts for much of this investment, but buildings, livestock, and machinery are also important. This large capital investment per farm requires a sound understanding of financial management principles in order to compete in today's agriculture.

Credit is important to capital acquisition and use. Credit is the capacity or ability to borrow money. Borrowing money is really the exchange of a borrower's credit for the use of the lender's money with a promise to return the money and pay interest. Borrowing money is like renting an asset such as land. In both cases rent (or interest) is paid for the use of another person's property with an agreement to return the property in good condition at the end of a specified period of time.

ECONOMICS OF CAPITAL USE

Broadly defined, capital is the money invested in the physical inputs used in agricultural production. It is needed to purchase or rent productive assets, pay for labor and other services, and for family living and other personal expenditures. Capital use can be analyzed using the economic principles discussed in Chapters 5 and 6. The basic questions to be answered are (1) how much total capital should be used and (2) how should limited capital be allocated among its many potential uses.

Total Capital Use

When unlimited capital is available, the problem is one of determining how much capital to use. In Chapter 5 the question of how much input to use was answered by finding the input level where the marginal value product (MVP) was equal to the marginal input cost (MIC). The same principal can be applied to the use of capital.

Figure 17-2 is a graphic presentation of MVP and MIC where MVP is declining, as occurs whenever diminishing marginal returns exist. Marginal input cost is equal to the additional dollar of capital plus the interest that must be paid to use it. Therefore, MIC is equal to $1 + i$, where i is the rate of interest or the opportunity cost of not investing the farms own capital elsewhere. In this example, profit will be maximized by using the amount of capital represented by a, where MVP is equal to MIC, that is, the additional return is equal to the additional cost.

In some cases the interest rate increases as more capital is used such as when a lender classifies a borrower as being in a higher risk category. At the point where this happens, the MIC curve would rise to a higher level. The optimum level of capital to use would be less than when the interest rate (and MIC) are constant.

Allocation of Limited Capital

Many businesses do not have sufficient capital of their own or may not be able to borrow enough to reach the point where MVP is equal to MIC for the total capital being used. In other words, capital is limited to something less than the amount which will maximize total profit. The problem now becomes one of allocating limited capital among its alternative uses. This can be accomplished by using the equal marginal principle, which was discussed in Chapter 5.

FIGURE 17-2   Using marginal principles to determine optimal capital use.

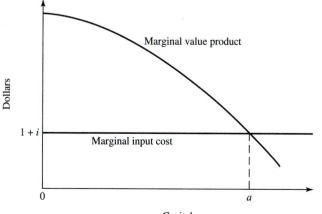

Use of the equal marginal principle results in capital being allocated among alternative uses in such a way that the marginal value product of the last dollar is equal in all uses. Even if no additional capital is available for investment, there may be opportunities for shifting capital between uses in order to more nearly equate the marginal value products.

This principle is often difficult to apply in an actual farm situation. First, there may be insufficient information available to calculate the marginal value products accurately. Prices and costs are constantly changing. Second, several alternative uses may require large lump-sum investments of an all-or-nothing nature. Third, capital invested in some assets such as land or buildings cannot easily be shifted to other uses. This makes it difficult to equate the marginal value products for these alternatives with others where capital can be invested dollar by dollar. Nonetheless, difficulties encountered in applying the equal marginal principle should not discourage its use. Whenever limited capital can be reallocated to make the marginal value products more nearly equal, profit will be increased.

SOURCES OF CAPITAL

Because capital consists of cash and assets purchased with cash, it is relatively easy to intermingle capital from different sources. An important part of farm financial management is the ability to obtain capital from several sources and use it in the proper proportions.

Owner's Equity

The operator's own capital is called equity or net worth. It is calculated as the difference between the total assets and the total liabilities of the business, as shown on the balance sheet statement discussed in Chapter 3. There are several ways the operator can secure or accumulate equity. Most farmers begin with a contribution of original capital acquired through savings. As the farm or ranch generates profits in excess of what is needed to pay personal expenses and taxes, retained earnings can be reinvested in the business. Some operators may have outside earnings such as a nonfarm job or other investment income which they can invest in their farming operation.

Assets that are already owned may increase in value through inflation. This does not increase the amount or productivity of the physical assets, but additional cash can be obtained by either selling the assets or using them as collateral for a loan.

Outside Equity

Some investors may be willing to contribute capital to a farm or ranch without being the operator. Under some types of share lease agreements, the landowner contributes operating capital to buy seed and fertilizer or even provides equipment and breeding livestock, as explained in Chapter 18. Larger agricultural operations may include limited or silent partners who contribute capital but do not engage in management. Farms that are incorporated may sell stock to outside investors. These arrangements increase the pool of capital available to the business but also obligate the business to share earnings with the investors.

Leasing

It is often cheaper to gain the use of capital assets by leasing or renting rather than owning them. Short-term leases make it easier for the operator to change the amount and kind of assets used from year to year. However, this also creates more uncertainty about the availability of the assets and discourages making long-term improvements on leased property.

Contracting

Farmers or ranchers who have very restricted access to capital or credit or who wish to limit their financial risk may contract their services to agricultural investors. Examples include custom feeding of livestock, contract broiler or egg production, and custom crop farming. The special skills of the operator can be leveraged over more units of production without increasing financial risk. However, potential returns per unit for contract operations may be lower than for a well-managed owner-operated business.

Credit

After owner's equity, capital obtained through credit is the second largest source of farm capital. Borrowed money can provide a means to more quickly increase business size, improve the efficiency of other resources, spread out the purchase of capital assets over time, and withstand temporary periods of negative cash flow.

Figure 17-3 shows the use of credit by U.S. farmers and ranchers by type of debt outstanding at the end of each year. Farm debt increased rapidly in the late 1970s and early 1980s as did the value of farm assets. However, both have since declined. Real estate debt or borrowing against land and buildings accounts for approximately one-half of the total debt. Borrowing against livestock, machinery, and grain inventories accounts for the other half. Comparing total farm debt against total farm assets indicates that U.S. agriculture is in sound financial condition (see Figure 17-1). However, this does not mean that every individual farmer and rancher is in sound financial condition. There are always individuals experiencing financial problems.

TYPES OF LOANS

Several systems are used to describe agricultural loans. Loans can be classified by length of repayment, use, and security pledged. All of them include certain terms used by lenders. A prospective borrower needs to be familiar with these terms to communicate effectively with lenders.

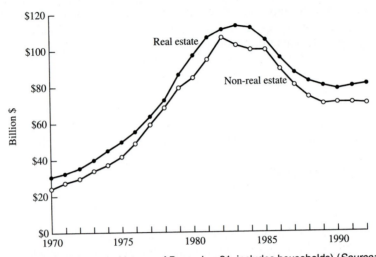

FIGURE 17-3   Total farm debt by type (data as of December 31; includes households) (*Source:* U.S. Department of Agriculture).

Length of Repayment

Classifying loans by the length of the repayment period is widely used in preparing balance sheets, as was discussed in Chapter 3. Three time classifications are commonly used in agricultural lending.

Short-Term Loans   Short-term loans include those with the full amount borrowed due within the next 12 months. They are usually used to purchase inputs needed to operate through the current production cycle, until income from this production is available. Money borrowed for the purchase of fertilizer, seed, feeder livestock, and feed would normally be a short-term loan, with repayment due when the crops are harvested and sold or when the feeder livestock are sold. Short-term loans are also called *production or operating* loans, and are listed under current liabilities on the farm balance sheet.

Intermediate-Term Loans   When the length of the loan is more than 1 year but less than 10 years, it is classified as an intermediate loan. Some payment is usually due each year. These loans are often used for the purchase of machinery, breeding and dairy livestock, and some buildings. A short-term loan is usually not appropriate for the purchase of these assets. They will be used in production for several years and cannot be expected to pay for themselves in one year or less.

Long-Term Loans   A loan with a term of 10 years or longer is classified as a long-term loan. Assets with a long or indefinite life such as land and buildings are often purchased with funds from long-term loans. Loans for the purchase of land may be made for a term of as long as 20 to 40 years, for example. Annual or semiannual payments are normally required throughout the term of the loan.

The Farm Financial Standards Task Force recommends that both intermediate and long-term loans be listed under noncurrent liabilities on the balance sheet.

Use

The use or purpose of the funds is another common criteria for distinguishing loans.

Real Estate Loans   This category includes loans for the purchase of real estate such as land and buildings or where real estate assets serve as security for the loan. Real estate loans are typically long-term loans.

Non-Real Estate Loans   All business loans other than real estate loans are included in this category, and are usually short-term or intermediate-term loans. Crops, livestock, machinery or other non-real estate assets may be pledged as security.

Personal Loans   These are nonbusiness loans used to purchase personal assets such as homes, automobiles, and appliances.

Security

The security for a loan refers to the asset or assets pledged or mortgaged to the lender to ensure loan repayment. If the borrower is unable to make the necessary principal and interest payments on the loan, the lender has the legal right to take possession of the mortgaged assets. These assets can be sold by the lender and the proceeds used to pay off the loan. Assets which are pledged or mortgaged as security are called loan *collateral*.

Secured Loans   With secured loans, some asset is mortgaged to provide collateral for the loan. Lenders obviously favor secured loans, as it gives them greater assurance that the loan will be repaid.

Unsecured Loans   A borrower with good credit and a history of prompt loan repayment may be able to borrow some money with only "a promise to repay" or without pledging any specific collateral. This would be an unsecured loan, also called a *signature loan,* as the borrower's signature is the only security provided the lender. Some short-term loans may be unsecured, although most lending practices and regulations discourage making unsecured loans.

Repayment plans

Loans can also be classified according to their repayment plan or schedule. There are many types and variations of repayment plans used as lenders try to fit repayment to the purpose of the loan, the type of collateral used to secure the loan, and the borrower's projected cash flow. In each case, the total interest paid will increase if the interest rate increases or if the money is borrowed for a longer period of time. The fundamental equation for calculating interest is

$$I = P \times i \times T$$

where I is the amount of interest to pay, P is the principal or amount of money borrowed or currently owed, i is the interest rate, and T is the time period over which interest accrues.

When a loan is negotiated the borrower and the lender should be in agreement about when it is to be repaid. The type of repayment plan used will depend on the purpose of the loan and the anticipated cash flow pattern of the business.

Single Payment   A single-payment loan has all the principal payable in one lump sum when the loan is due, plus interest. Short-term or operating loans are usually of this type. Single-payment loans require good cash flow planning to ensure that sufficient cash will be available when the loan is due.

The interest paid on a loan with a single payment is called *simple interest*. For example, if $10,000 is borrowed for exactly one year at 12 percent interest, the single payment would be $11,200, including the $1,200 interest. If the loan is repaid in less than or more than one year, interest would be computed for only the actual time the money was borrowed.

Line of Credit   The use of single-payment loans often means having more money borrowed than the operator really needs at one time, or having to take out several individual loans. As an alternative, some lenders allow a borrower to negotiate a line of credit. Loan funds are transferred into the farm account as needed, up to an approved maximum amount. When farm income is received the borrower pays the accumulated interest on the loan first, then applies the rest of the funds to the principal. There is no fixed repayment schedule or amount.

Table 17-1 contains an example of a line of credit. The amounts borrowed are $20,000 on February 1 and $10,000 on April 1. On September 1 a payment of $17,500 is made to the lender. The interest due is calculated as follows:

$$\$20,000 \times 12\% \times 2/12 = \$\ \ 400$$
$$\$30,000 \times 12\% \times 5/12 = \underline{\$1,500}$$
$$\$1,900$$

The remaining $15,600 is used to reduce the principal balance to $14,400.

On October 1 the interest rate is reduced to 11 percent. Another payment is made on December 1, for $12,000. The interest calculation for this payment is as follows:

$$\$14,400 \times 12\% \times 1/12 = \$144$$
$$\$14,400 \times 11\% \times 2/12 = \underline{\$264}$$
$$\$408$$

TABLE 17-1   ILLUSTRATION OF A LINE OF CREDIT

Date	Amount borrowed	Interest rate	Amount repaid	Interest paid	Principal paid	Outstanding balance
Feb. 1	20,000	12%	0	0	0	20,000
April 1	10,000	12%	0	0	0	30,000
Sept. 1	0	12%	17,500	1,900	15,600	14,400
Oct. 1	0	11%	0	0	0	14,400
Dec. 1	0	11%	12,000	408	11,592	2,808

After the interest is paid the remaining $11,592 goes toward reducing the outstanding balance.

A line of credit reduces the borrower's interest costs and results in less time spent in the loan approval process. However, the borrower must exercise more discipline in deciding when and how much to borrow or repay.

Amortized   An amortized loan is one which has periodic interest and principal payments. It may also be called an installment loan. As the principal is repaid, and the loan balance declines, the interest payments also decline. Assume $10,000 is borrowed and the repayment schedule is $5,000 in 6 months and the remaining $5,000 at the end of one year. The interest calculations would be as follows:

First payment: $10,000 at 12% for 1/2 year = $600
Second payment: $ 5,000 at 12% for 1/2 year = $300
                                 Total interest = $900

Notice that interest is paid only on the unpaid loan balance, and then only for the length of time that amount was still borrowed. The total interest is less than if the total amount was repaid at the end of the year, as only $5,000 was outstanding for the last half of the year.

There are two types of amortization plans, the *equal principal payment* and the *equal total payment.* An amortized loan with equal principal payments has the same amount of principal due on each payment date, plus interest on the unpaid balance. For example, a 10-year, 8 percent loan of $10,000 would have annual principal payments of $1,000. Since the loan balance decreases with each principal payment, the interest payments also decrease, as shown in Figure 17-4.

Borrowers often find the first loan payment the most difficult to make, as a new or expanded business may take some time to generate its maximum potential cash flow. For this reason, many long-term loans have an amortized repayment schedule with equal total payments, in which all payments are for the same amount. Figure 17-4 also shows the amount of principal and interest paid each year under this plan. A large portion of the total loan payment is interest in the early years, but the interest decreases and the principal increases with each payment, making the last payment mostly principal.

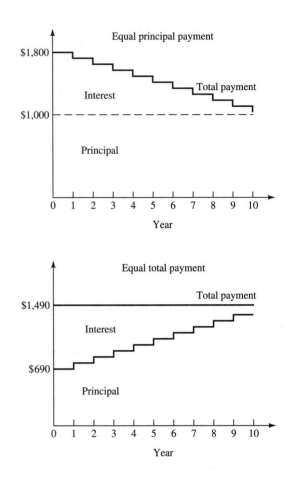

FIGURE 17-4   Loan repayment under two types of amortization.

To calculate the total loan payment for an amortized loan with equal total payments a table of amortization factors is needed. These factors are in Table 1 in the Appendix. The annual payment will depend on both the interest rate and the length of the loan. For example, the amortization factor for a 10-year loan at 8 percent interest is 0.1490. This factor is multiplied by the amount of the loan to find the total annual payment. A loan of $10,000 would have an annual payment of (0.1490 × $10,000), or $1,490.

Table 17-2 shows the actual principal and interest payments under each plan. The equal total payment method has a smaller total loan payment than the equal principal payment loan in the first 4 years. However, a total of $4,903 in interest is paid compared to only $4,400 under the first plan. The advantage of the lower initial payments is partially offset by more total interest being paid over the life of the loan, since the principal is being reduced at a slower rate.

Balloon Loan   Some amortization schedules are set up with lower periodic payments so not all of the principal is repaid by the end of the loan period. For ex-

TABLE 17-2 AMORTIZATION OF A $10,000 LOAN OVER 10 YEARS AT 8% INTEREST

| Year | Equal principal payments | | | | Equal total payments | | | |
	Principal paid	Interest paid	Total payment	Principal remaining	Total payment	Interest paid	Principal paid	Principal remaining
1	$1,000	$ 800	$1,800	$9,000	$1,490	$ 800	$ 690	$9,310
2	1,000	720	1,720	8,000	1,490	744	746	8,564
3	1,000	640	1,640	7,000	1,490	685	805	7,759
4	1,000	560	1,560	6,000	1,490	620	870	6,889
5	1,000	480	1,480	5,000	1,490	551	939	5,950
6	1,000	400	1,400	4,000	1,490	476	1,014	4,936
7	1,000	320	1,320	3,000	1,490	395	1,095	3,841
8	1,000	240	1,240	2,000	1,490	307	1,183	2,658
9	1,000	160	1,160	1,000	1,490	212	1,278	1,380
10	1,000	80	1,080	0	1,490	110	1,380	0
Total	$10,000	$4,400	$14,400		$14,900	$4,900	$10,000	

ample, half the principal may be paid with the periodic payments with the other half due at the end of the loan period. In some cases, the periodic payments may be interest only with all the principal due at the end of the loan period. These types of loans are called *balloon loans* as the last payment "balloons" in size. They have the advantage of smaller periodic payments but the disadvantage of a large final payment. Balloon loans often require some form of refinancing to make the final payment.

The payments for a balloon loan are calculated by amortizing the amount of principal to be repaid with periodic payments using either the equal principal or equal total payment method explained above and then adding the interest on the balloon principal to each periodic payment.

THE COST OF BORROWING

Lenders use several different methods of charging interest, which makes comparisons difficult. The true interest rate, or *annual percentage rate* (APR), should be stated in the loan agreement. Some lenders charge loan closing fees (sometimes called "points"), appraisal fees, or other fees for making a loan. These fees as well as interest rates affect the total cost of credit.

One way to compare the cost of various credit plans is to calculate the dollar amount to be repaid in each time period (principal, interest, and other fees) and find the discounted present value of the series of payments, as described in Chapter 15. If several financing alternatives are being compared, the same discount rate should be used for all of them. Subtracting the original loan from the present value of the payments leaves the net present value or true cost of the loan. The true cost can also be expressed in percentage terms, by calculating the internal rate of return (IRR) to the lender.

Fixed-rate loans have an interest rate that remains the same for the entire length of the loan. However, lenders do not like to make long-term loans at a fixed interest rate because the rate they must pay to obtain loan funds may change. Borrowers do not like to borrow long-term at a fixed rate if they think interest rates may decrease. Predicting future interest rates is difficult for both borrowers and lenders.

For this reason, variable-rate loans have been developed which allow for adjustment of the interest rate periodically, often annually. There may be limits on how often the rate can be changed, the maximum change on a single adjustment, and maximum and minimum rates. For example, the Farm Credit System offers a variable rate tied to the average interest rate on the bonds they issue to raise their loan capital. They also make loans with a fixed rate for a period of several years with this rate "adjusted" for the next time period.

Fixed interest rate loans will typically carry a higher initial interest rate than variable rate loans to protect the lender against future increases in rates. The borrower must weigh this higher but fixed rate against the possibility that the variable rate could eventually go even higher than the fixed rate. However, the variable rate could also decrease. Variable rate loans are one way to ensure that the interest rate is near the current market rate.

SOURCES OF LOAN FUNDS

Farmers and ranchers borrow money from many different sources. Some lending agencies specialize in certain types of loans and some provide other financial services in addition to lending money. The more important sources of funds for both real estate and non-real estate loans are shown in Figure 17-5.

Commercial Banks

Commercial banks are the single largest source of non-real estate loans for agriculture, and they also provide some real estate loans. Banks limit their long-term loans in order to maintain the liquidity needed to meet customers' cash requirements and unexpected withdrawals of deposits. However, the use of variable interest rates and balloon payments has allowed banks to increase their share of the farm real estate loan market.

The large share of non-real estate loans held by banks is due partially to the large number of local banks in rural communities. This proximity to their customers allows bank personnel to become well acquainted with customers and their individual needs. Banks also provide other financial services such as checking and savings accounts, which make it convenient for their farm and ranch customers to take care of all their financial business at one location.

Farm Credit System

The Farm Credit System was established by the U.S. Congress in 1916 to provide an additional source of funds for agricultural loans. Government funds were used initially

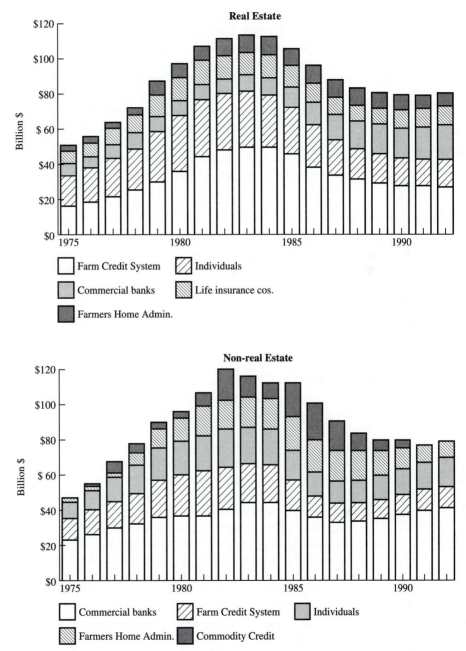

FIGURE 17-5 Sources of funds for real estate and non-real estate agricultural loans (data as of December 31) (*Source:* U.S. Department of Agriculture).

to organize and operate the system, but these funds have all been repaid. The Farm Credit System is now a private cooperative owned by its member/borrowers. However, the system is supervised, audited, and regulated by the Farm Credit Administration, which is an independent government agency.

The Farm Credit System obtains loan funds by selling bonds in the national money markets. Proceeds from these bond sales are made available to district Farm Credit Banks and Banks for Cooperatives located across the country. These district banks then either make direct loans to customers or provide funds to local associations, which then initiate and supervise loans to farmers and ranchers. Loans from the Farm Credit System may be used to purchase livestock, machinery, buildings, and land. Short-term operating credit is also available.

Life Insurance Companies

Life insurance companies acquire funds from the premiums paid on life insurance policies and from other earnings and reserve funds. They place these funds in various investments, including, for some companies, long-term agricultural real estate loans. The amount of money these companies make available for agricultural loans may vary from year to year depending on the rate of return from agricultural loans compared to alternative investments. Life insurance companies generally prefer large farm real estate loans, often over $500,000.

Farmers Home Administration

The Farmers Home Administration (FmHA) is an agency of the U.S. Department of Agriculture with offices in most agricultural counties. This agency makes farm ownership and operating loans as well as loans for rural development purposes. It also has authority to make emergency loans to qualified farmers and ranchers in officially declared disaster areas. These are temporary loans used to restore normal operations after a natural disaster such as flood or drought.

To be eligible for FmHA farm loans the borrower must operate a family-size farm or ranch, receive a substantial portion of total family income from farming or ranching, and be unable to obtain adequate loan funds from other lending institutions. The last requirement does not mean FmHA borrowers are always poor credit risks. Many are beginning farmers who simply do not have enough equity to borrow capital from other sources. As soon as FmHA borrowers improve their financial condition to the point where funds can be obtained from a commercial source, they must switch to another lender.

Farmers Home Administration loans usually carry a lower interest rate than loans from other sources. Over the years FmHA has moved away from direct loans made from congressional appropriations and toward more guaranteed loans in which a private lender provides the loan funds and FmHA guarantees repayment.

Individuals and Suppliers

Individuals, farm supply stores, dealers, and others are an important source of both real estate and non-real estate loans, as shown in Figure 17-5. Non-real estate loans can come from friends, relatives, or accounts payable at farm supply stores. Many suppliers allow customers 30, 60, or 90 days to pay their accounts before any interest is charged, and may finance a purchase for a longer period with interest. This policy is essentially a loan, and the total balances in these accounts can be large at certain times of the year. Farm equipment and automobile dealers also provide loans by financing purchases themselves or through an affiliated finance company.

The relatively large portion of real estate debt owed to individuals and others comes mostly from seller-financed land sales. Many land sales utilize a land purchase contract in which the seller provides the financing and the buyer makes periodic loan payments directly to the seller. This differs from a cash sale where the buyer borrows from a commercial lender, pays the seller cash for the full purchase price, and then makes periodic payments to the lender. A land purchase contract can have some income tax benefits for the seller, and the buyer may be able to negotiate a lower down payment, lower interest rate, and more flexible repayment terms.

Other Sources

The Commodity Credit Corporation (CCC) is a government entity that provides some non-real estate loans using stored grain or cotton as the collateral. For most commodities, a farmer must have participated in the latest government farm program to be eligible for a CCC loan. The loan can be repaid by forfeiting commodities to the CCC rather than paying in cash.

The Small Business Administration (SBA) also makes some agricultural loans and has an emergency loan program for farmers in designated disaster areas.

ESTABLISHING AND DEVELOPING CREDIT

When trying to establish or develop credit, it is useful to look at it from the lender's viewpoint. What does a lender consider when making a decision on a loan application? Why can one business borrow more money than another or receive different interest rates and repayment terms? A borrower should be aware of the need to demonstrate and communicate credit worthiness to lenders. Some of the more important factors that go into making loan decisions are the following:

1 Personal character
2 Management ability
3 Financial position and progress over time
4 Repayment capacity
5 Purpose of the loan
6 Collateral

When using these factors as a guide for establishing and developing credit, a prospective borrower should remember that lenders want to make loans. That is their business. However, they are looking for profitable and safe loans which will be repaid.

Personal Character   Honesty, integrity, judgment, reputation, and other personal characteristics of the loan applicant are always considered by lenders. Credit can be quickly lost by being untruthful in business dealings and slow to meet financial obligations. If the lender is not acquainted with the borrower, character references will usually be requested and checked. To maintain a good credit record borrowers should promptly inform lenders of any changes in their financial condition or farming operation which might affect loan repayment. An honest and open relationship with lenders is necessary to maintain credit.

Management Ability   A lender must try to evaluate a borrower's management ability. Established farmers and ranchers will be evaluated on their past records, but beginners can be judged only on their background, education, and training. These factors affect profitability and therefore the ability to repay a loan. Lenders often rate poor management ability as the number one reason for borrowers getting into financial difficulty.

Financial Position   Accurate, well-prepared balance sheets and income statements are needed to document the current financial position of the business and its profitability. Lenders can learn much about a business from these records. A record of good financial progress over time can be just as important as the current financial position.

Repayment Capacity   Having a profitable business does not guarantee repayment capacity. There must be enough cash income to meet family living expenses and income tax payments as well as the interest and principal payments on loans. Repayment capacity is best measured by the cash flow generated by the business. A cash flow budget projected for one or more years should be completed before borrowing large amounts and establishing rigid repayment schedules. Too often money is borrowed for a profitable business only to find that cash flow in the early years is not sufficient to make interest and principal payments. A longer loan or a more flexible repayment schedule may solve the problem if it is identified in time.

Purpose of the Loan   Loans which are *self-liquidating* may be easier to obtain. Self-liquidating loans are for items such as fertilizer, seed, and feeder livestock where the loan can be repaid from the sale of the crops or livestock. On the other hand, capital asset loans are those used to purchase long-term tangible assets such as land or machinery, which must generate additional revenue without being sold themselves. Capital asset loans may require extra collateral.

Collateral   Land, livestock, machinery, stored grain, and growing crops can all be used as loan collateral. The amount and type of collateral available are important factors in a loan request. Loans should not be made or requested unless repayment can be projected from farm income. However, lenders still ask for collateral to support a loan request. If the unexpected happens and the loan is in default, it may be the lender's only means of recovering the loan funds.

LIQUIDITY

The ability of a business to meet cash flow obligations as they come due is called liquidity. Maintaining liquidity in order to meet loan payments as they become due is an important part of establishing and maintaining credit.

Factors Affecting Liquidity

A farm or ranch that is profitable will usually have adequate liquidity in the long run. Chapter 16 discussed how to analyze a farm business and identify factors that affect profitability. However, even profitable businesses experience cash flow problems at times, due to several factors.

Business Growth   Holding back ever increasing inventories of young breeding livestock or feed reduces the volume of production sold in the short run. Construction of new buildings or purchases of land or machinery require large cash outlays up front, but may not generate additional cash income for months or years. Moreover, when new technology is involved several production cycles may pass before an efficient level of operation is achieved. All these factors produce temporary cash shortfalls and need to be considered when planning financing needs and repayment plans.

Nonbusiness Income and Expenditures   These are especially important for family farms and ranches. Family living expenses must be met even when agricultural profits are low. At certain stages of their lives farm families may have high educational or medical expenses. Generally, nonbusiness expenditures should be postponed until surplus cash is available. However, reinvesting every dollar earned into the farm business may eventually cause personal and family stress.

Over the years farm and ranch families have received more and more of their income from off-farm employment and investments. A regular and dependable source of outside income may not only stabilize resources for family living expenses but also help support the farm during periods of negative cash flow.

Debt Characteristics   The rates and terms of credit may affect cash flow as much as the amount of debt incurred. Seeking out the lowest available interest rate obviously reduces debt payments. Using longer term debt or balloon payment loans also reduces short-term financial obligations.

Debt Structure   Structure refers to the distribution of debt among current, intermediate, and long-term liabilities. Generally, loan repayment terms should correspond to the class of assets which they were used to acquire. Financing intermediate or long-term asset purchases with current debt often leads to repayment problems and cash flow problems.

Financial Contingency Plan

No matter how well the operator budgets or how efficiently the business is managed there will be periods in which cash flow is negative. Figure 17-3 illustrated how the amount of debt used by farmers and ranchers expanded dramatically from 1975 to 1983. However, interest rates climbed sharply in the early 1980s and reached levels that were not anticipated when many loans were made. This combined with generally lower agricultural prices produced what was known as the "farm financial crisis" of the 1980s.

The lessons learned during this period have made many farmers and lenders more conservative about the use of credit. Every operation should have a *financial contingency plan* to provide for unexpected cash flow shortfalls. In some cases actions that are not profitable in the long run may have to be taken in order to cover cash flow obligations in the short run. The following steps are possible financial contingency actions:

1 Maintain savings in a form that can be easily turned into cash and that carries a low risk of loss.

2 Maintain a credit reserve or some unused borrowing capacity for both current and long-term borrowing. A long-term credit reserve can be used to refinance excess current liabilities if the need arises.

3 Prepay debt in years when cash income is above average.

4 Reduce nonfarm expenditures or increase nonfarm earnings when farm cash flow is tight. Acquiring additional education or work experience may make obtaining off-farm employment easier.

5 Carry adequate insurance coverage against crop losses, casualties, medical problems, and civil liabilities.

6 Sell off less productive assets to raise cash. The marginal cost, marginal revenue principle should be employed to identify assets that will have the least effect on total farm profits. In some cases cash flow can be improved without reducing efficiency by selling assets and then leasing them back thereby maintaining the size of the operation.

7 Rely on relatives or other personal contacts for emergency financing or for the use of machinery or buildings at little or no cost.

8 Declare bankruptcy and work out a plan to gradually repay creditors while continuing to farm or conduct an orderly disposal of assets and cancellation of debts.

These actions are not substitutes for operating a profitable business. Some of them may even reduce the farm's profitability. But any of them can be applied, depending on the severity of the farm's financial condition, as a means to continue operating until profits increase.

SOLVENCY

While liquidity management focuses on cash flow, solvency refers to the amount of debt capital (loans) used relative to equity capital (net worth), and the security available to back it up. The farm debt/asset ratio is a common measure of solvency and can also be used to analyze leverage.

Leverage and the Use of Credit

Using a combination of equity capital and borrowed capital permits a larger business than would be otherwise possible. The degree to which borrowed capital is used to supplement and extend equity capital is called *leverage.* Leverage increases with increases in the debt/asset ratio. A debt/asset ratio of 0.50 indicates that one-half of the total capital used in the business is borrowed and one-half is equity capital.

When the return on borrowed capital is greater than the interest rate, profits will increase and equity will grow. Higher leverage will increase profits and equity even faster. For example, assume a firm has $50,000 in equity capital, as shown in the upper part of Table 17-3. If $50,000 more is borrowed at 10 percent interest, the debt/asset ratio is 0.50. If the total assets earn a net return of 15 percent (ROA), or $15,000, and the interest cost of $5,000 is paid, the remaining $10,000 is the return on equity. The rate of return on equity (ROE) is 20 percent. Increasing the leverage, or debt/asset ratio, to 0.67 results in an even higher return to equity, as shown in the right-hand column of the top portion of Table 17-3.

However, there is another side to the coin. If the return on assets is less than the interest rate on borrowed capital, return on equity is decreased by increased leverage, and can be negative. This is shown in the lower part of Table 17-3, where only a 5 percent ROA is assumed. Thus, high leverage can substantially increase the financial risk of the farm or ranch when interest rates are high or return on farm assets is low.

Maximum Debt/Asset Ratio

Most lenders use the debt/asset ratio or some variation of it to measure solvency. However, they do not always agree on what constitutes a "safe" ratio. Profitability of the business and the cost of borrowed funds also enter in. A simple relationship among these factors can be expressed as follows:

TABLE 17-3 ILLUSTRATION OF THE PRINCIPLE OF INCREASING LEVERAGE

	Debt/asset ratio			
	0.0	0.33	0.50	0.67
Equity capital ($)	50,000	50,000	50,000	50,000
Borrowed capital ($)	0	25,000	50,000	100,000
Total assets ($)	50,000	75,000	100,000	150,000
Return on assets (15%)	7,500	11,250	15,000	22,500
Interest paid (10%)	0	2,500	5,000	10,000
Return to equity ($)	7,500	8,750	10,000	12,500
Return to equity (%)	15.0	17.5	20.0	25.0
Return on assets (5%)	2,500	3,750	5,000	7,500
Interest paid (10%)	0	2,500	5,000	10,000
Return to equity ($)	2,500	1,250	0	−2,500
Return to equity (%)	5.0	2.5	0.0	−5.0

$$\text{Max debt/asset} = \frac{\text{ROA}}{\text{IR}}$$

where max debt/asset = maximum debt/asset ratio the business can sustain on its own
ROA = percent return on total assets after deducting the value of un-
paid labor
IR = average interest rate on debt

For example, a farm that earns a 6 percent average return on assets and pays an average interest rate of 12 percent on its debt could service a maximum debt/asset ratio
of 0.50 without reducing equity. In other words, the return on each $1 of assets pays
the interest on $.50 of debt.

The "maximum" debt/asset ratio is the level at which return on equity is just equal
to zero. Debt is serviced without reducing net worth or utilizing outside income.
Obviously farms that earn a higher return on assets or that can borrow money at a
lower interest rate can safely carry a higher debt load. The wise manager will maintain
a margin of safety, however.

Inflation

In the United States the rate of inflation has generally been maintained at a low level,
although it did surpass 10 percent for several years in the 1970s. Some countries have
experienced inflation rates of over 100 percent per year. With high inflation, managers
often prefer to hold tangible assets such as land that will increase in value, rather than
financial assets such as cash or bonds. However, tangible assets may also lose value
when economic conditions change as shown in Figure 17-1.

When farmers invest in intermediate or long-term assets, both their degree of solvency and growth in market value net worth become closely tied to changes in the values of these capital assets. However, changes in asset values alone have little effect on
liquidity or cash flow. Thus, during the land boom of the 1970s many farmland owners
found their net worth rising rapidly without a corresponding increase in cash income.
A few sold land and turned the asset appreciation into cash, although some capital was
lost to income tax payments. Others used their new net worth as collateral to borrow
money, but found repayment to be difficult.

In general, the longer the life of a farm asset, the lower its rate of cash return.
However, many operators have a goal to own land and buildings with the expectation
that they will increase in value over time. The prudent financial manager must carefully balance the need for short-term liquidity against the security and growth potential of
long-term investments.

SUMMARY

Capital includes money invested in machinery, livestock, buildings, and other assets as
well as cash and bank account balances. Sources of capital that are available to farmers include the operator's own equity, equity from outside investors, leased or con-

tracted assets, and borrowed funds. Today's farm managers must be skilled in acquiring, organizing, and using capital whether it is equity capital or borrowed capital. The economic principles discussed in Chapters 5 and 6 can be used to determine how much total capital can be profitably used and how to allocate limited capital among alternative uses.

Agricultural loans are available for the purchase of both real estate and non-real estate assets and for paying operating costs. They can be repaid over periods ranging from less than a year to as long as 40 years or more, in a single payment or with several types of amortized payments. Interest rates, loan terms, and repayment schedules vary from lender to lender and by loan type. Borrowers should compare the annual percentage rate of interest, loan fees, conditions under which the interest rate can vary, and other loan terms when shopping for credit.

Agricultural loans are available from commercial banks, Farm Credit Services, life insurance companies, Farmers Home Administration, individuals, and other sources. Some lenders specialize in certain types of loans, but all are interested in the quality and quantity of credit a prospective borrower possesses. Borrowers should work at improving their credit by maintaining good personal character, improving management skills, demonstrating adequate financial progress and repayment capacity, and providing sufficient collateral.

Liquidity or cash flow management is affected by business growth, nonbusiness income and expenses, and the characteristics and structure of the debt held. A financial contingency plan should be formulated to provide for unexpected cash flow shortages. Solvency refers to the degree to which farm liabilities are secured by assets. Increased leverage can increase the rate at which net worth grows but also increases the risk of losing equity. The amount of debt a farm can support depends on the rate of return earned on assets and the interest rate paid on debt capital. Inflation increases the market value of assets but does not contribute to cash flow unless the assets are sold.

QUESTIONS FOR REVIEW AND FURTHER THOUGHT

1 What economic principles are used to determine (a) how much capital to use and (b) how to allocate a limited amount of capital?

2 What is the single largest source of capital used in U.S. agriculture? What other sources are used?

3 Define the following terms:
   a Secured loan
   b Annual percentage rate
   c Amortized loan
   d Real estate loan
   e Collateral
   f Line of credit

4 What is the difference between short-term, intermediate-term, and long-term loans? List the type of assets which might serve as collateral for each.

5 Assume a $200,000 loan will be repaid in 30 annual payments at 12 percent annual interest on the outstanding balance. How much principal and interest will be due in the first payment if the loan is amortized with equal principal payments? If it is amortized with equal total payments? How would these figures change for the second payment in each case? (Use Appendix Table 1 to find the amortization factor for the equal total payment case.)

6 What are the advantages and disadvantages of a 10-year balloon payment loan versus a completely amortized loan of the same amount and interest rate?

7 Identify the different sources of agricultural loans in your home town or home county. Which types of loans does each lender specialize in? You might interview several lending institutions to learn more about their lending policies and procedures.

8 Select one agricultural lender and find out the rates and terms currently available for an intermediate or long-term loan. Are both fixed and variable interest rate loans available? What loan closing fees or other charges must be paid?

9 Assume you are a beginning farmer or rancher and need capital to purchase breeding livestock. What information and material would you need to provide a lender to improve your chances of getting a loan?

10 List several reasons why a request for a loan by one farm operator might be denied while a similar request from another operator is approved.

11 Explain the difference between liquidity and profitability. Give three reasons why a profitable farm could experience liquidity problems.

REFERENCES

Barry, Peter J., John A. Hopkin, and C. B. Baker: *Financial Management in Agriculture,* 4th ed., The Interstate Printers & Publishers, Danville, IL, 1988, chaps. 14, 15, 16.

Boehlje, Michael D., and Vernon R. Eidman: *Farm Management,* John Wiley & Sons, New York, 1984, chap. 15.

Klinefelter, Danny A., and Bruce Hottell: *Farmers' and Ranchers' Guide to Borrowing Money,* MP-1494, Texas Agricultural Experiment Station, October, 1981.

Lee, Warren F., Michael D. Boehlje, Aaron G. Nelson, and William G. Murray: *Agricultural Finance,* 8th ed., Iowa State University Press, Ames, IA, 1988, chaps. 6, 7, 18.

Penson, John B., Jr., Danny A. Klinefelter, and David A. Lins: *Farm Investment and Financial Analysis,* Prentice-Hall, Englewood Cliffs, NJ, 1982, chaps. 10, 11.

Recommendations of the Farm Financial Standards Task Force, American Bankers Association, Washington, D.C., 1991.

18

LAND—CONTROL AND USE

CHAPTER OBJECTIVES

1 To explore the unique characteristics of land and its use and management in agriculture

2 To outline the advantages and disadvantages of owning land

3 To note factors important in land purchase decisions including methods of land valuation and the legal aspects of a land purchase

4 To present the advantages and disadvantages of leasing land

5 To compare the characteristics of cash, crop share, livestock share, and other leasing arrangements

6 To demonstrate how an equitable share lease arrangement can be developed which will encourage efficient input use

7 To discuss considerations in conservation and sustainable agriculture that go beyond conventional budgeting analysis

Agriculture uses large areas of land, which distinguishes it from most other types of industries. Land is the basic resource which supports the production of all agricultural commodities including livestock, which depend on land to produce the forage and grain they consume. Land is the single most valuable asset in the balance sheet of U.S. agriculture, accounting for nearly three-fourths of the value of total assets. The index of land values shown in Figure 18-1a shows a steady upward trend during the 1950s and 1960s followed by a very sharp rise in the 1970s. Land values then dropped rapidly until 1987 but have recovered somewhat since then. Figure 18-1b indicates that land

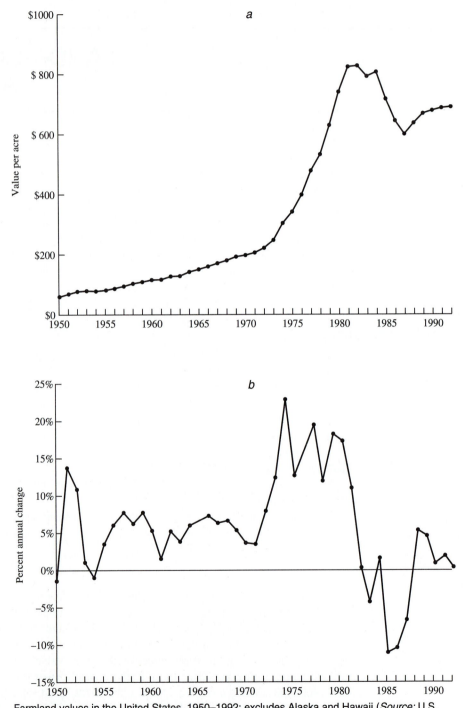

FIGURE 18-1   Farmland values in the United States, 1950–1992; excludes Alaska and Hawaii (*Source:* U.S. Department of Agriculture).

values increased in most of the years since 1950 except for the decline in the early 1980s.

A number of factors contribute to increasing land values. An increase in the profit per acre from crop production has been the driving force behind most land value increases. Increased crop yields due to many technological advances have contributed to this. During part of the 1970s a sharp rise in commodity prices also boosted profits. During most years land values have kept pace with or exceeded the rate of inflation in the United States, making ownership of land a good hedge against inflation. This factor attracts nonfarm investors as well as farmers into the land market.

While nonfarm investors have been important purchasers of land in some areas, the large majority of land buyers are still farmers and ranchers who desire to expand the size of their operation to take advantage of larger-scale technologies. They recognize the economies of size they can achieve with a larger operation, the opportunity for greater profits, and the effect of increasing land values on their net worth.

However, land prices do not always increase every year. In the 1980s high interest rates, drought, and low commodity prices brought financial difficulty for many farmers and ranchers. The result was a sudden decline in land values as buyers left the market. Many farmers and ranchers were forced to sell land at a loss or saw their net worth decline as the land on their balance sheets decreased in market value.

THE ECONOMICS OF LAND USE AND MANAGEMENT

Land has a number of unique characteristics not found in other agricultural or nonagricultural resources. These characteristics greatly influence the economics of land use and management.

Characteristics of Land

Land is a permanent resource which does not depreciate or wear out, provided soil fertility is maintained and appropriate conservation measures are used. Proper management not only will maintain the inherent productivity of land but can even improve it. Land is productive in its native state, producing stands of timber and native grasses, but the management efforts of farmers and ranchers have improved the agricultural productivity of many types of land. This has been accomplished through land clearing, drainage, good conservation practices, irrigation, the introduction of new and improved plant species, and the use of limestone and fertilizer. Land use often changes as a result of these improvements.

Each tract of land has a legal description which identifies its particular location, size, and shape. Land is immobile and cannot be moved to be combined with other resources. Machinery, seed, fertilizer, and water, for example, must be moved to the land to be combined with it in the production of crops and livestock.

Not only is land a unique resource in general, but each farm or specific parcel of land is unique. Any piece of land larger than several acres will often contain two or more distinct soil types, each with its own set of characteristics. Topography, different soil types, climatic features, and the existence of natural hazards such as flooding, wind and water erosion, and rock outcrops are other factors which combine to make land resources different from farm to farm.

The supply of land suitable for agricultural production is essentially fixed, although small amounts may be brought into production by clearing and draining, or are lost to nonfarm uses. This makes the price of land very sensitive to changes in the demand for it. Unlike other agricultural inputs, additional land cannot be manufactured when the demand increases. Therefore, changes in the profitability of agricultural production are eventually factored into land prices and rents, and the land owner receives the benefits or the losses.

Planning Land Use

The different land resource on each farm explains why one of the first steps in whole farm planning is a complete inventory of the land, including soil types, drainage, slope, and production potential. Without this inventory the most profitable farm plan cannot be developed. The potential livestock and crop enterprises, yields, fertility requirements, and necessary conservation practices are directly related to the nature of the land resources available. Whole farm planning was approached from the standpoint of maximizing the return to the most fixed resource. The fixed nature of land in the short run makes it the beginning of most planning efforts.

Land use is affected by regional differences in land productivity. However, the most profitable use for land also depends on relative commodity prices and production technology, which can change over time and bring about changes in land use. Cotton production moved westward out of the southeastern states and was often replaced by pasture and livestock production. Soybeans have become the second most important crop in the Midwest and Mississippi Delta region. The development of irrigation has transformed former livestock grazing areas into important crop production regions with interesting side effects. Irrigation greatly increased the supply of grain available in the Southern Plains region of Texas and Oklahoma, which in turn encouraged the development of large-scale cattle feeding operations. All these changes can be traced to changes in relative prices, new technology, and other economic factors.

LAND CONTROL—OWN OR LEASE?

How much land to acquire and how to acquire it are two important decisions to be made by any farmer or rancher. Errors made at this point may plague the business for many years. Too little land may mean the business is too small to utilize other resources fully. At the other extreme, too much land may require borrowing a large amount of money, cause serious cash flow problems, and overextend the operator's management capacity. Either situation can result in business failure.

Land acquisition should be thought of in terms of control and not just ownership. Control can be achieved by either ownership or leasing. Many farmers are finding a combination of ownership and leasing to be desirable, particularly when capital is limited. They often own some land and buildings to have a permanent "home base" and then lease additional land in the neighborhood to attain the desired farm size. These farmers are classified as part owners (as opposed to full owners) or full tenants if they own none of the land they farm. Part owners accounted for 44 percent of the U.S. farms with sales of over $10,000 in 1987. Another 43 percent were full owners, and 13 percent were full tenants or operators who owned no land.

As with many economic decisions, there can be no general recommendations about how much land should be owned and how much should be leased and the proportion of each. There are advantages and disadvantages to both methods of controlling land. Capital availability, personal preferences and goals, and the amount of land for sale or lease in the area are some of the more important decision factors.

Landownership

Landownership is an important goal for some individuals regardless of the economics involved. A certain amount of pride, satisfaction, and prestige is derived from owning land. It also provides a tangible estate to pass on to one's heirs.

Landownership has the following advantages:

1 *Security of Tenure* Landownership eliminates the uncertainty of losing a lease and having the size of the business reduced unexpectedly.

2 *Loan Collateral* Accumulated equity in land provides an excellent source of collateral when borrowing money. Increasing land values over time have provided substantial equity for landowners, although some of this equity can be lost with a decline in land values such as occurred during the early 1980s.

3 *Management Independence and Freedom* Landowners are free to make their own decisions about enterprise combinations, conservation measures, fertilizer levels, and other choices, without consulting with a landlord.

4 *Hedge against Inflation* Over the long run, land provides an excellent hedge against inflation, as increases in land values have tended to equal or exceed the rate of inflation. However, land values do not necessarily increase every year.

5 *Pride of Ownership* Owning and improving one's own property is a source of pride. It assures a future benefit from years of labor and investment.

Ownership of land controlled by the business can also have some disadvantages. These disadvantages are primarily related to the capital position of the business. Possible disadvantages are:

1 *Cash Flow* A large debt load associated with landownership can create serious cash flow problems. The cash earnings may not be sufficient to meet the required principal and interest payments as well as the other cash obligations of the business.

2 *Lower Return on Capital* Where capital is limited, there may be alternative uses with a higher return than investment in land. Limited capital often has a higher annual cash return when invested in machinery, livestock, and annual operating inputs such as fertilizer, seed, and feed.

3 *Less Working Capital* This disadvantage is related to the one above. A heavy debt load on land may restrict the amount of working capital available, severely limiting enterprise selection, input levels, and profit.

4 *Size Limits*   A combination of limited capital and a desire to own all the land to be operated will limit the size of the business. A small size may prevent the use of certain technologies and result in higher average costs.

The disadvantages of landownership are more likely to affect the beginning farmer with limited capital. With the accumulation of capital and borrowing capacity over time, they may become less and less important. Older farmers tend to own more land and have relatively less debt than younger farmers.

Land Leasing

The beginning farmer or rancher is often advised to lease land. With limited capital, leasing is a means of obtaining control over a larger number of acres. Other advantages of leasing land are:

1 *More Working Capital*   When capital is not tied up in land purchases, more is available to purchase machinery, livestock, and annual operating inputs.

2 *Additional Management*   A beginning farmer may be short of management skills. Management assistance can be provided by a knowledgeable landlord or professional farm manager employed by the landlord.

3 *More Flexible Size*   Lease contracts are often for only one year or at most several years. Year-to-year changes in business size can be easily accomplished by giving up old leases or leasing additional land.

4 *More Flexible Financial Obligations*   Lease payments are more flexible than mortgage payments which may be fixed for a long period. Share rents will vary with crop yields and prices. Cash rents are less flexible but can be negotiated each time the lease is renewed, taking into account current economic conditions.

Leasing land also has disadvantages which take on a particular significance when all the land operated is leased. These disadvantages are:

1 *Uncertainty*   Given the short term of most lease contracts, the danger always exists that all or part of the land being farmed can be lost with fairly short notice. This possibility discourages some good farming practices, may result in a rapid and substantial decrease in business size, and contributes to a general feeling of uncertainty about the future of the business.

2 *Poor Facilities*   Some landlords are reluctant to invest money in buildings and other improvements. Tenants cannot justify investing in improvements that are attached to someone else's property. Thus, family housing, livestock facilities, grain storage, and machinery housing may be less than desirable.

3 *Slow Equity Accumulation*   Without landownership, equity can be accumulated only in machinery, livestock, or cash savings. In periods of rising land values, landownership can generate substantial equity and increase the owner's borrowing capacity.

A review of ownership and leasing reveals no clear advantage for either method of acquiring control over land. Control is still the important factor, as income can be obtained from either owned or leased land. In the final analysis, the proper combination of owned and leased land is the one providing enough land to fully use the available labor, machinery, management, and working capital given individual capital limitations and personal preferences.

BUYING LAND

The purchase of a farm or ranch is an important decision often involving large sums of money. A land purchase will have long-run effects on both the liquidity and solvency of the business.

The first step in a land purchase decision should be a determination of its value. While income potential is one important determinant of land value, many other factors also contribute:

1 *Soil, Topography, and Climate*   These factors combine to determine the crop and livestock yield potential and therefore affect the expected income stream.

2 *Community*   A farm located in an area of well-managed, well-kept farms has a special attraction which increases its value.

3 *Location*   Location with respect to schools, churches, towns, paved roads, and farm input suppliers will affect value. A prospective purchaser who plans to live on the farm will pay particular attention to the community and services provided in the general area.

4 *Buildings and Improvements*   The number, size, condition, and usefulness of buildings, fences, and other land improvements will affect value. A neat, attractive farmstead with a modern house can add many dollars to a farm's value, while run-down, obsolete buildings will detract from it.

5 *Size*   Small farms may sell for a higher price per acre than large farms. A smaller total purchase price puts smaller farms within the financial reach of a larger number of buyers. Several neighboring farmers may consider the purchase of a small farm a good way to increase the size of their business, and bid up the price.

6 *Markets*   Proximity to a number of markets will reduce transportation costs, increase competition for the farm's products, and possibly raise selling prices.

7 *Urban Pressure*   Land that is close to urban or recreational areas may have a higher value than other agricultural land due to its other potential uses.

A thorough inspection of a property should be conducted to identify problem areas which might reduce its value. Buildings should be inspected for structural soundness, and soil types properly identified. Drainage and erosion problems, sand and gravel areas, rock outcroppings, and other soil hazards should be noted and yield expectations adjusted accordingly. It is also advisable to conduct an environmental audit. This is an inspection of the property looking for potential environmental problems such as chemical disposal sites, oil spills, and the existence of old underground fuel tanks which could be leaking. These problems could be costly to correct.

There is a tendency to overvalue poorer land, as the asking price may make it appear to be a bargain. If the land is basically productive but has been abused and the fertility level allowed to decline, good management practices may restore its productivity. This type of land may occasionally be found at a bargain price. However, land with a naturally low fertility level and/or many natural hazards must be carefully evaluated to determine its value. Labor and machinery costs for the primary field operations will be nearly the same as for more productive soils, and some other costs may be little different. With many costs nearly the same and a lower yield, the poor soil will have a smaller net income stream over time.

Land Appraisal

An appraisal is a systematic process which leads to an estimate of value for a given piece of real estate. There are business firms which provide appraisal services for a fee. They provide their clients with a detailed analysis of the factors which determine the value of the property and conclude the report with an estimate of current value. Two basic methods are used to determine the value of income-producing property such as land, the income capitalization method and the market data method.

Income Capitalization   This method uses investment analysis methods to estimate the present value of the future income stream from the land. It requires an estimate of the expected annual net income, the selection of a discount rate, and computation of the present value of an annuity, as outlined in Chapter 15. However, it is normally assumed that net income from land has an infinite life. The present value equation for a perpetual or infinite stream of income is

$$V = \frac{P}{i}$$

where P is the average annual net income, i is the discount rate, or capitalization rate as it is called in appraisal work, and V is the estimated land value.

The first step is estimating the net income stream P that can be obtained from the farm's physical production resources. This requires determining potential long-term yields, selling prices, and production costs for the most profitable enterprise combination. P is the net return to investment in land and is equal to the farm's estimated gross income minus the costs of all resources used to generate it including landownership costs such as property taxes, depreciation, and maintenance. The opportunity cost of the capital invested in the land, and any interest cost that might be incurred on a loan used to purchase it, is *not* subtracted.

The results of such a procedure are shown in the top portion of Table 18-1 for a hypothetical 160-acre farm. It has 150 tillable acres after deducting for roads, lanes, and waterways. The long-run crop plan is estimated to be 100 acres of corn and 50 acres of soybeans generating an annual gross income of $41,400. Annual expenses associated with the crop plan are also shown in Table 18-1. They total $27,700, which makes net income equal to $13,700 per year. Yield and price estimates are important. They strongly influence the final estimate of net income. Both items must be accurately esti-

TABLE 18-1 ESTIMATED ANNUAL INCOME AND EXPENSES FOR APPRAISAL PURPOSES
(Hypothetical 160-Acre Farm with 150 Tillable Acres)

	Acres	Yield (bu)	Price	Total
Income:				
Corn	100	120	$2.50	$30,000
Soybeans	50	38	$6.00	$11,400
Total income				$41,400
Expenses:				
Fertilizer				$5,000
Seed				2,700
Pesticides				3,200
Trucking				500
Drying				1,200
Labor and management				4,500
Machinery:				
Ownership costs				4,500
Operating costs				3,000
Property taxes and insurance				1,800
Repairs and maintenance				500
Depreciation of buildings, fences				800
Total expenses				$27,700
Annual net income				$13,700

Capitalization of income

Capitalization rate		Total value	Per acre
8% ($13,700 ÷ 0.08)	=	$171,250	$1,070
6% ($13,700 ÷ 0.06)	=	$228,333	$1,427
4% ($13,700 ÷ 0.04)	=	$342,500	$2,141

mated to arrive at an accurate estimate of value. County yield averages and farm yields assigned by the local Agricultural Stabilization and Conservation Service office are useful starting places for yield estimates. Prices should reflect the appraiser's best estimate of long-run prices after a careful review of past price levels.

The second step in the income capitalization method is to select the discount or capitalization rate. The effect of the capitalization rate on value is readily apparent, and the appropriate rate must be carefully selected. The estimate of the net income does not include any expectation of inflation. Therefore, as explained in Chapter 15, the discount rate should be based on the real interest rate, that is, the nominal or actual rate of interest minus the anticipated rate of inflation.

Actual appraisal practice is to estimate the average rate of return to land investment in the general area for similar farms at recent sale prices. Using this rate in the capitalization procedure will give a value which is comparable with the recent selling prices for other farms. This procedure often results in a capitalization rate of 3 to 6 percent, which is the range of historical rates of return to farmland based on current market values. Anticipation of long-term increases in land values helps explain why landowners have historically been willing to accept a current rate of return on land which is lower than other investments.

The third step is to divide the expected net income by the capitalization rate. This process is shown in the bottom portion of Table 18-1 for three different capitalization rates. The estimated value of the land in the example using a 6 percent capitalization rate is $13,700 divided by 0.06 = $228,333, or about $1,427 per acre.

Market Data   This appraisal approach compares the characteristics of land that has been sold recently with the tract of land being appraised. Prices of the *comparable sales* are adjusted for differences in factors such as soil type, productivity, size, the community, location, buildings and improvements, proximity to markets, and urban pressure, as discussed earlier. Three additional factors need to be considered when comparable sales values are being adjusted to reflect "fair market value" of the farm being appraised.

1 *Financing*   The method and terms of the financing arrangements for the purchase will affect selling price. Land may be sold with an installment purchase contract where the seller provides the financing. The terms of a contract may include a smaller down payment and/or a lower interest rate than conventional mortgage financing, which will increase the price a buyer is willing to pay for a given piece of land.

2 *Relationships*   If the buyer and seller are closely related, the selling price is often below the price that would be agreed on by unrelated parties.

3 *Time of Sale*   The more time that has passed since the comparable sale took place, the more likely the price must be adjusted to reflect current market conditions. It may be adjusted up or down, depending on whether land values have been rising or falling.

Financial Feasibility Analysis

A financial feasibility, or cash flow, analysis should be performed on any prospective land purchase. It is not a method for determining land value but, for a given purchase price, will show if there will be sufficient cash flow to meet the annual operating expenses as well as the interest and principal payments on a loan. Table 18-2 is a summarized cash flow analysis for the purchase of the example 160-acre farm for $1,600 per acre, or a total of $256,000.

The analysis assumes a 40 percent down payment ($102,400) and a 20-year loan of $153,600 with 9 percent interest and equal annual principal payments. Annual cash receipts and cash expenditures are based on the figures in Table 18-1, with 50 percent of the machinery ownership costs estimated to be cash expenses. Although depreciation is a noncash expense, some cash outflow must be included for machinery replacement. Labor costs should reflect the cost of hired labor and a portion of family living costs (estimated at $4,000 in this example), but no opportunity cost for unpaid labor. In the example, cash receipts and expenditures are assumed to inflate at a rate of 3 percent annually. Principal payments and the interest rate are fixed for the term of the loan, however.

TABLE 18-2 CASH FLOW ANALYSIS ON PURCHASE OF 160-ACRE FARM AT $1,600 PER ACRE*

	Year 1	Year 2	Year 3	Year 4	Year 5
Cash receipts ($)	41,400	42,642	43,921	45,239	46,596
Cash expenditures ($)					
Seed, fertilizer, pesticides, trucking, drying	12,600	12,978	13,367	13,768	14,181
Machinery	5,250	5,408	5,570	5,737	5,909
Family living	4,000	4,120	4,244	4,371	4,502
Taxes, repairs	2,300	2,369	2,440	2,513	2,589
Annual loan payment ($):					
Principal	7,680	7,680	7,680	7,680	7,680
Interest	13,824	13,133	12,442	11,750	11,059
Total cash outflow ($)	45,654	45,688	45,743	45,819	45,920
Net cash flow ($)	−4,254	−3,046	−1,822	−580	676

*Assumes a down payment of $102,400 with the balance financed by a 20-year loan of $153,600 at 9% interest with equal principal payments made annually.

For the first four years, the cash operating income is not enough to meet both the cash operating expenses and the required principal and interest payments. Because of inflation and the declining interest payment each year, net cash flow becomes positive in the fifth year of ownership. In the meantime, the cash flow deficits in the first four years must be met with cash from other sources, such as other land being farmed, livestock, or off-farm employment.

The situation shown in Table 18-2 is typical of many land purchases in recent years. Land prices include expectations of appreciation in value, growth in net income, security, and pride of ownership. These higher prices cause negative cash flows from a debt-financed land purchase to be the rule rather than the exception. Earnings will seldom be sufficient to meet both operating expenses and debt repayment. Several things would reduce the cash flow deficit, including a lower purchase price, a lower interest rate, a larger down payment, or a longer term on the loan. An amortized loan with equal total payments would also help reduce the cash outflow the first few years. Obviously a cash flow projection for the entire business should be done to determine if cash will be available from other parts of the business to make the loan payments on a new land purchase.

Legal Aspects

Purchasing land is a legal as well as a financial transaction. The legal description of the property should be checked for accuracy and the title examined for any potential problems. For example, there may be unpaid property taxes and mortgages on the property. The buyer should also be aware of any easements for roads, pipelines, or power lines which might interfere with the use of the land. Zoning and other local land use restrictions should also be checked.

Any water rights and mineral rights passing to the new owner should be carefully identified and understood. Rights to underground minerals do not automatically transfer with the rights to the land surface. Where oil, gas, coal, or other minerals are important, the fraction of the mineral rights being received can have a large impact on land values. In the irrigated areas of the western United States the right to use water for irrigation purposes is often limited and allocated on an individual farm basis. Obviously, the extent and duration of these water rights should be carefully determined. An existing but undetected environmental problem could result in anything from difficulty obtaining a loan to the expense of eliminating the problem.

This is only a partial list of the many ways land buyers can receive less than they expected because of some unknown restriction or problem. For these and other reasons it is advisable for land buyers to retain the services of an attorney with experience in land transactions before any verbal or written offer is made on a property.

LEASING LAND

Obtaining control of land through leasing has a long history in the United States and other countries. Not all this history has been good, as there has been exploitation of tenants and sharecroppers and poor land use. Full landownership has been advocated by some as a means of eliminating these problems. However, it is not likely we will ever have an agriculture where all farmers own all the land they operate. The capital requirements are too large, and improvements in leasing arrangements have reduced many of the problems and inefficiencies.

Leases on agricultural land are strongly influenced by local custom and tradition. The type, terms, and length of leases tend to be fairly uniform within a given area or community. This reliance on custom and tradition results in fairly stable leasing arrangements over time, which is desirable. However, it can also mean lease terms are slow to change in response to changing economic conditions and new technology. Inefficient land use can result from outdated leasing arrangements.

A lease is a legal contract whereby the landowner gives the tenant the possession and use of an asset such as land for a period of time in return for a specified payment. The payment may be cash, a share of the production, or a combination of the two. Oral leases are legal in most states but are not recommended. It is too easy for memories to fail, causing disagreements over the terms of the original agreement, and because documentation may be needed for an estate settlement or a tax audit.

A lease should contain the following information: (1) the legal description of the land, (2) the term of the lease, (3) the rent to be paid and the time and place of payment, (4) the names of the landlord (lessor) and the tenant (lessee), and (5) the signatures of all parties to the lease. These are only the minimum requirements of a lease. A good lease will contain other provisions spelling out the rights and obligations of both the landlord and tenant. Many good leases contain a clause describing the arbitration procedure to be followed in case of unresolved disputes. Dates and procedures for notification of lease renewal and cancellation should also be included, particularly if they differ from state law requirements.

An example of a lease form is included at the end of this chapter, but this is only one example. Blank lease forms are available through the Agricultural Extension Service in many states, as well as from attorneys and professional farm managers. It is important to modify the language in a lease form to fit the characteristics of the particular property as well as the laws of the particular state. As with all legal documents, the contracting parties may wish to obtain the advice of an attorney before signing a lease.

The three basic types of leases commonly used in the leasing of agricultural land are cash rent, crop share, and livestock share. Each has some advantages and disadvantages to both landlord and tenant.

Cash Rent

A cash rent lease has the rent specified as a cash payment of either a fixed amount per acre or a fixed lump sum. Rent may be due in advance, at the end of the crop season, or as some combination of these. If the rent is set per acre, the number of acres should be specified in the lease. Some cash leases show rent for land and for buildings separately. Under a cash lease the landlord furnishes the land and buildings while the tenant receives all the income and typically pays all expenses except property taxes, property insurance, and major repairs to buildings and improvements. The lease may contain restrictions on land use and require the tenant to use certain fertility practices, control weeds, and maintain fences, waterways, terraces, and other improvements in their present conditions.

The characteristics of a cash lease create both advantages and disadvantages. Some of the more important are:

1 *Simplicity*   There is less likelihood of disagreements, as the rent can be easily spelled out and understood by both parties. A cash lease is also easy for a landlord to supervise, as there are few management decisions to be made. For this reason, landlords who live a long distance from their farm or who have little knowledge of agricultural production often prefer a cash lease.

2 *Managerial Freedom*   Tenants are free to make their own decisions regarding crop and livestock programs and other management options. Tenants who are above-average managers often favor a cash lease, as they receive the total benefits from their management decisions.

3 *Risk*   A cash lease provides the landlord with a known, steady, and dependable rental income. The tenant stands all risk from yield, price, and cost variability. This avoidance of risk is one reason cash rents tend to yield lower average returns to the owner than a crop share lease on comparable land.

4 *Capital Requirements*   The landlord has fewer capital requirements under a cash lease as there is no sharing of the annual operating inputs. Conversely, the tenant has a larger capital requirement including all operating inputs and the cash rent.

5 *Land Use*   With all income accruing to the tenant, some may be tempted to maximize short-term profits from the land at the expense of long-term productivity, particularly under short-term leases. This can be prevented by including fair and proper land use restrictions in the lease.

6 *Improvements*   Landlords may be reluctant to invest in buildings and other improvements when using a cash lease, as they do not share in any additional income they provide.

7 *Rigid Rents* Cash rents tend to be inflexible and slow to change. Unless they are renegotiated each year to reflect changes in prices, land values, and technology, inequities can soon develop.

8 *Investment Income* Some landlords may favor cash leases for another reason. Cash rental income where the landlord is not closely involved in management of the business may qualify as investment income rather than self-employment income. There is no self-employment tax on investment income.

Setting a Fair Cash Rent

Cash rental rates for particular parcels of farmland ultimately depend on the productivity of the land, the supply of and demand for farmland in the area, and the bargaining positions of the owner and the operator. However, there are several different approaches to estimating a fair rent.

1 *Landowner's Costs* In the long run, the landowner wants enough rent to cover both the cash and opportunity costs of owning the land. Assume that the owner of the example farm in Table 18-1 feels that it has a current market value of $1,500 per acre, or $240,000, and the opportunity cost rate for similar investments is 4 percent. The owner's total costs would be 4 percent of $240,000, or $9,600, for opportunity cost on the investment, $1,800 for property taxes, $500 for repairs and maintenance, and $800 for depreciation for a total cost of $12,700, or $79 per acre. This would be the minimum rent needed to pay all the owner's costs.

2 *Tenant's Residual* A second approach is to estimate how much income the tenant will have left after paying all other costs. Using the example in Table 18-1 again, the tenant would have gross income of $41,400 and expenses of $24,600 for fertilizer, seed, pesticides, trucking, drying, labor, and machinery. The difference is $16,800, or $105 per acre. Note that the tenant would not include any land ownership costs. If the rented land could be farmed with no extra machinery investment, then machinery ownership costs could be excluded also. This approach estimates the maximum rent the tenant could pay without allowing for any return to management, risk, or profit.

3 *Crop Share Equivalent* A partial budgeting approach can be used to estimate how much cash rent the tenant could pay and receive the same return as from a crop share lease. Suppose that by shifting from a crop share to a cash lease the tenant would receive 100 percent of the crop instead of 50 percent. Using the Table 18-1 example, this would increase income by $20,700 (half of $41,400). However, the landowner would no longer pay half of the fertilizer, seed, pesticide, trucking, and drying costs, so the tenant would have $6,300 in additional costs. The net gain would be $14,400, or $90 per acre. This is how much cash rent could be paid and still receive the same net return as under a crop share lease.

4 *Share of Gross Income* The last approach estimates a cash rental rate as a percent of the expected gross income from the land. For example, if land costs generally represent about 35 percent of the total cost of production for particular crops in a region, rent could be estimated at 35 percent of the expected gross income. In the example this would be equal to $41,400 × 0.35 = $14,490, or $91 per acre.

These various approaches will not give identical answers. However, they can help define a range within which the owner and operator can negotiate.

Pasture Rent   A fair cash rent for pasture land is more difficult to determine because the potential income is very uncertain. Factors such as the quality of the established pasture, water supply, condition of fences, buildings, and location must all be considered.

Pasture can be rented for a fixed rate per acre or by the month. One common method is to establish a charge per animal unit month (AUM). One AUM is equivalent to one mature cow grazing for one month. With this method the rent is proportional to the stocking capacity of the pasture.

Crop Share

Crop share leases are popular in the Midwest and other areas where cash grain farms dominate. These leases specify that the landlord will receive a certain share of the crops produced, with the proceeds from their sale becoming the rent. The tenant typically provides all the labor and machinery. Fertilizer, seed, chemicals, and irrigation costs may be shared, along with harvesting and other costs in some areas. Along with sharing production and expenses, landlords often participate in management decisions.

The landlord's share will vary from one-half down to one-quarter or less, depending on the type of crop, local custom, and whether or not any or how many variable costs are shared. Soil productivity is an important factor, as many of the tenant's costs such as labor and machinery will be nearly the same regardless of soil type. Therefore, tenants receive a larger share of the production from poorer soils because the value of the land contribution is less. Figure 18-2 shows how crop shares tend to vary by soil productivity level. Landlords whose farms contain poorer soils may find

FIGURE 18-2   The relationship between divisions of crop shares and soil productivity rating.

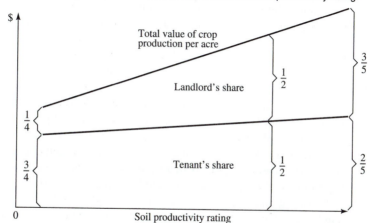

they have to take a smaller share and/or pay more of the variable costs to attract a good tenant. A typical crop share lease is shown at the end of this chapter.

The advantages and disadvantages of crop share leases can be summarized as follows:

1 *Risk*   The rent will vary with changes in yields and prices, so the risk is shared by the tenant and owner. This may be a disadvantage to the owner if the rent is an important part of the owner's total income as it may be for some retired persons. Variable rent is an advantage to the tenant, as the rental payment varies with the ability to pay.

2 *Management*   Landlords using a crop share lease usually maintain some direct or indirect control over crop selection and other management decisions. This may be an advantage to an inexperienced tenant and gives the landlord some control over land use.

3 *Capital Requirements*   Because some crop production expenses are shared and no cash rent must be paid, the landlord's capital requirements increase while the tenant's decrease when compared with a cash lease.

4 *Expense Sharing*   A problem with crop share leases is determining a fair and equitable sharing of expenses. As a general rule, variable expenses should be shared in the same proportion as the production, but this rule is little help in determining which expenses should be shared. The adoption of new technology often creates new problems with the proper share of its costs and benefits.

5 *Buildings and Pasture*   Since the landlord receives no direct benefits from buildings (except grain storage) and livestock pasture, problems can arise concerning a fair rent for these items. A cash rent supplementing the crop share is often paid by the tenant for use of buildings and livestock pasture.

6 *Marketing*   The owner is generally free to sell his/her share of the crops wherever or whenever desired. This requires some additional expertise by the owner. Some owners let the tenant or a professional farm manager make all the marketing decisions.

7 *Material Participation*   Landowners are usually more involved in management decisions under a crop share lease. This management participation often qualifies share rent income as self-employment income, which helps build social security earnings. Also, it may possibly qualify the farm for a lower valuation for estate tax purposes.

Livestock Share

A livestock share lease is much like a crop share lease except livestock income is also shared between landlord and tenant. The tenant typically furnishes all labor and machinery and a share of the livestock and operating inputs, with the landlord furnishing land, buildings, and the remaining share of livestock and operating inputs. Most livestock share leases are 50–50 shares, although other share arrangements are possible depending on the type of livestock and how expenses are shared.

Livestock share leases can be complex, and there is considerable latitude in the number and type of expenses to be shared. In addition to livestock expenses such as feed, and health expenses, the landlord may share in the cost of machinery and equipment related to livestock production. Examples include milkers, feed grinders, feeders, waterers, and forage harvesting equipment. The lease should contain a full and detailed list of the expenses to be shared and the sharing ratio.

The advantages, disadvantages, and potential problems with a livestock share lease are similar to those of a crop share lease, with the following additional considerations:

1 *Rent*   A livestock share lease provides the landlord with some rent on buildings and pasture by sharing in the livestock production. The landlord's return will generally be higher than for a crop share lease. Landlords are more likely to furnish and maintain a good set of buildings. However, tenants may desire additional building improvements to reduce their labor requirements in the shared livestock production.

2 *Records*   The sharing of both crop and livestock income and expenses requires good records to ensure a proper division. There should be a periodic accounting of income and expenses with compensating payments made to balance the accounts.

3 *Lease Termination*   Terminating a livestock share lease can be complex and time-consuming. All livestock and shared equipment must be divided in a fair and equitable manner. The lease should contain a method for making the division and an arbitration procedure for any unresolved disputes.

4 *Management*   There is a greater need and opportunity for sharing management decisions under a livestock share lease, and this requires a good working relationship between the landlord and tenant.

Other Lease Types

There are several other types of leases and variations of the above types that are in use.

Labor Share Lease   In a labor share lease the landlord provides all the land, machinery, and other variable inputs. The tenant provides only the labor and receives a share of the production. However, this share would be smaller than with the other share leases. This arrangement works well for a tenant who would like to begin farming but has very limited capital and for a landlord who has a complete set of farming resources but is ready to retire.

Variable Cash Lease   Rigidity of cash rents was identified as one of the disadvantages of cash leases. Variable cash leases are sometimes used to overcome this problem and divide some of the risk between the landlord and tenant. Annual cash rent under a variable cash lease can be tied to the actual yield, the price received, or to both of these factors. For example, the cash rent may increase or decrease a specified amount for each bushel the actual yield is above or below some base or average yield. The same could be done for price or for gross income. Variable cash leases maintain most of the properties of fixed cash leases but allow the landlord and tenant to share at least part of the price and/or production risk.

One type of variable cash lease sets a rent based on the most likely price and yield, and then adjusts it in proportion to the amount by which the actual price and yield exceed the base values. In the example below the base rent is $90 per acre for a corn yield of 110 bushels per acre and a price of $2.50. Assume the actual yield turns out to be 125 bushels per acre, but the price is only $2.10. The actual rent is then calculated at $85.91 per acre, as shown below:

$$\text{Actual rent} = \$90 \times \frac{\text{actual yield}}{110} \times \frac{\text{actual price}}{2.50}$$

$$= \$90 \times \frac{125}{110} \times \frac{2.10}{2.50} = \$85.91$$

Another common cash rent formula is to simply pay a fixed percent of the actual gross income earned from the crops on the rented land. Since the tenant's ability to pay is affected by both price and yield, a variable cash rent formula that depends on both of these factors provides the most risk reduction. The tenant and landowner should agree in advance on how the actual yield and price used will be determined.

Bushel Lease   Another type of lease is the bushel lease, or standing rent lease. With this lease the landlord typically pays no crop production expenses but receives a specified number of bushels or quantity of the product as rent. Production risk falls entirely on the tenant as a fixed quantity of the product must be delivered to the landlord regardless of the actual yield obtained. Part of the price risk is shared with the landlord as the actual rent value will depend on the price received for the product.

Custom Farming   The practice of custom farming is not a true leasing agreement but represents another alternative arrangement between owner and operator. Typically the operator provides all machinery and labor for doing all the field operations and hauling in exchange for a fixed payment. There may be a bonus for above-average yields. The owner supplies all the operating inputs, receives all the income, and stands all the price and yield risks.

Table 18-3 summarizes the important characteristics of the lease types discussed in this section.

TABLE 18-3   COMPARISON OF LEASE TYPES

	Fixed cash	Variable cash	Fixed bushel	Crop or livestock share	Custom farming
Price risk borne by:	Tenant	Both	Both	Both	Owner
Production risk borne by:	Tenant	Both	Tenant	Both	Owner
Operating capital supplied by:	Tenant	Tenant	Tenant	Both	Owner
Management decisions made by:	Tenant	Tenant	Tenant	Both	Both
Marketing done by:	Tenant	Tenant	Both	Both	Owner
Terms adjust:	Slowly	Quickly	Medium	Quickly	Slowly

Efficiency and Equity in Leases

Because many leases are based on local custom and tradition, inefficiencies, poor land use, and a less than equitable division of income and expenses can occur. There have been many critics of a land tenure system based on leasing because of the existence of these problems. It may not be possible to write a perfect lease, but improvements can be made. There are two broad areas of concern in improving the efficiency and equity of a lease. The first is the length of the lease, and the second is the cost sharing arrangements.

Farm leases are typically written for a term of one year, although some livestock share leases are written for three to five years. Most one-year leases contain a clause providing for automatic renewal on a year-to-year basis if neither party gives notice of termination before a certain date. Many such leases have been in effect for long periods of time, but the possibility always exists that the lease may be canceled on short notice if the landlord sells the farm or finds a better tenant. This places the tenant in a state of insecurity. It also tempts the tenant to use practices and plant crops that maximize immediate profits rather than conserve or build up the property over time.

Short leases also discourage a tenant from making improvements as the lease may be terminated before the cost can be recovered. These problems can be at least partially solved by longer-term leases and agreements to reimburse the tenant for the unrecovered cost. Improvements at the tenant's expense such as application of limestone and erection of soil conservation structures can be covered under an agreement of this type.

Inefficiencies can also arise from poor cost-sharing arrangements if the costs of those inputs which directly affect yields are not shared in the same proportion as the income or production. Seed, fertilizer, chemicals, and irrigation water are examples. Table 18-4 shows what can happen. The profit-maximizing level of fertilizer use is where its marginal value product is equal to its marginal input cost, which occurs at 140 pounds of fertilizer per acre. However, if the tenant receives only one-half the crop but pays all the fertilizer cost, the marginal value product for the tenant is only one-half the total. This is shown in the last column of Table 18-4. Under these conditions the tenant will use only 100 pounds of fertilizer per acre to maximize profit. While this amount will maximize profit for the tenant under these conditions, it reduces the total profit per acre from what it would be at 140 pounds of fertilizer.

On the other hand, fertilizer has a zero marginal input cost to the landlord. Additional fertilizer use does not increase the landlord's costs but does increase yield, which is shared. The landlord would like to fertilize for maximum yield as the way to maximize profit. This would be at 160 pounds of fertilizer per acre in the example in Table 18-4.

These types of conflicts can be eliminated by sharing the cost of yield-determining inputs in the same proportion that the production is shared. If fertilizer cost were equally shared in this example, both parties would pay one-half the marginal input cost and receive one-half the marginal value product. They would both agree on 140 pounds of fertilizer as the profit-maximizing level. Tenants are understandably reluctant to adopt any new yield-increasing technique or technology if they must pay all the additional cost but receive only a share of the yield increase.

TABLE 18-4   EXAMPLE OF INEFFICIENT FERTILIZER USE UNDER A CROP SHARE LEASE
(Landlord Receives One-Half of Crop but Pays No Fertilizer Cost)

Fertilizer (lb)	Yield (bu)	Marginal input cost at $0.20/lb ($)	Total marginal value product, corn at $2.20 per bu ($)	Tenant's marginal value product ($)
60	95			
80	100	$4.00	$11.00	$5.50
100	104	4.00	8.80	4.40
120	107	4.00	6.60	3.30
140	109	4.00	4.40	2.20
160	110	4.00	2.20	1.10

Similar problems can arise when changing technology allows the substitution of shared inputs for nonshared inputs. An example is the use of herbicides in place of mechanical or hand weed control. If the tenant pays only one-half the cost of the herbicide and all the cost of labor and machinery, the profit-maximizing combination of inputs for the tenant will include more herbicide and less labor and machinery than in a cash lease or owner-operator situation.

Determining Lease Shares

The objective of any lease should be to provide a fair and equitable return to both parties for the inputs they contribute to the total farm operation. Lease terms and rental rates should be designed to encourage maximum profit given the productivity potential for the farm. This is equally as important as the sharing rate, as it is obvious that either party is better off with 40 percent of a $50,000 income than with 50 percent of a $30,000 income.

Most people agree that a fair and equitable sharing arrangement exists when each party is paid for the use of their inputs according to the contribution those inputs make toward generating income. Application of this principle requires the identification and valuation of all resources contributed by both the landlord and tenant.

One method for determining the proper shares for a crop share lease is shown in Table 18-5. This same procedure could be used for a livestock share lease, but it would contain more detail. The example in Table 18-5 starts from an assumed 50–50 sharing of expenses for seed, fertilizer, lime, and other crop inputs. Other costs are allocated to the landlord or tenant based on who owns the asset or provides the service.

The most difficult part of this procedure is placing a value on the services of fixed assets such as labor, land, and machinery. The same procedures that are used for estimating cash and opportunity costs for farm budgets should be applied. Land costs can include both cash costs and an opportunity interest charge, or they can be estimated

TABLE 18-5 DETERMINING INCOME SHARES UNDER A CROP SHARE LEASE

Cost item	Whole farm	Owner	Tenant
Fertilizer	$5,000	$2,500	$2,500
Seed	2,700	1,350	1,350
Pesticides	3,200	1,600	1,600
Trucking	500	250	250
Drying	1,200	600	600
Labor, management	4,500	500	4,000
Machinery ownership	3,000	0	3,000
Machinery operating	4,500	0	4,500
Property taxes, insurance	1,800	1,800	0
Repairs and maintenance	500	500	0
Depreciation of buildings, fences	800	800	0
Opportunity cost for land	9,600	9,600	0
Total costs	$37,300	$19,500	$17,800
Percent contributed		52%	48%

from current cash rental rates. Do not use principal and interest payments, however, since they represent debt repayment rather than economic costs.

This example has a cost sharing very close to a 50–50 proportion. It may be so close that both parties feel that a 50–50 division of production is fair without any adjustments. If the cost shares are substantially different, one of two methods can be used to make adjustments. First, the party contributing the smaller part of the costs can agree to furnish more of the capital items or pay for more of the variable costs, to make the cost sharing closer to 50–50. The second method would be to change the production shares to 60–40 or some other division to more nearly match the cost shares. This should include changing the cost sharing on the variable inputs to the same ratio to prevent the inefficiencies discussed earlier. Whichever alternative is used to match the cost and production shares, the final result should have the tenant and landlord sharing farm income in the same proportion as they contribute to the total costs of production.

CONSERVATION AND ENVIRONMENTAL CONCERNS

Conservation can be defined as the use of those farming practices that will maximize the present value of the long-run social and economic benefits from land use. This definition does not prevent land from being used but does require following practices which will maintain soil productivity and water quality over time. Farming systems that accomplish these goals are sometimes called *sustainable agriculture.*

Ordinary budgeting techniques are often inadequate for deciding how best to achieve the goals of conservation. Nevertheless, farm and ranch managers need to be conscious of how their decisions affect the long-term quality of life for themselves and society in general. There are three general areas in which sustainable agriculture considerations go beyond conventional analysis.

Long-Run versus Short-Run Consequences

Narrow profit margins and tight cash flows make it tempting to "mine" the soil in order to maximize short-run returns. Most conservation practices require some extra cash expenditures. They may also reduce crop yields temporarily as soil and cropping patterns are disturbed. This short-run reduction in profit may be necessary to achieve higher profits in the future or to prevent a long-run decline in production if no conservation practices are used.

Figure 18-3 illustrates the possible income trends with and without conservation practices. Since conservation practices are a form of investment, the methods discussed in Chapter 15 can be used to evaluate the profitability of a particular practice. Unfortunately, the long-term effects of soil depletion and erosion on productivity are not always well known nor understood. Likewise, it takes years of study to determine the long-term effects of continued use of high rates of fertilizer and pesticides on soil, water, and wildlife.

Some conservation practices require large initial investments such as terraces, drainage ditches, and diversion structures. The opportunity cost of capital and the planning horizon of the landowner become very important in the adoption of conservation practices. Higher discount rates reduce the present value of the larger future incomes resulting from conservation. Shorter planning horizons also discourage conservation. In Figure 18-3, a farmer with a 5-year planning horizon is not likely to adopt the conservation practices, as the net present value of the income flow would be less than that with no conservation. In this example, it may take a planning horizon of 15 years or more before the net present value of the income flow with conservation exceeds that from no conservation. For this reason, many farmers are looking at changes in tillage practices and crop rotations as lower cost alternatives to achieve conservation goals.

FIGURE 18-3   Hypothetical illustration of income flows over time with and without conservation practices.

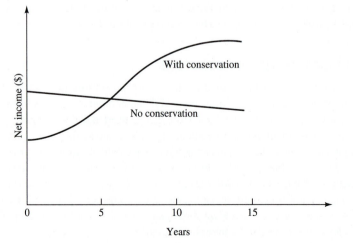

Leasing arrangements also affect the type of conservation practices followed. One-year leases discourage tenants from considering the long-term effects of their farming practices. On the other hand, a landlord may be reluctant to make large conservation investments if they feel the tenant will receive all or a share of the benefits. Tenants and landlords need to fully discuss the farming practices needed to meet long-run conservation and environmental considerations. Share leases should attempt to divide the costs and benefits of such practices whenever possible.

Systems Analysis

Because most farms and ranches carry out more than one type of crop or livestock activity, it is important to understand how they affect each other. This is called *systems analysis*. For example, crops grown in rotation or in combination with other crops have different fertility and pest control needs than the same crops grown alone or continuously. Tillage practices and soil and water runoff will also differ. Livestock enterprises can complement crop enterprises by efficiently utilizing feedstuffs with a low market value and returning fertility to the land through manure disposal. There can be interactions among production practices used on a single crop, such as the placement of fertilizer and pesticides and the type of tillage practices followed. These interactions are difficult to quantify and difficult to incorporate into standard enterprise budgets. However, they should be considered in whole farm planning and budgeting.

Off-Farm Effects

Many of the decisions farmers make with regard to land use and production practices have consequences that go far beyond the boundaries of the farm. Buildup of silt in rivers and lakes, pollution and contamination of groundwater, destruction of wildlife habitat, and the presence of chemical residues in livestock products are just a few examples. These effects are very difficult to evaluate in monetary terms, and often difficult to relate back to specific agricultural practices and locations. Nevertheless, the research into causes and effects is continuing. Agriculture must consider more than just farm input costs when making decisions about input use. The total societal costs of various technologies is quickly becoming an important factor in choosing production practices.

Regulations and Incentives

Both the federal and state government have enacted laws to promote and sometimes require land-use practices that preserve and enhance soil, water, and air resources. As more and more of these regulations are enacted, future conservation efforts may increasingly become a matter of selecting the least-cost combination of practices to meet the relevant requirements.

Long-term land retirement programs such as the Conservation Reserve Program offer guaranteed annual payments in return for taking erodible land out of production. The opportunity cost of the lost production must be evaluated against the incentive payments and the long-run conservation benefits. Other regulations such as "swampbuster" and "sodbuster" rules restrict farmers from cropping certain areas with a history of pasture or wetland use. Farmers are also now required to develop and follow an approved conservation plan in order to participate in many government farm programs.

Restrictions on the use of pesticides and some chemical fertilizers vary among states. Regular soil testing and careful scouting for pest problems will ensure that such products are used only at environmentally safe and economically profitable levels. The principles of marginal cost and marginal revenue can be used to determine the economic threshold at which the potential losses from a pest problem are greater than the cost of treating it.

Some farms have been found to have environmental hazards such as leaking underground fuel storage tanks or accumulations of discarded pesticide containers. The costs of cleaning up such problems can significantly reduce the value of a farm. Before closing a sale of a farm or ranch property, a prudent buyer will arrange for an *environmental audit* to identify any potential problems of this type and the estimated costs of correcting them.

A high opportunity cost of capital, short planning horizons, and limited direct evidence of environmental effects combine to explain why many landowners are reluctant to adopt sound environmental practices. However, society has an interest in conservation in order to maintain and expand our nation's long-run food production potential as well as to ensure food safety. Societal planning horizons are typically longer and broader than those for an individual farmer, with society wanting more conservation practices than farmers are willing or able to adopt. In recognition of this and the limited capital position of many farmers, society has devised means to encourage more conservation practices. Technical assistance is available at no cost through local offices of the Soil Conservation Service. Financial assistance to pay for part of the costs of certain conservation practices has been made available through the Agricultural Conservation Program and from some states.

Today's farmers and ranchers must think beyond simply complying with regulations and maximizing current profits. The ethic of conservation tells us that owning or using land for agricultural purposes carries with it the responsibility to follow practices that will support society well into the future.

SUMMARY

Land is an essential resource for agricultural production. It is a permanent resource that is fixed in supply and location. The decision to buy or lease land is an important one which will affect the production capacity and financial condition of the business for many years. Land ownership has many advantages. However, purchasing land requires a strong capital position and adequate cash flow potential if credit is used. Land can be valued based on its net earnings or by comparing sale prices of similar land.

A mixture of owned and leased land exists in many farm and ranch businesses. Cash rent, crop share, and livestock share are the most common types of lease arrangements. Each lease type has advantages and disadvantages to both the tenant and landlord in terms of the capital contribution required, the amount of price and yield risk to be shared, and the making of management decisions. Share leases should provide for the sharing of income in the same proportion as each party contributes to the total cost of production. Variable inputs should also be shared in this proportion in order to allocate resources efficiently.

Land-use decisions that promote conservation and environmental concerns need to consider long-run effects, interactions among enterprises, and consequences that occur beyond the borders of the farm.

QUESTIONS FOR REVIEW AND FURTHER THOUGHT

1 List as many reasons as you can to explain why people buy farmland. How would your list, or the importance of each reason, be different for an operating farmer than for a non-farm investor?

2 What are the advantages and disadvantages of owning land compared to renting land?

3 Using the capitalization method and a 6 percent discount rate, how much could you afford to pay for an acre of land that would have a net return of $81 per year? What would the answer be with a discount rate of 4 percent?

4 Why is a cash flow analysis important in a land purchase decision?

5 List three advantages and disadvantages of each type of lease for both the tenant and the landlord.

6 What are the typical terms for crop share leases in your community? Analyze them using Table 18-5 as a guide. Are the terms fair? What problems did you encounter in the analysis?

7 Develop a flexible cash lease for your area using both price and yield as the variable factors upon which the rent is based.

8 What additional considerations must a farmer take into account when choosing practices that promote conservation? Give two examples.

REFERENCES

Davis, K. C., and Cecil D. Maynard: *The Oklahoma Farm Lease Agreement,* Oklahoma State University Extension Facts No. 121.

Erickson, Duane E., and John T. Scott, Jr.: *Farm Real Estate,* North Central Regional Extension Publication No. 51, University of Illinois, Champaign-Urbana, IL, 1990.

Henderson, Philip A.: *Fixed and Flexible Cash Rental Arrangements for Your Farm,* North Central Regional Extension Publication 75, Kansas State University, Manhattan, KS, 1979.

Kadlec, John E.: *Farm Management: Decisions, Operation, Control,* Prentice-Hall, Inc., Englewood Cliffs, NJ, 1985, chaps. 14, 15.

Murray, William, Duane G. Harris, Gerald A. Miller, and Neill S. Thompson: *Farm Appraisal and Valuation,* 6th ed., Iowa State University Press, Ames, IA, 1983.

Pretzer, Don D.: *Crop Share or Crop Share-Cash Rental Arrangements for Your Farm,* North Central Regional Extension Publication No. 105, Kansas State University, Manhattan, KS, 1989.

Reiss, Franklin J.: *Farm Leases for Illinois,* Illinois Cooperative Extension Service Circular 1199, 1982.

Stoneberg, E. G.: *Improving Your Farm Lease Contract,* FM 1564 (Rev.), Iowa State University Extension Service, Ames, IA, 1984.

Wise, Murray R.: *Investing in Farmland,* Probos Publishing Company, Chicago, 1989.

Crop-Share or Crop-Share-Cash Farm Lease

North Central Regional
Publication No. 77

This form can provide the landlord and tenant with a guide for developing an agreement to fit their individual situation. This form is not intended to take the place of legal advice pertaining to contractual relationships between the two parties. Because of the possibility that a farm operating agreement may be legally considered a partnership under certain conditions, seeking proper legal advice is recommended when developing such an agreement.

This lease is entered into this _____ day of _____, 19_____, between

_____, landlord, of _____

_____
(address)

_____, spouse, of _____

_____
(address)

hereafter known as "the landlord," and

_____, tenant, of _____

_____
(address)

_____, spouse, of _____

_____
(address)

hereafter known as "the tenant."

I. PROPERTY DESCRIPTION

The landlord hereby leases to the tenant, to occupy and use for agricultural and related purposes, the following described property:

_____

_____

_____

consisting of approximately _____ acres situated in _____ County (Counties), _____ (State) with all improvements thereon except as follows:

_____

_____

_____

II. GENERAL TERMS OF LEASE

A. Time period covered. The provisions of this agreement shall be in effect for _____ year(s), commencing on the _____ day of _____, 19_____. This lease shall continue in effect from year to year thereafter unless written notice of termination is given by either party to the other at least _____ days prior to expiration of this lease or the end of any year of continuation.

B. Review of lease. A written request is required for a general review of the lease or for consideration of proposed changes by either party, at least _____ days prior to the final date for giving notice to terminate the lease as specified in IIA.

C. Amendments and alterations. Amendments and alterations to this lease shall be in writing and shall be signed by both the landlord and tenant.

D. No partnership intended. It is particularly understood and agreed that this lease shall not be deemed to be nor intended to give rise to a partnership relation.

E. Transfer of property. If the landlord should sell or otherwise transfer title to the farm, he will do so subject to the provisions of this lease.

F. Right of entry. The landlord reserves the right for himself, his agents, his employees, or his assigns to enter the farm at any reasonable time to: a) consult with the tenant; b) make repairs, improvements, and inspections; and c) (after notice of termination of the lease is given) do plowing, seeding, fertilizing, and any other customary seasonal work, none of which is to interfere with the tenant in carrying out regular farm operations.

G. No right to sublease. The landlord does not convey to the tenant the right to lease or sublet any part of the farm or to assign the lease to any person or persons whomsoever.

H. Binding on heirs. The provisions of this lease shall be binding upon the heirs, executors, administrators, and successors of both landlord and tenant in like manner as upon the original parties, except as provided by mutual written agreement.

I. Landlord's lien for rent and performance. The landlord's lien provided by law on crops grown or growing shall be the security for the rent herein specified and for the faithful performance of the terms of the lease. If the tenant fails to pay the rent due or fails to keep the agreements of this lease, all costs and attorney fees of the landlord in enforcing collection or performance shall be added to and become a part of the obligations payable by the tenant hereunder.

J. Additional provisions.

_____

_____

III. LAND USE

A. General provisions. The land described in Section I will be used in approximately the following manner. If it is impractical in any year to follow such a land-use plan, appropriate adjustments will be made by mutual written agreement between the parties.

1. Cropland

   a) Row crops   _____ acres

   b) Small grains   _____ acres

   c) Legumes   _____ acres

   d) Rotation pasture   _____ acres

2. Permanent pasture: _____ acres

3. Other: _____ _____ acres

     _____ _____ acres

4. Total   _____ acres

B. Restrictions. The maximum acres harvested as silage shall be _____ acres unless it is mutually decided otherwise.

The pasture stocking rate shall not exceed:

PASTURE IDENTIF.	
_____	_____ acres/animal unit
_____	_____ acres/animal unit
_____	_____ acres/animal unit

(1000 pound mature cow is equivalent to one animal unit.)

Other restrictions are:

_____

_____

_____

_____

_____

C. Government programs. The extent of participation in government programs will be discussed and decided on an annual basis. The course of action agreed upon shall be placed in writing and be signed by both parties. A copy of the course of action so agreed upon shall be made available to each party.

IV. CROP-SHARE-CASH RENT AND RELATED PROVISIONS

A. General agreement. The tenant agrees to pay as rent for the use of the land the share of crops shown in Table 1 of this section. The tenant also agrees to furnish all labor, machinery, and cash operating expenses except for landlord's share (percent and/or dollar charge per unit) indicated in Table 1.

Table 1—Landlord's Share (% and/or $) of Crops and Crop Expenses

	Corn example	Corn	Grain sorghum	Small grain	Soybeans	Hay, _____		
SHARE OF CROPS	50%							
SHARE OF CROP EXPENSES:								
Fertilizer:								
Materials	50%							
Application	50%							
Herbicide:								
Materials	50%							
Application								
Insecticide:								
Materials	50%							
Application								
Seed	50%							
Lime, rock phosphate*	100%							
Harvesting (per ac.)	$7.50							
Drying	50%							
Baling								
Delivery to:								
Storage/bu.								
Market/bu.	$ .07							

* Lime, rock phosphate, and other fertilizers having more than one year life paid by the tenant should be recorded in the compensation table in Section V-C-2.

B. Other crop-share-cash agreements.

   1. Operating expenses. Additional agreements relative to the sharing of expenses are as follows:

     _____

     _____

     _____

   2. Storage, landlord's crop. At the landlord's request, the tenant agrees to store as much of the landlord's share of the crops as possible, using storage space reserved by the landlord and not to exceed _____ percent of the storage space not specifically reserved.

   3. Delivery of grain. The tenant agrees to deliver the landlord's share of crops at a place and at a time the landlord shall designate, not over _____ miles distant at the charge shown in Table 1 of this section. Additional agreements are:

     _____

     _____

     _____

   4. Cash rent on non-shared items. The tenant agrees to pay cash rent annually for the use of the following non-shared items.

Table 2—Amount of Annual Cash Rent
(Complete at beginning of lease)

	Total
Pasture	$_____
Hayland: _____	$_____
_____	$_____
Farmstead: Dwelling	$_____
Service bldgs.	$_____
Timber and waste	$_____
Total cash rent	$_____

Payment of cash rent: The tenant agrees to pay cash rent as follows:

$_____ on or before _____ day of _____ (month)

$_____ on or before _____ day of _____ (month)

$_____ on or before _____ day of _____ (month)

$_____ on or before _____ day of _____ (month)

If rent is not paid when due, the tenant agrees to pay interest on the amount of unpaid rent at the rate of _____ percent per annum from due date until paid.

5. Pasturing. The tenant will prevent damage to cropland and growing crops by livestock.

6. Home use. The tenant and landlord may take for home use the following kinds and quantities of jointly owned crops:

_____

_____

_____

_____

7. Buying and selling. The landlord and tenant will buy and sell jointly owned property according to the following agreement:

_____

_____

_____

8. Division of property. At the termination of this lease, all jointly owned property will be divided or disposed of as follows:

_____

_____

_____

V. OPERATION AND MAINTENANCE OF FARM

In order to operate this farm efficiently and to maintain it in a high state of productivity, the parties agree as follows:

A. The tenant agrees:

1. General maintenance. To provide the unskilled labor necessary to maintain the farm and its improvements during his tenancy in as good condition as it was at the beginning. Normal wear and depreciation and damage from causes beyond the tenant's control are excepted.

2. Land use. Not to: a) plow pasture or meadowland, b) cut live trees for sale or personal use, or c) pasture new seedings of legumes and grasses in the year they are seeded without consent of the landlord.

3. Insurance. Not to house automobiles, motor trucks, or tractors in barns, or otherwise violate restrictions in the landlord's insurance policies without written consent from the landlord. Restrictions to be observed are as follows:

_____

_____

_____

_____

4. Noxious weeds. To use diligence to prevent noxious weeds from going to seed on the farm. Treatment of the noxious weed infestation and cost thereof shall be handled as follows:

_____

_____

_____

5. Addition of improvements. Not to: a) erect or permit to be erected on the farm any nonremovable structure or building, b) incur any expense to the landlord for such purposes, or c) add electrical wiring, plumbing, or heating to any building without written consent of the landlord.

6. Conservation. Control soil erosion as completely as practicable; keep in good repair all terraces, open ditches, inlets and outlets of tile drains; preserve all established watercourses or ditches including grassed waterways; and refrain from any operation or practice that will injure such structures.

7. Damages. When he leaves the farm, to pay the landlord reasonable compensation for any damages to the farm for which he, the tenant, is responsible. Any decrease in value due to ordinary wear and depreciation or damages outside the control of the tenant are excepted.

8. Costs of operation. To pay all costs of operation except those specifically referred to in Sections IV, V-A-4, and V-B.

9. Repairs. Not to buy materials for maintenance and repairs in an amount in excess of $_____ within a single year without written consent of the landlord.

B. The landlord agrees:

1. Loss replacement. To replace or repair as promptly as possible the dwelling or any other building regularly used by the tenant that may be destroyed or damaged by fire, flood, or other cause beyond the control of the tenant or to make rental adjustments in lieu of replacements.

2. Materials for repairs. To furnish all material needed for normal maintenance and repairs.

3. Skilled labor. To furnish any skilled labor tasks which the tenant himself is unable to perform satisfactorily. Additional agreements regarding materials and labor are:

_____

_____

4. Reimbursement. To pay for materials purchased by the tenant for purposes of repair and maintenance in an amount not to exceed $_____ in any one year, except as otherwise agreed upon. Reimbursement shall be made within _____ days after the tenant submits the bill.

5. Removable improvements. Let the tenant make minor improvements of a temporary or removable nature, which do not mar the condition or appearance of the farm, at the tenant's expense. He further agrees to let the tenant remove such improvements even though they are legally fixtures at any time this lease is in effect or within _____ days thereafter, provided the tenant leaves in good condition that part of the farm from which such improvements are removed. The tenant shall have no right to compensation for improvements that are not removed except as mutually agreed.

6. Compensation for crop expenses. To reimburse the tenant at the termination of this lease for field work done and for other crop costs incurred for crops to be harvested during the following year. Unless otherwise agreed, current custom rates for the operations involved will be used as a basis of settlement.

C. Both agree:

1. Not to obligate other party. Neither party hereto shall pledge the credit of the other party hereto for any purpose whatsoever without the consent of the other party. Neither party shall be responsible for debts or liabilities incurred, or for damages caused by the other party.

2. Capital improvements. Costs of establishing hay or pasture seedings, new conservation structures, improvements (except as provided in Section V-B-5), or of applying lime and other longlived fertilizers shall be divided between landlord and tenant as set forth in the following table. The tenant will be reimbursed by the landlord either when the improvement is completed, or the tenant will be compen-

sated for his share of the depreciated cost of his contribution when he leaves the farm based on the value of the tenant's contribution and depreciation rate shown in the following table. (Cross out the portion of the preceding sentence which does not apply.)

Rates for labor, power, and machinery contributed by the tenant shall be agreed upon before construction is started.

3. Mineral rights. Nothing in this lease shall confer upon the tenant any right to minerals underlying said land, but same are hereby reserved by the landlord together with the full right to enter upon the premises and to bore, search, and excavate for same, to work and remove same, and to deposit excavated rubbish, and with full liberty to pass over said premises with vehicles and lay down and work any railroad track or tracks, tanks, pipelines, power lines, and structures as may be necessary or convenient for the above purpose. The landlord agrees to reimburse the tenant for any actual damage he may suffer for crops destroyed by these activities and to release the tenant from obligation to continue farming this property when development of mineral resources interferes materially with the tenant's opportunity to make a satisfactory return.

VI. ARBITRATION OF DIFFERENCES

Any differences between the parties as to their several rights or obligations under this lease that are not settled by mutual agreement after thorough discussion, shall be submitted for arbitration to a committee of three disinterested persons, one selected by each party hereto and the third by the two thus selected. The committee's decision shall be accepted by both parties.

Compensation for Improvements Table

| Type of improvement | Date to be completed | Estimated total cost (dollars) | Proportion to be contributed by tenant Unskilled | | | Total value of tenant's contrib. (dollars) * | Rate of annual depreciation |
			Material	labor	Mach.		
			%	%	%		%
		$				$	
		$				$	
		$				$	
		$				$	
		$				$	

* To be recorded when improvement is completed.

Executed in duplicate on the date first above written:

_____
(tenant)

_____
(tenant spouse)

_____
(landlord)

_____
(landlord spouse)

STATE OF _____
COUNTY OF _____ } SS:

On this _____ day of _____ A.D., 19_____, before me, the undersigned, a Notary Public in said State, personally appeared _____

_____, _____, _____,

and _____, to me known to be the identical persons named in and who executed the foregoing instrument, and acknowledged that they executed the same as their voluntary act and deed.

_____
Notary Public

Programs and activities of the Cooperative Extension Service are available to all potential clientele without regard to race, color, sex, national origin, or handicap.

In cooperation with NCR Educational Materials Project

Issued in furtherance of Cooperative Extension work, Acts of Congress of May 8 and June 30, 1914, in cooperation with the U.S. Department of Agriculture and Cooperative Extension Services of Illinois, Indiana, Iowa, Michigan, Minnesota, Missouri, Nebraska, North Dakota, Ohio, South Dakota, and Wisconsin. Fred D. Sobering, Director of Cooperative Extension Service, Kansas State University, Manhattan, Kansas 66506.

KSU-EXT-5-85-11.25M

LABOR MANAGEMENT

CHAPTER OBJECTIVES

    1 To outline the trends in agricultural labor use and the factors which affect its productivity

    2 To discuss planning for the quantity and quality of labor needed for different situations

    3 To illustrate methods for measuring and improving labor efficiency

    4 To suggest ways to improve the management of agricultural employees, including selection, compensation, and retention

    5 To summarize laws regulating agricultural workers and employers

Labor is one of the few agricultural inputs whose use has diminished substantially over time. There has been a dramatic decline in farm labor use since 1950, as shown in Table 19-1. The introduction of mechanization and new technology in many forms has allowed agricultural production to increase in spite of the decrease in labor, however. Energy in the form of electrical and mechanical devices has replaced much of the physical energy provided by humans in the past. More of the labor input on today's farm is spent operating, supervising, and monitoring these mechanical activities and less is expended on physical effort. This change in the tasks performed by agricultural labor has required increased education, skills, and training.

    The mere availability of new technology such as larger machinery, mechanical feed and manure handling systems, and computers does not explain its rapid and widespread adoption. There has to be an economic justification before farmers will use any

TABLE 19-1   LABOR USE ON UNITED STATES FARMS (1950–1990)

Year	Persons employed (million)	Hours worked (million)	Persons fed per farm worker
1950	9.9	19,300	15.5
1960	7.1	12,500	25.8
1970	4.5	7,800	47.7
1980	3.7	6,600	75.7
1990	2.9	5,550	96.0

Source: Economic Indicators of the Farm Sector, Production and Efficiency Statistics, 1990, Economic Research Service, U.S. Department of Agriculture, 1992. Based on data from the U.S. Census of Agriculture.

new technology or it will simply "sit on the shelf." Most labor-saving technology has been adopted for one or more of the following reasons:

1 It is less expensive than the labor it replaced.

2 It allows farmers to increase their volume of production and total profit.

3 It makes the farmers' work easier and more pleasant.

4 It allows certain operations such as planting and harvesting to be completed on time even when weather is unfavorable or labor is in short supply.

Input substitution occurs because of a change in the marginal physical rates of substitution and/or a change in the relative prices of inputs, as was discussed in Chapter 6. Both factors have been important in the substitution of capital-using technology for labor in agriculture. Marginal rates of substitution have changed as new technology altered the shape of the relevant isoquants, making it profitable to use less labor and more capital.

Both interest rates and wage rates have increased since 1950, but wage rates have increased by a higher percentage, making labor relatively more expensive than capital. This change in wage rates affects the price ratio in the substitution problem, also making it profitable to use more capital and less labor. The increased amount of capital per worker in agriculture has caused a substantial increase in the productivity of agricultural labor (see Table 19-1), making it feasible to pay higher wages.

CHARACTERISTICS OF AGRICULTURAL LABOR

A discussion of agricultural labor must recognize the unique characteristics which affect its use and management on farms. Labor is a continuous flow input, meaning the service it provides is available hour by hour and day by day. It cannot be stored for later use; it must be used as it becomes available or it is lost.

Full-time labor is also a "lumpy" input, meaning it is available only in whole, indivisible units. Part-time and hourly labor is often used, but the majority of agricultural labor is provided by full-time, year-round employees. If labor is available only on a full-time basis, the addition or loss of an employee is a major change in the labor supply on a farm with only a few employees. For example, a farmer hiring the first employee is increasing the labor supply by 100 percent, and the second employee represents a 50 percent increase. A problem facing a growing business is when and how to acquire the additional resources needed to keep a new worker fully employed. When other resources such as land and machinery also come in "lumpy" units, it becomes very difficult to avoid a shortage or excess of one or more resources.

The operator and other family members provide all or a large part of the labor used on most farms and ranches. This labor does not generally receive a direct cash wage, so its cost and value can be easily overlooked or ignored. However, as with all resources, there is an opportunity cost on operator and family labor which can be a large part of the farm's total fixed cost. Compensation for operator and family labor is received indirectly through expenditures for family living expenses and other withdrawals. This indirect wage or salary may vary widely, particularly for nonessential items, as net farm income varies from year to year. Cash farm expenses have a high priority claim on any cash income causing nonessential living expenses to fluctuate with farm income.

The human factor is another characteristic distinguishing labor from other resources. If an individual is treated as an inanimate object, productivity and efficiency suffer. The hopes, fears, ambitions, likes, dislikes, worries, and personal problems of both the operator and employees must be considered in any labor management plan.

PLANNING FARM LABOR RESOURCES

Planning the farm's labor resources carefully will help avoid costly and painful mistakes. Figure 19-1 illustrates the process. The first step is to assess the farm's labor needs, both quantity and quality, and the conditions under which work will be performed.

Quantity of Labor Needed

Most farm managers judge the quantity of labor needed by observation and experience. When new enterprises are being introduced, typical labor requirements from published enterprise budgets can be used. A worksheet such as the one illustrated in Table 19-2 is useful for summarizing labor needs.

The seasonality of labor needs must also be considered. For example, labor requirements may exceed available labor during the months when planting, harvesting, calving, and farrowing take place. The chart in Figure 19-2 shows an example of the total monthly labor requirements for all farm enterprises and the monthly labor provided by the farm operator and a full-time employee. The farm operator in this example has a problem not uncommon on many farms and ranches. Operator labor will not meet the requirements in several months, but the addition of a full-time employee results in large amounts of excess labor during several months. Longer workdays, temporary help, or hiring a custom operator may be necessary to perform the required tasks on

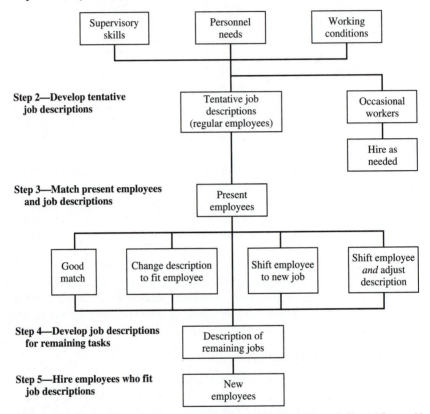

Step 1—Assess your situation

Step 2—Develop tentative job descriptions

Step 3—Match present employees and job descriptions

Step 4—Develop job descriptions for remaining tasks

Step 5—Hire employees who fit job descriptions

FIGURE 19-1   Flowchart of the farm labor planning process (*Source:* Thomas, Kenneth H. and Bernard L. Erven, *Farm Personnel Management,* North Central Regional Extension Publication 329).

time. A more permanent solution may be to increase the capacity of field machinery or processing equipment, or to shift to different enterprises.

How much labor should be utilized to maximize profits depends on its availability, its cost, and whether it is a fixed or variable input.

Labor Fixed but Plentiful   The total labor supply of the operator and/or full-time employees may be fixed but not fully utilized. If this labor is being paid a fixed sum regardless of the hours worked, there is no additional or marginal cost for utilizing another hour. In this situation, labor use should be increased in each activity or enterprise until its marginal value product is zero, to equate it with the marginal input cost. However, even permanent labor may have an opportunity cost greater than zero, from either leisure or off-farm employment. This becomes the minimum acceptable earning rate or marginal value product, and whenever marginal earnings are below this value the individual will choose not to work additional hours.

TABLE 19-2   LABOR ESTIMATE WORKSHEET

		Total hours for year	Distribution of hours			
			Dec. thru March	April May June	July Aug.	Sept. Oct. Nov.
	(1)	(2)	(3)	(4)	(5)	(6)
1	Suggested hours for full-time worker	2,400	600	675	450	675
2	My estimate for full-time worker					
3	Labor hours available					
4	Operator (or Partner No. 1)					
5	Partner No. 2					
6	Family labor					
7	Hired labor					
8	Custom machine operators					
9	Total labor hours available					
10	Direct labor hours needed by crop and animal enterprises					
11	Crop enterprises Acres Hr/Acre					
12						
13						
14						
15						
16						
17						
18						
19						
20	Total labor hours needed for crops					
21	Animal enterprises No. Units Hr/Unit					
22						
23						
24						
25						
26	Total labor hours needed for animals					
27	Total hours needed for crops and animals					
28	Total hours of indirect labor needed					
29	Total labor hours needed (lines 27 & 28)					
30	Total available (line 9)					
31	Additional labor hours required (line 29 minus line 30)					
32	Excess labor hours available (line 30 minus line 29)					

Source: Missouri Farm Planning Handbook, Manual 75; Feb. 1986. Department of Agricultural Economics, University of Missouri, Columbia, MO.

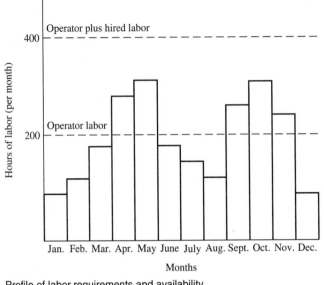

FIGURE 19-2 Profile of labor requirements and availability.

Labor Variable but Plentiful   When the manager can hire an unlimited amount of labor on a hourly or monthly basis, it should be hired until its marginal value product equals its marginal input cost, which is the wage rate plus the cost of any extras or fringe benefits.

Limited Labor   If the total labor available is less than the amount necessary to equate the marginal value product and marginal input cost in all uses, the equal marginal principle should be used to allocate labor among alternative and competing uses. Profit from the limited labor supply will be maximized when labor is allocated among alternative uses in such a way that the marginal value products of the last hour used in each alternative are all equal.

Quality of Labor Needed

Not all farm labor is equal. New agricultural technology requires more specialized and sophisticated skills. Some activities such as application of certain pesticides may even require special training and certification. An assessment of labor needs requires identification of special skills such as operating certain types of machines, performing livestock health chores, balancing feed rations, using computers or electronic control devices, or performing mechanical repairs and maintenance. If the operator or family members do not possess all the special skills needed, then employees must be hired who do, or certain jobs may be contracted to outside consultants, repair shops, or custom operators. Training programs may also be available to help farm workers acquire new skills.

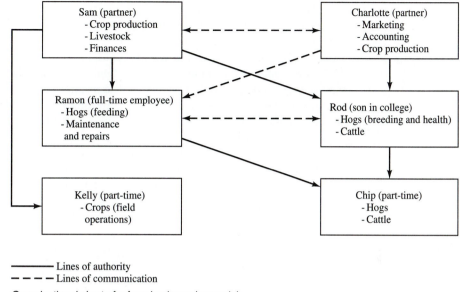

─────── Lines of authority
─ ─ ─ ─ Lines of communication

FIGURE 19-3   Organizational chart of a farm business (example).

Management Style

Some farm and ranch managers utilize hired labor more effectively than others. Many operators are accustomed to working alone, and prefer employees who can work independently with a minimum amount of supervision and instruction. Other employers prefer to work closely with workers, and give specific instructions about how a job is to be performed. Both management styles can be effective, but a good manager recognizes his or her own style and seeks out workers who can function effectively under it.

Once the quantity and quality of labor needed by the farm or ranch operation has been analyzed, tentative job descriptions should be developed. Larger operations will allow for much more specialization than smaller ones, of course. Then compare the skills of the workers currently available to the job descriptions. Some duties may have to be rearranged to match the skills and interests of certain employees, family members or partners. Needs that cannot be met by the current work force will have to be filled by providing job training, securing additional workers, or contracting with outside services.

Developing an *organizational chart* is especially useful where several employees and employers are involved. In particular, it should be made clear if some employees are expected to take directions from other employees or members of the operator's family. Figure 19-3 shows an example of an organizational chart for a medium-size operation.

MEASURING LABOR EFFICIENCY

Labor efficiency depends not only on the skills and training of the labor used but also on the size of the business, enterprises, degree of mechanization, type of organization, and many other factors. Measures of labor efficiency should be used to compare and evaluate results only on farm businesses of approximately the same size and type.

Labor efficiency measures often use the concept of person-years of labor employed. This is a procedure for combining operator, family, and hired labor into a total labor figure which is comparable across farms. The example below shows 21 months of labor provided from three sources. Dividing this total by 12 converts it into 1.75 person-years, or the equivalent of 1.75 persons working full time during the year.

Operator labor	12 months
Family labor	4 months
Hired labor	5 months
Total	21 months

21 ÷ 12 = 1.75 person-years equivalent

Labor efficiency measures convert some physical output, cost, or income total into a value per person-year. The following measures are commonly used.

Value of Farm Production per Person   This measures the total value of agricultural products produced on the farm per person-year equivalent. It is affected by business size, type of enterprise, and the amount of machinery and other labor-saving equipment used.

Labor Cost per Tillable Acre   The cost of labor per tillable acre is found by dividing the total labor cost for the year by the number of tillable acres. The opportunity cost of operator and family labor is included in total labor cost. Values will be affected by machinery size, type of crops grown, and whether or not livestock are part of the farm operation.

Tillable Acres per Person   The number of tillable acres per person is found by dividing by the number of person-years of labor used.

Table 19-3 contains some labor efficiency values for farms in Iowa. The data show that productivity per person usually increases as farm size increases, mostly because of more investment in larger machinery and equipment.

IMPROVING LABOR EFFICIENCY

Labor efficiency can be improved by increasing the capital investment per worker, through the use of larger machinery and other forms of mechanization. However, the objective should be to maximize profit and not just to increase labor efficiency at any cost. Marginal rates of substitution and the prices of labor and capital determine the proper combination. Increasing the capital investment per worker will increase profit only if (1) total cost is reduced while output remains constant or (2) the labor which is saved can be used to increase output value elsewhere by more than the cost of the investment.

TABLE 19-3 LABOR EFFICIENCY BY FARM SIZE FOR IOWA FARMS, 1991

Item	Farm size (sales)			
	$10,000 to $39,999	$40,000 to $99,999	$100,000 to $249,999	$250,000 or more
Person-years of labor per farm	0.75	0.92	1.25	2.08
Value of farm production per person ($)	47,894	71,957	110,981	134,048
Labor cost per tillable acre ($)	47.69	42.27	35.44	37.15
Tillable acres per person	260	320	389	391

Source: 1991 Iowa Farm Costs and Returns, Iowa State University Extension Publication FM-1789, August 1992.

When the labor supply is increased by adding a full-time worker, some additional capital investment may be needed to fully utilize the available labor. Figure 19-4 shows how costs per cow may go up when a dairy farm adds employees if the number of cows is not increased at the same time. This illustrates the "lumpy" nature of full-time labor as an input.

Simplifying working procedures and routines can have a high payoff in increased labor efficiency. Considerable time can be saved by having all necessary tools and other supplies at the work area, not having to stop to open and close gates, having equipment well maintained, and having spare parts on hand. Making changes in the farmstead layout, building designs, field size and shape, and the storage sites of materials relative to where they will be used can also save time and increase labor efficien-

FIGURE 19-4 Labor cost per dairy cow for different herd sizes.

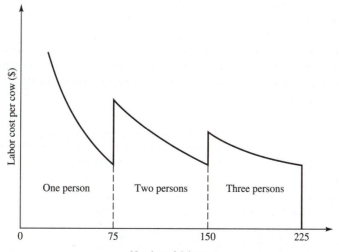

cy. When materials must be moved, consider using conveyers, carts, small vehicles, augers, and other labor-saving devices. Automatic switches and timers can eliminate time spent waiting. As always, the additional cost of any changes must be weighed against the value of the labor saved.

Labor efficiency can also be improved by making sure workers have safe and comfortable working conditions whenever possible. Although most agricultural work is still performed outside, modern machinery cabs and vehicles help reduce fatigue from environmental conditions. Making sure that machinery guards are in place, livestock buildings are well ventilated, and all safety measures are followed will prevent lost work time from injuries or illness. Workers should be provided with suitable clothing and other safety equipment when working with agricultural chemicals or performing other hazardous jobs.

The simple procedure of planning and scheduling work in advance will help reduce wasted time. Tasks which must be done at a specific time should be scheduled first, and those that are not time-specific, such as building repairs and painting, can be planned for months of slack labor. Keep a list of jobs to be done, and assign a priority and deadline to each one. Keep it where all workers can see it and add to it, and cross off tasks that have been completed. From this a daily work schedule can be planned in a few minutes each morning or evening. Time spent organizing the next day's tasks and their order of importance is time well spent.

OBTAINING AND MANAGING HIRED LABOR

Hired labor accounts for over one-fourth of the total labor supply on farms and ranches, as shown in Table 19-4. Acquisition, training, and retention of hired workers are common subjects of conversation whenever managers of larger farms get together. Farm managers are finding that skills in human relations and personnel management are valuable assets.

Recruiting

The process of hiring labor starts with recruiting, including the announcement of the job opening and the receiving of applications. Placing newspaper ads, talking with other farmers, relatives, agribusiness people, or farm consultants in the community, or

TABLE 19-4   THE AGRICULTURAL WORK FORCE IN THE UNITED STATES

Type of worker	Number	Percent
Farm operator	2,674,000	35
Unpaid farm worker	2,853,000	37
Paid farm worker	2,160,000	28
	7,687,000	100

Source: Characteristics of Agricultural Work Force Households, 1987, Agricultural Information Bulletin 612, United States Department of Agriculture.

utilizing college placement offices and employment agencies are ways to inform individuals about a job opening and identify potential candidates to fill it.

The job announcement should clearly state the skills and experience desired. In addition it should make the position sound desirable. Emphasize reasons why the applicant would want to work for this operation instead of the one down the road. In some areas the employer may have to negotiate with a labor contractor to supply a large number of temporary workers for harvesting or other labor-intensive activities.

It is generally helpful to provide an application form to each applicant. Basic information should be obtained about the applicant's background, work experience, training, personal goals, references, and other factors. Remember that questions about race, religion, age, or marital status are illegal to ask in most states.

Interviewing and Selecting

The completed application forms can be used to select a small number of candidates for the next step, interviewing. The interview should be carefully planned to efficiently acquire more information. Sufficient time and opportunity should be allowed for the applicant to ask questions about the job, its duties, and responsibilities. Interviewing involves not only obtaining information about applicants but also providing information to them so they can assess their own interest in and qualifications for the job. A skills test may be needed for certain technical jobs. A tour of the work location should be provided, and the candidate should have a chance to visit with other employees.

The information obtained about each job applicant through the application form, the interview, and references must now be evaluated. Many factors must be considered in selecting a candidate, including personal compatibility. Farm employers often work more closely with their employees on a day-to-day basis than other employers do, sometimes under stressful conditions. This close working relationship increases the chance for friction if the individuals are not compatible.

The Employment Agreement

Once a job offer is made and accepted, a written employment agreement should be developed. An example of one is shown in Table 19-5. The purpose of the agreement is to clarify the work expectations of both the employer and employee, and to serve as a reference for evaluation of performance later on.

The employment agreement should start with the job description, including duties and responsibilities, lines of authority, and the job title. Other important information includes wages and benefits, working hours and days, vacation, sick leave and personal leave, safety rules, allowable uses of farm property, training opportunities, bonus or incentive plans, and procedures for evaluation and promotion or termination. A review of the employment agreement should be made once or twice a year as part of the evaluation process.

TABLE 19-5 EMPLOYMENT AGREEMENT EXAMPLE

Farm Employer–Employee Agreement of Employment

I, _____, agree to employ _____ to work on
my farm located: _____ beginning (date) _____ and continuing until such
time as either wishes to terminate this agreement by a _____ day notice. _____, the
employer, and _____, the employee, agree to comply with the following conditions
and actions:

1. To pay _____ $_____ per _____ from
   which the employee's Income Tax (Yes/No) and Social Security taxes will be withheld. Wages will be paid on
   _____ (day) of (Week/Bi-weekly/Monthly).

2. To provide a house with utilities including heat and electricity. The maintenance is to be done by _____
   and paid for by _____. Any other agreements pertaining to the employee's house will be noted on the
   back of this page.

3. The normal working hours are from _____ A.M. to _____ P.M. with one hour off for breakfast and one hour
   off for lunch. Overtime will be paid for any work done after 7:00 P.M. at the rate of 1½ times the normal wage rate.
   Overtime will be paid after _____ hours are worked in any one week, Sunday through Saturday.

4. Time off shall be every other Sunday and holidays. The holidays for purposes of this agreement are New Year's Day,
   Easter, Memorial Day, Labor Day, Thanksgiving Day, and Christmas Day. On Sundays and holidays only the chore
   work will be done. The employer, _____, shall notify the employee,
   _____, at least 45 days before the holiday of what the time-off
   arrangements will be.

5. The employee is entitled to _____ weeks vacation with pay annually which shall be taken during the nonheavy
   work season and agreed upon with the employer 30 days prior to beginning of vacation.

6. The employee is entitled to _____ days sick leave with pay annually for the time off due to actual illness.

7. The employee is entitled to _____ quarts of milk per day.

8. The employee is entitled to _____ pounds of beef and _____ pounds of pork per year.

9. The employee is entitled to a 15-minute break in midmorning and midafternoon.

10. The following insurance plans will be carried on the employee:
    _____
    _____
    _____

11. A bonus or incentive plan (is, is not) included. If included, the provisions are noted on Form G, attached.

12. Other provisions not included above are listed on the reverse side of this form.

_____   Date_____
          Employer Signature

_____   Social Security No. _____
          Employee Signature

Source: Thomas, Kenneth H. and Bernard L. Erven: *Farm Personnel Management,* North Central Regional Extension Publication 329.

Compensation

A competitive compensation package is essential in a successful labor hiring program. The actual cash wage or salary paid is the most important item. Whether to pay a fixed or variable wage is the first decision. Positions in which the duties and hours worked will be fairly constant throughout the year usually receive a fixed weekly or monthly salary. Positions with highly variable hours, such as found on a cash grain farm, are often paid by the hour, as are most part-time positions. Workers who are employed in harvesting activities are sometimes paid on a piece-rate basis.

Salaries depend on the position and the particular employee filling it. The size of the farming operation, the skill level of duties performed, and the number of years the employee has worked on the farm are factors that influence the level of compensation.

Fringe Benefits   Fringe benefits are often a large part of the total compensation for farm employees. Table 19-6 shows the employee benefits that were reported in a farm labor study. They accounted for 12 percent of the average employee's compensation.

TABLE 19-6   BENEFITS RECEIVED BY FARM EMPLOYEES

Type	Percent of employees receiving	Average cost per employee
Personal use of vehicle	19	$ 966
Farm produce	43	380
Farm commodities	15	988
Meals	25	675
Clothing	14	258
Insurance (total)	47	$2,298
Health, single	18	
Health, family	27	
Dental, optical	7	
Life	6	
Unemployment	10	
Housing (total)	45	$1,979
Mobile home	3	
Apartment	6	
House	36	
Utilities (total)		$1,326
Heat	21	
Electric	30	
Telephone	11	
Water, sewer	25	
Continuing education	25	$ 243
Recreation	4	168
Retirement plans	5	2,588
Other	14	1,349

Source: Farm Employee Management in Iowa, Iowa State University Extension Service Publication FM-1841, June 1991.

Prospective employees should be made aware of the value of benefits so they can fairly evaluate a job offer. It is the spendable income left after food, shelter, and salary deductions that should be compared against comparable figures in other employment. Fringe benefits such as housing, utilities, garden space, meat, milk, and the use of a vehicle will make a lower cash wage for farm work competitive with nonfarm employment that has a higher cash salary.

Fringe benefits are most useful when the employer can provide them for less than it would cost the employee to obtain them elsewhere. Examples include the use of existing housing, vehicles, or livestock facilities. Some benefits such as health insurance are tax deductible for the employer but not taxable to the employee.

Incentive Programs and Bonuses Bonuses are often used to supplement base wages, improve labor productivity, and increase retention. However, bonuses may do little to increase labor efficiency if they are not closely tied to performance. Employees soon come to expect the bonus and consider it part of their basic cash salary. If the size of the bonus is tied to annual profit, an employer may find it difficult to decrease the bonus in a poor year after employees have experienced several years with a larger bonus.

Most bonus plans are based on one of four factors: volume, performance, longevity, or profitability.

1 *Volume* can be measured by the number of pigs weaned, calves born, or acres harvested. The employee's wages increase when the work load increases, and a modest incentive for efficiency is provided. Care should be taken that higher costs are not incurred simply to produce more, however.

2 *Performance* can be measured by the number of pigs weaned per sow, calving percentage, milk production per cow, or crop yields per acre. The bonus is often based on how much the actual performance exceeds a certain base level. This type of bonus can be an effective work incentive as long as it is based on factors over which the employee has at least some control.

3 *Longevity* is rewarded by paying a bonus based on the number of years the employee has worked for the business. This recognizes the value of experience and worker continuity to the employer.

4 *Profitability* bonuses are usually based on a percent of the gross or net income of the business. They allow the employee to share in the risks and rewards of the business, but may depend on many factors not under the control of the employee. They also require the employer to divulge some financial information to the employee.

There are several basic principles which will increase the effectiveness of any incentive program.

1 The program should be simple and easily understood by the employee.

2 The program should be based on factors largely within the control of the employee.

3 The program should aim at rewarding work that is in the best interest of the employer.

4 The program should provide a cash return large enough to provide motivation for improved performance.

5 The incentive payment should be made promptly or as soon after the completion of work as possible.

6 An example of how the bonus will be computed should be provided in writing, using typical performance levels.

7 The incentive payment should not be considered as a substitute for a competitive base wage and good labor relations.

Training Hired Labor

Farm and ranch managers sometimes hire unskilled workers and then expect them to perform highly skilled tasks in livestock management or with expensive machinery. They may also expect employees to automatically do things exactly the way they would do them. The result is disappointment, frustration, high repair bills, poor labor productivity, and employee dissatisfaction.

Studies of employment practices on farms and ranches show little evidence of formalized training programs for new employees. Even a skilled employee will need some instruction on the practices and routines to be followed. Employees with lesser skills should receive complete instructions and proper supervision during a training period. Employers need to develop the patience, understanding, and time necessary to train and supervise new employees. Unfortunately, in production agriculture the training period may need to last as long as a year or until the new employee has had a chance to perform all the tasks in a complete cycle of crop and livestock production.

Periodic training may be necessary for even long-time employees. The adoption of new technology in the form of different machinery, new chemicals, feed additives, seed varieties and electronics, or the introduction of a new enterprise may require additional training for all employees. Extension short courses, bulletins, video tapes, field days, and other short-term educational programs available in the community can be used for employee training. Participation in these activities will not only improve employees' skills but also their self-esteem.

Motivation and Communication

Hiring and training new employees is costly in terms of both time and money. If labor turnover is high, these costs can become excessive, and labor efficiency will be low. Employers should be aware of the reasons for poor employee retention and take actions to improve it.

Farm employees often report they are attracted to farm work by previous farm work experience, the chance to work outdoors, and their interest in working with crops and livestock. Disadvantages cited are long hours, little time off, early morning or late evening work, uncomfortable working conditions, and poor personal relationships with their employer. Low pay is seldom at the top of the list of disadvantages, indicating that they have personal goals other than obtaining the highest pay.

A set policy for adequate vacations and time off is important to employees, as is the opportunity to work with good buildings and equipment. Job titles can also be important. The title "hired hand" contributes little to an employee's personal satisfaction and self-image. Herd superintendent, crop manager, group leader, and machine operator are examples of titles most people associate with a higher status than that of a hired hand.

Good human relations may be the most important factor in labor management. This includes such things as a friendly attitude, loyalty, trust, mutual respect, the ability to delegate authority, and the willingness to listen to employee suggestions and complaints. Work instructions should be given in sufficient detail that both parties know what is to be done, when, and how. Everyone responds positively to public praise for a job well done, but criticism and suggestions for improvement should be communicated in private. Unpleasant jobs should be shared by all employees, and all should be treated with equal respect. Employers who reserve the larger air-conditioned tractor for their own use while employees drive smaller tractors without cabs, for example, are more likely to have labor problems.

As employees grow in their abilities and experience, they should be given added responsibilities along with the chance to make more decisions. Likewise, the employer must be willing to live with the results of those decisions or tactfully suggest changes.

Evaluation

All employers are constantly evaluating their employees' performances. Too often, however, communication takes place only when serious problems arise. Regular times for communication and coordination should be scheduled. If meetings are held only when there is a problem the employee will immediately be put on the defensive. Some managers take their employees to breakfast once every week or month. The employer should listen carefully to each employee's concerns and ideas, even if not all of them can be acted upon.

Operations with a sizable work force should use written evaluations, with performance measured against job descriptions. Employees should be warned if their performance is unsatisfactory, first orally and then in writing, and given a chance to improve. If termination is necessary, document the reasons carefully and give written notice in advance. All this takes time, but can prevent costly complaints or lawsuits later. If an employee quits, discuss the reasons and decide if changes in employee recruitment or management practices are needed.

GOVERNMENT REGULATIONS AND LABOR USE

Federal and state regulations affecting the employment of agricultural labor have become an important factor in labor management in the United States. Their thrust is to extend to farm workers much of the same protection enjoyed by nonfarm workers. For the employer the effects of these regulations may be increased costs for higher wages and more benefits, more labor records to keep, and additional investments for safety and environmental protection, but also a more satisfied and productive work force.

It is not possible to list and describe all the federal and state regulations pertaining to agricultural labor in a few pages. Only a brief discussion of some of the more important and general regulations will be included here.

Minimum Wage Law Agricultural employers who used more than 500 person-days of labor in any calendar quarter of the preceding year must pay at least the federal minimum wage to all employees except immediate family members, certain piece-rate hand harvesters, and cowboys and shepherds employed in range production of livestock. The minimum wage is subject to increases over time. This law also requires employers to keep detailed payroll records to prove compliance with the minimum wage legislation. Agricultural employers are not required to pay extra wages to employees who work more than 40 hours per week, however.

Social Security Employers must withhold social security, or FICA (Federal Insurance Contributions Act), taxes from their employee's cash wages and match the employee tax with an equal amount. This regulation applies to farmers with one or more employees if the employee was paid $150 or more in cash wages during the year *or* the employer paid $2,500 or more in total wages to all employees. Social security taxes are not withheld for children under 18 years of age employed by their father or mother.

The tax rate and the maximum earnings subject to this tax increase from year to year. Employers should contact the local Social Security Administration office or obtain Internal Revenue Service (IRS) Circular E, *Employer's Tax Guide,* for the current rates. It is important to remember that the tax rate and maximum tax figures represent the amounts withheld from the employee's pay. The employer must match them with an equal amount for remittance to the Internal Revenue Service.

Federal Income Tax Withholding Farmers are now required to withhold federal income taxes from wages paid to agricultural workers if they meet the same criteria described for social security withholding. IRS Circular E or Circular A, *Agricultural Employee's Tax Guide,* provide details. Records should be kept of each employee's name, age, social security number, cash and noncash payments received, and amounts withheld or deducted.

Workers' Compensation This is an insurance system to provide protection for workers who suffer job-related injuries or illnesses. It also frees the employer from liability for such injuries. Laws regarding workers' compensation insurance for farm employees vary from state to state, but it is required in many states. Where required, the employer pays a premium based on a percentage of the total employee payroll.

Unemployment Insurance This insurance temporarily replaces part of a person's income lost due to unemployment. Farm employers are included if, during the current or previous calendar year, they (1) employed 10 or more workers in each of 20 or more weeks or (2) paid $20,000 or more in cash wages in any calendar quarter. Benefits are financed by an employer-paid payroll tax.

Child Labor Regulations   These regulations require an employee to be at least 16 years of age to be employed in agriculture during school hours. The minimum age is 14 for employment after school hours with two exceptions: children 12 or 13 can be employed with written parental consent, and children under 12 can work on their parents' farm.

There are also age restrictions which apply to certain jobs that are classified as hazardous. Examples of these jobs are working with certain chemicals, operating most farm machinery, and working with breeding livestock. Employees under the age of 14 cannot be employed in hazardous jobs, and those who are 14 or 15 must be certified for these jobs. Certification can be obtained by taking qualified safety and machinery operation courses. Child Labor Bulletin No. 102 of the U.S. Department of Labor contains details.

Occupational Safety and Health Act (OSHA)   OSHA is a government agency charged with ensuring the health and safety of employees by providing safe working conditions. Examples are the requirements for slow-moving farm equipment using public roads to display a slow-moving vehicle (SMV) sign on the rear and for roll bars to be on most farm tractors. Employers of seven or more workers must keep records of accidents. Farm employers should check the latest OSHA regulations to be sure they are meeting all current requirements.

Immigration Reform and Control Act   This act requires employers to check documents that establish the identity and eligibility of all workers hired after November 7, 1986. Both employer and employee must complete and retain a Form I-9, available from the Immigration Service or Department of Labor, which certifies that the employee is a U.S. citizen, a permanent alien, or an alien with permission to work in the United States.

Migrant Labor Laws   The U.S. Migrant and Seasonal Agricultural Worker Protection Act prescribes housing, safety, and record-keeping requirements for employers of migrant or seasonal agricultural workers. State laws may also stipulate minimum housing, safety, and health standards.

SUMMARY

Although the quantity of labor used in agriculture has fallen dramatically in this century, its productivity and skill level have increased even more rapidly. Planning farm labor needs involves assessing both the quantity and quality of labor needed. Marginal principles of input substitution should be used to plan the most profitable combination of labor and capital for the farm. Good labor management techniques can improve labor efficiency.

Effective management of hired labor starts with the recruiting, interviewing, and selection process. An employment agreement should be drawn up which specifies work rules and compensation, including salary, benefits, and bonuses. Bonuses should be carefully planned to provide the desired work incentives. Farm employees must also be trained, motivated, and evaluated.

A manager of farm labor must be familiar with the many state and federal laws that protect and regulate agricultural workers.

QUESTIONS FOR REVIEW AND FURTHER THOUGHT

1 Why has the amount of labor used in agriculture decreased?
2 Why are the training and skills needed by agricultural workers greater today than in the past?
3 Under what conditions would it be profitable to substitute labor for capital?
4 Why are there wide variations in the monthly labor requirements on a crop farm and less variation on a dairy farm?
5 Why is the labor cost per acre generally less for larger farms than for smaller ones?
6 Observe an actual routine farm chore such as milking or feeding livestock. What suggestions would you have for simplifying the chore to save labor? Would your suggestions increase costs? Would they be profitable?
7 Write an advertisement to place in a newspaper or magazine for recruiting a livestock manager for a large farm.
8 Write a detailed job description for the same position.
9 Discuss what you would include in a training program for a new farm employee.
10 Discuss some procedures that could be used to minimize poor personal relationships between a farm employer and farm employees.
11 What type of incentive program would be most effective on a dairy farm if the objective was to increase milk production per cow? If the objective was to retain the better employees for a longer time?
12 Which labor laws and regulations have to do with withholding taxes? Safety and health? Employing minors?

REFERENCES

Boehlje, Michael D., and Vernon R. Eidman: *Farm Management,* John Wiley & Sons, New York, 1984, chap. 12.

Edwards, Richard A.: *Texas Farm Labor Handbook,* Texas Agricultural Extension Service, B-1250, College Station, TX, 1992.

Edwards, William M., and Corinne Heetland: *Farm Employee Management in Iowa,* Iowa State University Extension Publication FM 1841, Ames, IA, 1991.

"Employee Relations," *National Hog Farmer,* Vol. 36, No. 12, Fall 1991, Intertec Publishing Company, Minneapolis, MN.

Erven, Bernard, and Russell Coltman: *Ohio Farm Labor Handbook,* The Ohio State University Cooperative Extension Service Publication MM 371, ESS 537, Columbus, OH, 1992.

Maloney, Thomas R., C. Arthur Bratton, Kay Embry, and Joan S. Petzen: *Human Resource Management on the Farm: A Management Letter Series,* Publication A.E. Ext. 88-22, Cornell University, Ithaca, NY, 1988.

Thomas, Kenneth H. and Bernard L. Erven: *Farm Personnel Management,* NCR Extension Publication 329, University of Minnesota, St. Paul, MN, 1989.

20

MACHINERY
MANAGEMENT

CHAPTER OBJECTIVES

1 To illustrate the importance of good machinery management on farms and ranches

2 To identify the costs associated with owning and operating machinery

3 To demonstrate procedures for calculating machinery costs

4 To show the relationships between machinery costs per hour or per acre and the amount of annual use

5 To discuss some factors important in machinery selection including size, total costs, and timeliness of operations

6 To compare owning, renting, leasing, and custom hiring as different means of acquiring machinery use

7 To outline methods for increasing the efficiency of machinery use

8 To discuss factors that influence when machinery should be replaced

Mechanization has had a dramatic effect on costs, production levels, production techniques, energy use, and labor requirements in agriculture all over the world. The decline in farm labor and the increase in tractor and horsepower use was discussed in Chapter 19. As tractor size has increased, the size of other machinery has also increased to keep pace. These increases have contributed to higher machinery investment per farm, larger farm sizes, and more efficient use of labor.

Discussions on the increased use of mechanization in agriculture are often in terms of tractors and other crop production machinery. Perhaps no less dramatic has been the rise in the use of power and equipment in livestock production and materials handling. Physical labor requirements have been reduced in many areas by the use of small engines, electric motors, augers, elevators, conveyors, and so forth. Grain handling, manure collection and disposal, livestock feeding, feed grinding and mixing, and hay handling have all been greatly mechanized to reduce labor requirements and costs.

IMPORTANCE OF MACHINERY MANAGEMENT

Machinery and motor vehicles represent a large investment on commercial farms and ranches, as much as $150 to $200 per acre on many cash grain farms. The manager must be careful to control the size of this investment and the related operating costs. Machinery costs can be controlled by four general methods: (1) avoid unnecessary machinery investment to reduce annual total fixed costs, (2) control the variable or operating costs, (3) increase annual machine use to reduce the average ownership cost per unit, and (4) utilize alternatives to ownership for acquiring machinery use.

Machinery management is the application of decision-making principles to a specific resource, machinery, with the objective of increasing profits. However, machinery management and use are related to enterprise combinations and the other resources used in the farm business. This makes it difficult to separate a discussion of machinery management completely from an analysis of the total business. The remainder of this chapter will concentrate on the use of economic principles and budgeting in machinery management, but the relationships between machinery, other resources, and the production enterprises will also be included.

ESTIMATING MACHINERY COSTS

Machinery is costly to purchase, own, and operate. A manager must be able to calculate the costs of owning and operating a machine and understand how they are related to machinery use. It is easy to underestimate machinery costs, as many of them may involve large, but infrequent, cash outlays.

Both ownership and operating costs are important in machinery management. Machinery ownership costs are also called overhead or fixed costs, because they are fixed with respect to the amount of annual use. Operating costs are also referred to as variable or direct costs, because they vary directly with the amount of machine use. Fixed and variable costs were discussed in Chapter 7 but will be reviewed in the following sections as they apply to farm machinery.

Ownership Costs

Ownership or fixed costs begin with the purchase of the machine, continue for as long as it is owned, and cannot be avoided by the manager except by selling the machine. For this reason it is important to estimate what the resulting ownership costs will be before a machine is purchased.

Ownership costs can be as much as 60 to 80 percent of the total annual costs for a machine. As a rule of thumb, annual ownership costs will be about 15 to 20 percent of the original cost of the machine, depending on the type of machine, its age and expected useful life, and the cost of capital.

Depreciation   Depreciation is a noncash expense that reflects a loss in value from age, wear, and obsolescence. It is also an accounting procedure to recover the initial purchase cost of an asset by spreading this cost over its entire ownership period. Since most depreciation is caused by age and obsolescence and is not affected very much by annual use, it is considered a fixed cost once the machine is purchased.

Annual depreciation can be estimated using the straight-line, declining balance, or sum-of-the-year's digits methods. However, if only the *average* annual depreciation is needed, it can be found from the equation

$$\text{Depreciation} = \frac{\text{cost} - \text{salvage value}}{\text{ownership life}}$$

Salvage value can be estimated as a percent of the new list price of a similar machine, as shown in Table 20-1.

Internal Revenue Service regulations specify methods to be used when calculating depreciation for tax purposes. However, the results may not reflect the actual annual depreciation of the machine. Ownership costs should be based on the concept of economic depreciation or the actual decline in value and not on tax depreciation. The commonly used tax depreciation method results in high depreciation in the early years of ownership and little or no depreciation in later years (see Chapter 14).

TABLE 20-1   ESTIMATED SALVAGE VALUE AS A PERCENTAGE OF THE NEW LIST PRICE FOR A SIMILAR MACHINE

Age of machine	Tractors	Self-propelled combines	Other harvesting equipment	Tillage and other equipment
1	62.6%	56.6%	49.6%	53.1%
2	57.6	50.1	43.9	47.0
3	53.0	44.4	38.8	41.6
4	48.7	39.3	34.4	36.8
5	44.8	34.7	30.4	32.6
6	41.2	30.7	26.9	28.8
7	37.9	27.2	23.8	25.5
8	34.9	24.1	21.1	22.6
9	32.1	21.3	18.6	20.0
10	29.5	18.9	16.5	17.7
11	27.2	16.7	14.6	15.7
12	25.0	14.8	12.9	13.9

Source: Edwards, William: *Estimating Farm Machinery Costs,* Iowa State University Extension, Ames, Iowa, Pm-710, based on *Agricultural Engineers Yearbook* Data 230.4.

Interest   Investing in a machine ties up capital and prevents it from being used for an alternative investment. Capital has an opportunity cost, and this cost is part of the true cost of machine ownership. The opportunity cost used for machinery capital should reflect the expected return from investing the capital in its next best alternative use. When borrowed capital is used to purchase machinery, the interest cost is the loan interest rate. Depending on the source of capital, the proper interest rate to use is the rate of return on alternative investments, the expected interest rate on borrowed capital, or a weighted average of the two.

The interest component of average annual fixed costs is calculated from the equation

$$\text{Interest} = \frac{\text{cost} + \text{salvage value}}{2} \times \text{interest rate}$$

The first part of the equation gives the *average value* of the machine over its ownership period, or its value at midlife. Since the machine is declining in value over time, its average value is used to determine the *average annual* interest charge. This procedure assumes that the capital tied up in the machinery investment decreases over the life of the machine, just as the balance owed on a loan used to purchase a machine declines as it is paid off.

An alternative to calculating depreciation and interest is to compute the annual *capital recovery* charge, as discussed in Chapter 8. The capital recovery charge includes both annual depreciation and interest costs. It can be found by

$$\text{Capital recovery} = \text{amortization factor}$$
$$\times \text{(beginning value minus salvage value)}$$
$$+ \text{(interest rate} \times \text{salvage value)}$$

The amortization factor is the value from Appendix Table 1, which corresponds to the ownership life and the interest rate for the machine. The capital recovery amount is the annual payment which will recover the capital lost through depreciation, plus interest on the unrecovered amount. This value is generally slightly higher than the sum of the average annual depreciation and interest values calculated from the previous equations as capital recovery assumes the interest charge is compounded annually.

Taxes   A few states levy a property tax on farm machinery. The actual charge will depend on the evaluation procedure and tax rate in a particular location. Machinery cost studies often use a charge of about 1 percent of the average machine value as an estimate of annual property taxes. A higher rate should be used for pickups and trucks to cover the cost of the license and any other road use fees.

Insurance   Another ownership cost is the annual charge for insurance to cover the loss of the machine from fire, theft, or windstorm damage, and for any liability coverage. A charge for insurance should be included in ownership costs even if the owner carries no formal insurance and personally stands the risk. Some loss can be expected over time, and some recognition of this personal risk needs to be included in the total ownership costs. The proper charge for insurance will depend on the amount and type of coverage and insurance rates for a given location. Machinery cost studies estimate the annual insurance charge at 0.4 to 0.6 percent of the machine's average value. This charge will need to be higher for pickups and trucks because of their higher premiums for property damage, collision, and liability coverage.

Housing   Some machinery cost studies include an annual cost for housing the machine. A properly housed machine will have fewer repair bills and may last longer. Cost studies have found annual housing costs to be about 1.5 percent of the average value of the machine, so this rule is often used to estimate machinery housing costs. A housing charge can also be estimated by calculating the annual cost per square foot for the machine shed (possibly including a prorated charge for a shop area) and multiplying this by the number of square feet the machine occupies.

Leasing   Some machinery is acquired under long-term leasing agreements. Although the machine is not actually owned, the annual lease payment should be included with fixed or ownership costs. Typically the operator also pays the insurance premiums for leased machines and provides housing.

Operating Costs

Operating costs are directly related to use. They are zero if the machine is not used but increase directly with the amount of annual use. Unlike ownership costs, they can be controlled by varying the amount of annual use, by improving efficiency, and by following a proper maintenance program.

Repairs   Annual repair costs will vary with use, machine type, age, preventive maintenance programs, and other factors. Table 20-2 shows some average repair costs per hour of use, for each $1,000 of new list price for various types of machines. These rates are based on average costs over the life of the machine. However, average annual repair costs usually increase at an increasing rate over time. If a new machine is purchased under warranty, repair costs to the operator may be very low at first. Any rule of thumb for estimating repair costs needs to be used with caution. The best source of information is detailed records of actual repair costs for each machine under the existing level of use, cropping pattern, and maintenance program.

Fuel and Lubrication   Gasoline, diesel fuel, oil and other lubricants, and filters are included in this category. These costs are minor for nonpowered equipment but are important for self-powered machinery. Fuel use per hour will depend on engine size, load, speed, and field conditions. Farm records can be used to estimate average fuel use. Data from the University of Nebraska tractor tests indicate average fuel consumption in gallons per hour can be estimated from power takeoff horsepower as follows:

TABLE 20-2 AVERAGE REPAIR COSTS PER HOUR OF USE, FOR EACH $1,000 OF NEW LIST PRICE

Item	Repair costs per hour of use, per $1,000 of list price
Two-wheel drive tractors	$0.12
Four-wheel drive tractor	0.10
Combine	0.26
Disk	0.29
Chisel plow	0.50
Field cultivator	0.40
Row crop cultivator	0.51
Planter, grain drill	0.66
Mower	0.75
Mower-conditioner	0.39
Rake	0.50
Baler	0.40
Forage harvester (pull type)	0.40
Forage harvester (self-propelled)	0.25
Wagon	0.26

Source: Schlender, John R., and David A. Pacey: *A Look at Machinery Cost,* Kansas State University Farm Management Guide MF-842, Kansas Cooperative Extension Service, 1986.

$$\text{Gasoline} = 0.060 \times \text{maximum rated pto horsepower}$$
$$\text{Diesel} = 0.044 \times \text{maximum rated pto horsepower}$$

Costs for lubricants and filters average about 15 percent of fuel costs for self-powered machines. The cost of lubricants for nonpowered machines is generally small enough that it can be ignored when estimating operating costs.

Labor   The amount of labor needed for machinery depends on the operation being performed and the size of machinery used. Labor costs are generally estimated separately from machinery costs but need to be included in any estimate of the total cost of performing a given machine operation. Machinery-related labor costs will be underestimated if only the time actually spent in operating the machine in the field is considered. The total labor charge should also include time spent fueling, lubricating, repairing, adjusting, and moving machinery between fields.

Custom Hire or Rental   When a custom operator is hired to perform certain machinery operations or a machine is rented by the acre, hour, or day, these costs should also be a part of the machinery variable or operating costs for the farm.

Other Operating Costs   Some specialized machines have additional operating costs associated with their use. Items such as twine, baling wire, and bags may need to be included in variable costs.

Cash and Noncash Machinery Costs

Machinery costs are often underestimated as a relatively large part of them can be noncash expenses. Depreciation is always a noncash expense. Interest may be a cash expense to the extent that interest is being paid on borrowed money, or it may be entirely a noncash expense if the machine was purchased with the owner's equity capital. Insurance would be a cash expense when using commercial insurance, or a noncash expense when the owner assumes the risk. Likewise, labor is a cash expense if hired labor is employed, but a noncash expense if it is furnished by the owner. A complete accounting of all noncash as well as cash expenses is necessary for an accurate estimation of total machinery costs.

The following is a breakdown of the various machinery costs into their possible cash and/or noncash categories:

Type of cost	Cash	Noncash
Depreciation		×
Interest on the investment	× (loan)	× (opportunity)
Property taxes	×	
Insurance	× (policy)	× (personal risk)
Housing	×	×
Leasing	×	
Repairs and maintenance	× (hired)	× (operator labor)
Fuel and lubricants	×	
Labor	× (wages)	× (operator labor)
Custom hire or rental	×	

Machinery Costs and Use

Annual ownership or fixed costs are assumed to be constant regardless of the amount of machine use during the year. Operating or variable costs increase with the amount of use, generally at a constant rate per acre or per hour. The result is an annual total cost which also increases at a constant rate. These relationships are shown in Figure 20-1a.

For decision-making purposes it is often necessary to express machine costs in terms of the average cost per acre, per hour, or per unit of output. Average fixed costs will decline as the acres, hours, or units of output increase while average variable costs will be constant *if* total variable costs increase at a constant rate. Because average total cost is the sum of average fixed and average variable costs, it also declines with added use. These relationships are shown in Figure 20-1b, where the vertical distance between the average fixed cost and average total cost curves is constant and equal to average variable cost.

Machinery cost estimates are only as accurate as the estimate of annual use. As much effort should go into estimating this value as in estimating the cost components. Using actual field records and tractor-hour meters will improve the accuracy of the estimates.

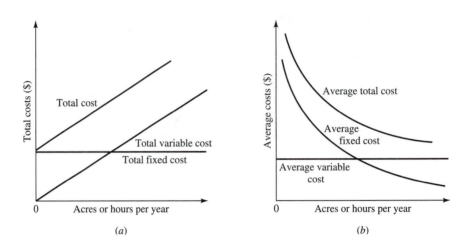

FIGURE 20-1 Relationships between total and average machinery costs.

EXAMPLES OF MACHINERY COST CALCULATIONS

Table 20-3 shows an example of how to calculate total annual machinery costs and average costs per hour and per acre for a new combine. The first step is to list the basic data, such as purchase cost, salvage value, useful life, annual use, and cost of capital. Next, total fixed costs are calculated and converted into average fixed cost per hour of use. Based on the assumed 250 hours of annual use, the average fixed cost per hour for the combine in the example is estimated to be $64.02. Remember that this value will change if the actual hours of use are different.

If the capital recovery method had been used, the estimate of fixed costs would have been slightly higher. The capital recovery factor from Appendix Table 1 for an 8-year life and 9 percent interest rate is 0.18067, so the capital recovery amount would be

$$[0.18067 \times (100,000 - 24,100)] + (0.09 \times 24,100) = \$15,882$$

Total fixed costs would be $16,813 per year, or $67.25 per hour, using this method.

Total variable costs and variable costs per hour, including labor, are calculated in step 3 of Table 20-3. The labor requirement was increased by 20 percent over the estimated machine use to account for time required to service, adjust, and transport the combine. Average variable cost per hour is estimated to be $38.36.

The total cost per hour is estimated at $102.38, based on 250 hours of annual use (step 4). This figure can be converted to a cost per acre or per unit of output if the field capacity per hour is known. In step 5 of the example in Table 20-3 three different performance rates are used to show how the cost per acre is affected.

When one tractor is used to pull several implements, or more than one harvesting head is used on the same self-propelled unit, ownership and operating costs must be cal-

TABLE 20-3   CALCULATING MACHINERY COSTS FOR A NEW COMBINE

Step 1: List basic data	
New combine, 18-foot header, 120-hp diesel engine:	
Cost	$100,000
Salvage value (24.1% of new cost)	$ 24,100
Average value ($100,000 + 24,100)/2	$ 62,050
Useful life	8 years
Estimated annual use	250 hours
Cost of capital	9 %
Step 2: Calculate fixed costs	
Depreciation ($100,000 – 24,100)/8 years	$ 9,488
Interest (9% of average value) = 9% × $62,050	5,585
Taxes, insurance and housing (1.5% of average value)	931
Total annual fixed costs	$ 16,004
Fixed costs per hour ($16,004/250 hours)	$ 64.02
Step 3: Calculate variable costs	
Repairs ($0.26 per $1,000 of new cost × 250 hours)	$ 6,500
Diesel fuel (120 horsepower × 0.044 gallon/horsepower-hour	
× $.85 per gallon × 250 hours)	1,122
Lubrication and filters (15% of fuel cost)	168
Labor (300 hours × $6 per hour)*	1,800
Total annual variable costs	$ 9,590
Variable cost per hour ($9,590/250 hours)	$ 38.36
Step 4: Calculate total cost per hour	
Fixed cost per hour	$ 64.02
Variable cost per hour	$ 38.36
Total cost per hour	$ 102.38
Step 5: Calculate cost per acre	
Performance rate:	
4 acres per hour ($102.38 /4 acres)	$ 25.60
5 acres per hour ($102.38 /5 acres)	20.48
6 acres per hour ($102.38 /6 acres)	17.06

*Labor requirements are increased by 20% to account for time spent servicing, adjusting, repairing, and traveling between fields.

culated separately for both the power unit and for the attachment. They can then be added together to find the combined cost of performing the operation. In Table 20-4, ownership costs and operating costs per hour have been estimated for a 125-horsepower tractor and a 17-foot tandem disk, using the methods explained previously. Notice that fuel and lubrication costs and the labor charge were assigned only to the tractor. The costs per hour for the tractor and disk are then combined to find the total cost of disking per hour, and divided by the field capacity to find the cost per acre. Note that it is not correct to combine the *annual* costs of two machines when only part of the annual use of the power unit is with this implement. Instead, the *hourly* costs should be combined.

The importance of an accurate estimate of annual use and performance rate must be stressed. Increasing the hours of annual use will decrease the total cost per hour, and improving the performance rate will decrease the total cost per acre. The manager may

TABLE 20-4   COMBINED COST OF A TRACTOR AND IMPLEMENT

	125-horsepower tractor		17-foot tandem disk
Annual ownership costs	$5,700		$600
Annual hours of use	600		100
Ownership cost per hour	$ 9.50		$6.00
Operating costs per hour:			
Fuel and lubrication	5.38		
Repairs	3.58		1.87
Labor	6.00		
Total cost per hour	$24.46		$7.87
Combined cost per hour		$32.33	
Field capacity, acres per hour		9.0	
Combined cost per acre		$ 3.59	

wish to calculate costs for several levels of annual use or performance rate to find the range of costs which might occur.

Sometimes cost estimates are desired for a number of machines of different sizes or with various amounts of annual use. Computer programs are now widely used to compute machinery costs for enterprise budgets and other uses. Once the basic data are stored in the program, estimates of machinery costs can be quickly and easily computed for a wide range of machine operations, performance rates, and annual use levels.

FACTORS IN MACHINERY SELECTION

One of the more difficult problems in farm management is proper machinery selection. This process is complicated not only by the wide range of types and sizes available, but also by capital availability, labor requirements, the particular crop and livestock enterprises in the farm plan, tillage practices, and climatic factors. The objective in selecting machinery is to purchase the machine which will perform the required task within the time available for the lowest possible total cost. This does not necessarily result in the purchase of the smallest machine available, however, because labor and timeliness costs must also be considered.

Machinery Size

The first step in analyzing machinery sizes is to determine the field capacity in acres per hour for each size that is available. The formula for finding capacity in acres per hour is

$$\text{Field capacity} = \frac{\text{speed (mph)} \times \text{width (ft)} \times \text{field efficiency}}{8.25}$$

For example, a 12-foot wide windrower operated at 8 miles per hour with a field efficiency of 82 percent would have an effective field capacity of (8 mph × 12 feet × 82%)/8.25 = 9.54 acres per hour.

Field efficiency is included in the equation to recognize that a machine is not used at 100 percent of its theoretical capacity because of work overlap and time spent turning, adjusting, repairing, and handling materials. Operations such as planting, which require frequent stops to refill seed, chemical, and fertilizer boxes, may have field efficiencies as low as 50 percent. Field efficiencies as high as 85 to 90 percent are common for some tillage operations, particularly in large fields when turning time and work overlap is minimized. Larger machines typically have higher field efficiencies due to less overlapping and less time needed for adjustments relative to the area covered.[1]

The next step is to compute the minimum field capacity (acres per hour) needed to complete the task in the time available. This value is found by dividing the number of acres the machine must cover by the number of field hours available to complete the operation. In turn, the number of field hours available depends on how many days the weather is suitable for performing the operation and the number of labor or field hours available each day. The formula for finding the minimum field capacity in acres per hour is

$$\text{Minimum field capacity} = \frac{\text{acres to cover}}{\text{hours per day} \times \text{days available}}$$

Suppose the operator wants to be able to windrow 150 acres in 2 days and can operate the windrower 8 hours per day. The minimum capacity needed is [150 acres/ (8 hours × 2 days)] = 9.38 acres per hour. The 12-foot windrower, with a capacity of 9.54 acres per hour, should be large enough.

When several field operations must be completed within a certain number of days, it may be more convenient to calculate the days needed for each operation and add them together to test whether or not a proposed machinery complement has sufficient capacity. The formula for each operation is

$$\text{Field days needed} = \frac{\text{acres to cover}}{\text{hours per day} \times \text{acres per hour}}$$

Using the same example, the operator also owns a baler with an effective field capacity of 5 acres per hour. If baling can be performed only 6 hours per day, then 150 acres requires [150/(6 hours × 5 acres per hour)] = 5.0 days of field work. Thus, a total of 7 suitable field days are required each time the 150 acres are both windrowed and

---

[1]Average field efficiencies may be found in Donnell Hunt, *Farm Power and Machinery Management,* Iowa State University Press, Ames, Iowa, 1977.

baled. The operator must decide if typical weather patterns will permit this many days to be available without a risk of significant crop losses.

The size of machinery required to complete field operations in a timely manner can be reduced by (1) increasing the labor supply so machinery can be operated more hours per day or two operations can be performed at the same time, (2) reducing the number of field operations performed, or (3) producing several crops with different critical planting and harvesting dates instead of just one crop.

Machinery selection may also involve the choice of one large machine or two smaller ones. The purchase cost and annual fixed costs will be higher for two machines, because the same capacity can usually be obtained at a lower cost in one large machine. Two tractors and two drivers will also be needed if tractor-drawn equipment is involved. The primary advantage of owning two machines is an increase in reliability. If one machine breaks down, work is not completely stopped but can continue at half speed utilizing the remaining machine.

Timeliness

Some field operations do not have to be completed within a fixed time period, but the later they are performed the lower the harvested yield is likely to be. The yield reduction may be in terms of quality, such as for hay that is allowed to mature or dry too long, or in quantity, such as for grain that suffers from too short a growing season or has excessive field losses during harvesting. Figure 20-2 shows a typical relationship between planting time and potential yield for winter wheat where planting too early or too late reduces yield.

FIGURE 20-2   Hard red winter wheat yields as a function of planting date at Stillwater, Oklahoma (*Source:* F. M. Epplin, D. E. Beck, and E. G. Krenzer, *Current Farm Economics,* Vol. 64, No. 3, September, 1991).

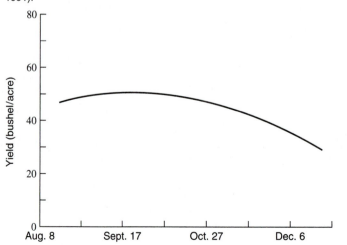

A yield reduction reduces profit and should be included as part of the cost of utilizing a smaller machine. This cost is referred to as timeliness cost. The dollar cost of poor timeliness is difficult to estimate, as it will vary from year to year depending on weather conditions. However, some estimate should be included when comparing the cost of owning machines of different sizes. Figure 20-3 is a hypothetical example of how timeliness and other machinery costs change as larger machinery is used to perform the same amount of work. At first, larger machinery reduces timeliness costs and lowers total costs. After some point, no more gains in timeliness are available and increasing ownership costs cause total costs to rise. Timeliness concerns must be balanced against higher ownership costs to select the least-cost machine size.

Partial budgeting is a useful tool to use when making decisions about machinery sizes, and timeliness can be included. Table 20-5 is an example which emphasizes the importance of considering all costs. Changing from a 6-row planter to a 12-row planter will double the annual ownership costs. However, the manager estimates that improved timeliness of planting will result in an average yield increase of 3 bushels per acre each year. The value of this added yield, combined with slightly lower labor and repair costs, results in a projected increase in net income of $1,505 per year from purchasing the larger planter. Of course, the date of planting also depends on how long it takes to complete tillage and other operations. Thus, the total machinery complement should be analyzed when timeliness factors are considered.

ALTERNATIVES FOR ACQUIRING MACHINERY USE

Most farm and ranch managers prefer to own their own machinery. Ownership gives them complete control over the use and disposal of each machine. However, alterna-

FIGURE 20-3 Hypothetical effect of timeliness and machine size on cost.

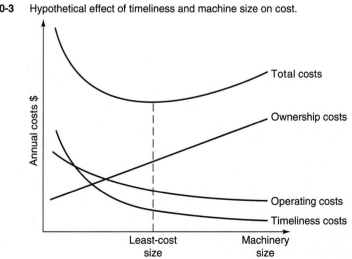

TABLE 20-5 EXAMPLE OF A PARTIAL BUDGET FOR SELECTING THE MOST PROFITABLE MACHINE SIZE

Adjustment: Change from a 6-row planter ($12,000 cost) to a 12-row planter ($24,000 cost). Assumes a 20% salvage value, 8-year ownership period, 10% interest rate. The 6-row machine requires 75 hours to plant 500 acres, the 12-row machine requires 40 hours.

Additional revenue:

3 bushels per acre increase in yield × 500 acres × $2.25 per bushel

Total additional revenue		$3,375

Reduced costs:

	Ownership	Operating	
Depreciation	$1,200		
Interest	720		
Taxes and insurance	120		
Repairs		$594	
Labor (35 hours less at $6/hour)		210	
Subtotal	2,040	804	
Total reduced costs			$2,844
Total of additional revenue and reduced costs			$6,219

Additional costs:

	Ownership	Operating	
Depreciation	$2,400		
Interest	1,440		
Taxes and insurance	240		
Repairs		$634	
Subtotal	$4,080	$634	
Total additional costs			$4,714
Reduced revenue:			0
Total additional costs and reduced revenue			$4,714
Net change in farm income:		($6,219 – $4,714) =	$1,505

tives such as leasing and custom hiring may be less costly in some situations. Even when machinery is owned, it can be purchased either new or used.

New versus Used Machines

Farm machines, particularly tractors and other self-powered machines, decline in market value most rapidly during the first few years of their useful lives. Buying used machinery can be an economical way to lower machinery costs. Purchasing a machine is essentially purchasing the service it can provide over time. If a used machine can provide the same service at a lower total cost than a new one, it is a more economical alternative.

Used machinery requires a lower initial investment and therefore lower annual ownership costs for depreciation and interest. Offsetting some of the lower ownership costs may be higher repair costs and decreases in reliability and timeliness. A used machine may also become obsolete sooner than a new one. The owner's ability as a mechanic and the facilities and time to do major repair work at home are often crucial for utilizing used machinery.

Used machinery should be considered when capital is limited, interest rates are high, the machine will have a relatively small annual use, and reliability and timeliness are not critical. The importance of the lower capital investment and interest cost is shown in Table 20-6. Ignoring interest, the new machine has the lower cost. However, the smaller investment in the used machine results in a lower annual interest cost and a lower total cost, making it the least-cost alternative.

A little experimenting with different interest rates for capital shows that a rate of about 7 percent is the break-even point in this example. The used machine has a lower annual cost when the cost of capital is above 7 percent, and the new machine has the advantage when it is 7 percent or lower. As in this example, the cost of capital combined with differences in repair costs and reliability is important in the decision.

Rental and Leasing

When investment capital is very limited or interest rates are high, leasing or renting a machine may be preferable to owning it. Short-term rental arrangements usually involve a period of from a few days to a whole season. The operator pays a rental fee plus the cost of insurance and daily maintenance, but not major repairs. Machinery rental is especially attractive when (1) a specialized machine is needed for relatively low use, (2) extra capacity or a replacement machine is needed for a short time, or (3) the operator wants to experiment with a new machine or production practice without making a capital investment.

A lease is a long-term formal agreement whereby the machine owner (the lessor), which may be a machinery dealer or leasing company, grants control and use of the machine to the user (the lessee) for a specified period of time for a regular lease payment. Most machinery leases are for three to five years or longer. As with any formal agreement, the lease should be in writing and should cover such items as payment rates and dates, payment of repairs, taxes, and insurance, responsibilities for loss or damage, and provisions for cancellation.

Some leases allow the lessee to purchase the machine at the end of the lease period for a specified price. To maintain the income tax deductibility of the lease payments,

TABLE 20-6   EXAMPLE OF COST COMPARISON FOR A NEW AND USED MACHINE

	New machine	Used machine
Purchase price	$10,000	$5,500
Useful life	10 years	6 years
Annual costs:		
Depreciation (no salvage value)	$ 1,000	$ 917
Taxes and insurance	200	150
Repairs	500	800
Total annual cost without interest	$ 1,700	$1,867
Interest at 10%	500	275
Total annual cost	$2,200	$2,142

the purchase must be optional and for a value approximately equal to the market value of the machine at the end of the lease. Otherwise, the lease may be termed a sales contract and depreciation and interest become tax deductible for the lessee instead of the lease payments.

Leasing machinery may help operators reduce the amount of capital they have tied up in intermediate assets. However, lease payments represent a cash flow obligation just like a loan payment. Leasing reduces the risk of obsolescence because the lessee is not obligated to keep the machine beyond the term of the lease agreement. Some operators who have very little taxable farm income and cannot utilize depreciation deductions from owning machinery may find leasing to have a lower after-tax cost than owning.

There are also some disadvantages to leasing farm machinery. The practice is not well established in many areas, and the desired model may not be available for lease. Lease rates may also be relatively high compared with loan payments, and the lessee may not be allowed to cancel the lease early without paying a substantial penalty.

Custom Hire

Custom hiring is an important practice in some areas for operations such as applying chemicals and harvesting grain and forages. The decision to own a machine or custom hire the service depends on the costs involved, and the amount of work to be done. For machines which will be used very little it is often more economical to hire the work done on a custom basis. However, the availability and dependability of custom operators must be considered. A manager may not want to rely on a custom operator for a task such as planting, where timeliness is critical.

Total costs per acre or per unit of output should be compared when deciding between machine ownership or custom hiring. Custom charges are typically a fixed rate per acre or per unit of output while ownership costs will decline with increased use. These relationships are shown in Figure 20-4. At low levels of use, hiring a custom operator is less expensive, while for higher usage the cost is lower if the machine is owned. The point where the cost advantage changes, or the break-even point, is shown as output level a in Figure 20-4.

When the necessary cost data are available, the break-even point can be found from the following equation:

$$\text{Break-even units} = \frac{\text{total annual fixed costs}}{\text{custom rate} - \text{variable costs per unit}}$$

For example, the ownership or fixed costs for the combine in Table 20-3 were $16,004 per year, and the variable costs for operating it were $7.67 per acre for a performance rate of 5 acres per hour ($38.36 per hour divided by 5 acres per hour). If the custom hire rate for a similar combine is $24 per acre, the break-even point would be

$$\frac{\$16,004}{\$24.00 - \$7.67} = 980 \text{ acres}$$

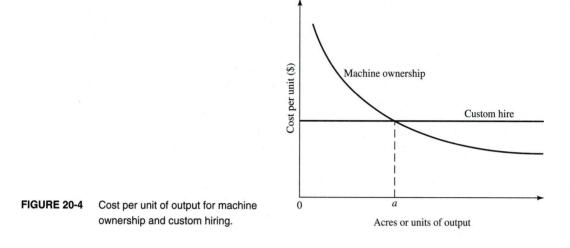

FIGURE 20-4   Cost per unit of output for machine ownership and custom hiring.

If the machine would be used on less than 980 acres, it would be less costly to hire the work done, while above 980 acres it would be less expensive to own the machine. A determination of the break-even point provides a useful guide to help managers choose between machine ownership and custom hiring.

Labor use is another consideration in custom hiring. The custom operator typically provides the labor necessary to operate the machine, which frees the farm operator's labor for other uses. This can be an advantage if it reduces the amount of hired labor needed or if the owner's labor has a high opportunity cost at the time the custom work is being performed. It can also be an advantage for some operations which take special skills to perform, such as the application of chemicals.

IMPROVING MACHINERY EFFICIENCY

Two values are often used to measure the efficiency of machinery use. The first is *investment per tillable acre,* which is calculated by dividing the current value of all crop machinery by the number of tillable acres. The current value of all machinery for a given year can be found by taking the average of the beginning and ending machinery inventory values for the year. Ideally, this should be the current market value, but current book or depreciated value (cost less accumulated depreciation) can be used if market values are not available.

The second measure of machinery efficiency is *machinery cost per tillable acre.* It is found by dividing the total annual machinery costs, both ownership and operating, by the number of tillable acres. Some farm record analyses also include pickup and truck expenses, the farm share of automobile expenses, the farm share of utilities, machinery lease payments, and custom hiring expenses in machinery costs. When possible, the cost of machinery used for livestock purposes should be excluded, in order to make a fair comparison among farms.

Lower values are desirable for both these efficiency measures, but they should be used and compared with caution. Numerous studies have shown that cost per acre and investment per acre decline with increases in farm size, and also vary with farm type. It is important to compare values only with those calculated in the same manner for farms of the same approximate size and type. Table 20-7 shows recent machinery costs per acre, excluding labor, for soybean producers in central Iowa. Low-profit producers incurred over $20 per acre higher machinery costs than high-profit producers and had over $40 per acre more invested in machinery.

Values which are too high or too low compared with similar farms or past records indicate machinery investment and cost problems. Investment per acre can be kept low by leasing, by custom hiring, and by using older, smaller machinery, but these practices may increase total annual machinery costs.

A number of techniques can be employed to improve machinery efficiency. Choosing the optimum size of machine, and comparing alternatives for acquiring machinery services have already been discussed. There are three other areas which can also have a large impact on efficiency: maintenance and operation, extending machinery use, and replacement decisions.

Maintenance and Operation

Repairs are a large part of machinery variable costs, but they are a cost which can be controlled by proper use and maintenance. Agricultural engineers report that excessive repair costs can generally be traced to (1) overloading or exceeding the rated machine capacity, (2) excessive speed, (3) poor daily and periodic maintenance, and (4) improper adjustment. All these items can be corrected by constant attention and proper training of machine operators.

A system of scheduling and recording repairs and maintenance is essential for controlling repair costs. Adherence to the manufacturer's recommended maintenance

TABLE 20-7   MACHINERY COSTS FOR CENTRAL IOWA SOYBEAN PRODUCERS ($ PER ACRE)

	Low profit third	High profit third
Depreciation	$ 20.13	$ 13.28
Interest	13.94	7.19
Lease payments	3.55	4.49
Housing	.56	.97
Insurance	.88	.80
Fuel and lubrication	8.89	6.44
Repairs	15.65	9.35
Custom hire	1.70	1.98
Total machinery cost	$ 65.30	$ 44.50
Machinery investment	$176.68	$136.26

Source: Crop Enterprise Record Summary, Central Iowa 1990. Iowa State University Extension Publication FM 1835, 1991.

schedule will keep warranties in effect, prevent unnecessary breakdowns, and reduce lifetime repair costs. Complete records of repairs on individual machines will help identify machines with higher than average repair costs. When these machines are identified, they should be considered for early replacement.

How a machine is operated affects both repair costs and field efficiency. Speed should be adjusted to load a machine to capacity without overloading it or lowering the quality of the work being done. Practices to improve field efficiency reduce costs by allowing more work to be done in a given time period, by permitting the same work to be done in less time, or by allowing a smaller machine to be used. Small and irregular shaped fields requiring frequent turns, frequent stops, and work overlap reduce field efficiency. For example, a 20-foot disk operated with a 2-foot overlap loses 10 percent of its potential efficiency from this factor alone.

Extending Machinery Use

Low annual use of an expensive specialized machine also contributes to high machinery costs. Ownership costs per unit of output remain high unless the annual use can be increased. Some farmers purchase low-use machines jointly with other operators. This not only decreases the fixed costs per unit but also decreases the investment required from each individual. Jointly owned machinery can reduce machinery costs for each individual owner, but it is important that the partners are compatible and can agree on the details of the machine's use. Whose work will be done first and the division of expenses such as repairs, insurance, and taxes should be agreed to before the machine is purchased. Normally, operators provide their own fuel and labor, then pay a fixed rate per acre into a fund used to pay ownership costs and repairs.

Some owners exchange the use of specialized machinery. If the use or value of the machines is unequal, some payment may be exchanged as well. Many operations such as harvesting are accomplished more efficiently when two or more people work together, anyway.

Machine cost per unit of output can also be reduced by performing custom work for other operators. If custom work does not interfere with timely completion of the owner's work, it will provide additional income to pay ownership costs. The custom rate charged should reflect the opportunity cost of labor in the operator's own business as well as the costs of owning and operating the machine. Some skilled operators do custom machinery work as a part-time or full-time alternative to farming their own or rented land.

Replacement

One of the more difficult decisions in machinery management is when to replace or trade a machine. There is no simple, easy decision rule applicable to all types of machines and conditions. Besides costs and reliability, replacement decisions must also consider the effects of income taxes and cash flow.

Replacement decisions can be made for any of the following reasons:

1 The present machine is worn out—the age and accumulated use are such that the machine is no longer capable of performing the required task reliably.

2 The machine is obsolete—new developments in machinery technology or changes in cropping patterns allow a newer machine to perform the job better.

3 Costs are increasing with the present machine—repair and timeliness costs are increasing rapidly, both total and per unit of output.

4 The capacity is too small—the area in production has increased or timeliness has become so critical that the old machine cannot complete the job on time.

5 Income taxes—in a high-profit year, machines may be replaced to take advantage of the tax-reducing benefits of depreciation deductions. However, replacement decisions should not be made on this basis alone.

6 Cash flow—many machines are replaced in years of above-average cash income in order to avoid borrowing funds later. Likewise, replacement of machinery is often postponed when cash flow is tight.

7 Pride and prestige—there is a certain "pride of ownership" involved in the purchase of new and larger machinery. While this may be important to some individuals, it can be a costly reason and one which is difficult to include in an economic analysis.

These reasons can be used individually or in combination to determine the replacement age for a specific machine. Cost and repair records on each machine can greatly aid the replacement decision. Income tax savings, pride, and prestige should not be overemphasized, as they can result in higher than necessary costs in the long run.

SUMMARY

Machinery investment is the second largest investment on most farms and ranches after real estate. Annual machinery costs are a large part of a farm's total annual costs. Ownership costs include depreciation, interest, taxes, insurance, housing, and lease payments. Repairs, fuel and lubrication, labor, custom hire, and rental payments are included in operating costs.

Selection of the optimum machine size should consider both total costs and the effects of timeliness in completing operations. Operators who are short of capital or who are skilled in machinery repair can benefit from investing in used machinery. Rental and custom hiring are other alternatives for acquiring machinery use and should be considered for some types and sizes of farms, particularly for specialized machines with low annual use.

Machinery efficiency can be improved by proper maintenance and operation, by extending machinery use through joint ownership with other operators, or exchanging the use of individually owned machines. The proper time to trade machinery depends on repair costs, reliability, obsolescence, cash flow, tax considerations, and personal pride.

QUESTIONS FOR REVIEW AND FURTHER THOUGHT

1 What is the total annual fixed cost for a $45,000 tractor with a 12-year life and a $9,000 salvage value when insurance and taxes are 2 percent of average value and there is a 9 percent opportunity cost on capital? What is the average fixed cost per hour if the tractor is used 500 hours per year? If it is used 800 hours per year?

2 What is the field capacity in acres per hour for a 28-foot-wide tandem disk operated at 4 miles per hour with 80 percent field efficiency? How much does it change if the field efficiency could be increased to 90 percent? How much time would be saved on 800 acres?

3 Owning and operating a certain machine used on 250 acres would have an ownership cost of $2,000 per year plus operating costs of $5.00 per acre including labor. Leasing a machine with a field capacity of 2.5 acres per hour to do the same work would cost $25 per hour plus the same operating costs. Hiring the work done on a custom basis would cost $12 per acre. Which alternative has the lowest total cost?

4 Assume a self-propelled windrower would have annual fixed costs of $3,850 and variable costs of $3 per acre. A custom operator charges $7.50 per acre. What is the breakeven point in acres per year?

5 List ways to improve the field efficiency of machine operations such as planting, disking, and combining grain.

6 How does the cost of capital enter into the decision to purchase, lease, or custom hire a machine?

7 What factors are important in a machinery replacement decisions? How would you rank them in order of importance? Might your ranking be different for different types of machines?

8 If you decide to use your combine to do custom work for others, what would happen to total ownership cost? Average ownership cost per acre harvested? Total operating cost? Average operating cost per acre harvested?

9 List and discuss factors that affect a machine's operating cost per acre.

REFERENCES

Boehlje, Michael D., and Vernon R. Eidman: *Farm Management,* John Wiley & Sons, New York, 1984, chap. 14.

Bowers, Wendell: *Machinery Management,* John Deere Service Training, Moline, IL, 1987.

Edwards, William, and Vernon M. Meyer: *Acquiring Farm Machinery Services: Ownership, Custom Hire, Rental, Lease,* Iowa State University Extension, Pm-787, Ames, IA, rev. 1986.

Kadlec, John E.: *Farm Management,* Prentice-Hall, Inc., Englewood Cliffs, NJ, 1985, chap. 16.

LaDue, Eddy L., Ronald Durst, Glenn D. Pederson, James S. Plaxico, and Anne O'Mara McDonley: *Financial Leasing in Agriculture: An Economic Analysis,* North Central Regional Research Publication 303, North Dakota State University, Fargo, ND, 1985.

Schlender, John R., and David A. Pacey: *A Look at Machinery Cost,* KSU Farm Management Guide MF-842, Kansas Cooperative Extension Service, Manhattan, KS, 1986.

Waters, William K., and Donald R. Daum: *Farm Machinery Management Guide,* Pennsylvania Cooperative Extension Service Special Circular 192, University Park, PA.

Willett, Gayle S.: *A Method for Analyzing Farm Machinery Purchase, Lease, Rent, or Custom-Hire Decisions,* Washington Cooperative Extension Service, EM 4359, Pullman, WA, rev. 1980.

21

FARM MANAGEMENT IN THE TWENTY-FIRST CENTURY

CHAPTER OBJECTIVES

1 To discuss changes that will be found in the structure and technology of agriculture in the twenty-first century and how they might affect farm managers

2 To assess the skills that farm managers will need in the twenty-first century in order to respond to these changes

What will farm managers be doing in the twenty-first century? They will be doing what they are doing now, making decisions. They will still be using economic principles, the various budgeting methods, record analysis, investment analysis, and other management techniques to help make those decisions. What kind of decisions will managers be making in the twenty-first century? Most will be the same type they are making now. Managers will still be making decisions about input and output levels and combinations, and when and how to acquire additional resources. They will continue to analyze the risk and returns from adopting new technology, making new capital investments, adjusting farm size, and changing enterprises.

Will anything about management decisions in the twenty-first century be different? Yes. While the broad types of decisions being made will be the same, the details and information used will change. Technology will continue to provide new inputs to consider and new and more specialized products for possible production and marketing. Electronic innovations will provide more accurate and timely information for use in management decision making. Farmers and ranchers will have to compete more aggressively with nonagricultural businesses for use of the land, labor, and capital resources needed. As usual, the better managers will be able to adapt to these changes and will continue to leverage their management skills over larger and larger units of production.

STRUCTURE OF FARMS AND RANCHES

The number of farms in the United States has been decreasing since about 1935. Since the amount of land in farms and ranches has been relatively constant, this means the average farm size has increased considerably. Several factors have contributed to this change.

First, labor-saving technology in the form of larger agricultural machinery, automated equipment, and specialized livestock buildings has made it possible for fewer workers to produce more. Second, higher wages and salaries have made it more expensive for farmers to hire labor, and employment opportunities outside of agriculture have become more attractive and plentiful. This encouraged labor to move out of agriculture. Also during this period of change, the cost of labor has increased faster than the cost of capital, making it profitable for farm managers to substitute capital for labor in many areas of production.

Third, farm and ranch operators have tried to attain higher levels of income, comparable to those of nonfarm families. One way to achieve this has been for each farm family to control more resources and produce more output, while holding costs per unit level or even decreasing them wherever economies of size exist. The desire for improved standards of living has provided much of the motivation for increasing farm size, and new technology has provided the means.

Fourth, some new technology is available only in a minimum size or scale. This encourages a larger farm size in order to spread the fixed costs of the technology over enough units to be economically efficient. Examples include grain drying and handling systems, large four-wheel drive tractors, large combines and cotton pickers, confinement livestock buildings, and large cattle feedlots. Perhaps even more important is the time and effort required for a manager to learn new skills in production, marketing, and finance. These also represent a fixed investment and generate a larger return to the operator when they are applied to more units of production.

All of these forces are likely to continue into the next century, meaning that farm sizes will continue to increase. Future managers will find themselves making more use of hired labor or machines to perform less skilled tasks, while they spend their time applying more sophisticated management skills to producing, financing, and marketing more and more units of production.

THE INFORMATION AGE

Many decision-making principles and budgeting tools have been difficult to use and were underused in the past. The individual farm data needed to use them were often not available or the process for analyzing it was too complex. The coming century will see rapid changes in methods for data collection, analysis, and interpretation.

Electronic sensors and processors now used in large-scale industries will become accessible and affordable to farms and ranches.

Not only will more whole farm data be available but data specific to very small land areas or to individual animals will become more common. This very specific data will help managers customize the treatment of each acre or each head. Yields may be monitored and recorded as harvesting machines move across the field. Automated machines may be able to take a soil sample every few yards which is instantly analyzed and the results recorded by field location. Satellite photographs and other techniques may provide information on the specific location of weed and insect infestations permitting a limited, pinpoint application of pesticide.

Miniature electronic sensors could collect and record information from livestock with a continuous monitoring of individual animal performance levels, feed intake, and health status. When undesirable changes are detected, there could be an automatic adjustment in environmental conditions and feed rations. This information could also be related back to genetic background, physical facilities, feed rations, and other management factors to improve and fine-tune animal performance.

Financial transactions may be recorded and automatically transferred to accounts through the use of debit cards and bar code symbols whenever purchases and sales occur. These transactions might also be posted automatically to the accounting system for an individual farm. This would mean the information in a farmer's accounting system would be accurate and up to date at the end of each day.

Personal computers will have greatly enhanced capacities to receive, process, and store this information, and to communicate with outside data sources. Portable computers will allow precise decisions to be made in the pickup or on the tractor as well as in the office. Fiber optic systems and electronic transmission technology will increase the speed and accuracy of data sharing.

Managers in the twentieth century have often found the lack of accurate, timely, and complete information to be frustrating. The managers of the twenty-first century may also be frustrated by information, only it will be by the large quantity and continual flow of information available to them. A continual and vital task of twenty-first century managers will be to determine which information is critical to their decision making, which is useful, and which is irrelevant. Even when this is done, the critical and useful information must be analyzed and stored in an easily accessible manner for future reference.

FINANCIAL MANAGEMENT

Outside capital will continue to be needed to finance larger scale operations. Management of traditional sources of farm credit, such as rural banks, will be more vertically integrated, and funds will come from national money markets. Farm managers will increasingly have to compete with nonfarm businesses for access to capital, as the rural and urban financial markets become more and more closely tied together. This will require more detailed documentation of financial performance and credit needs, and more conformity to generally accepted accounting principles and performance measures. The standardized financial ratios and farm accounting practices recommended by the Farm Financial Standards Task Force are a step in this direction.

Some day farmers may lose the option to use cash accounting for income tax preparation. This, combined with farm lenders requesting audited financial statements prepared according to nonfarm accounting standards, will place continued pressure on farm managers to improve their record keeping. Standardized records will help make comparative analysis with similar farms more meaningful. The farm manager may have to decide whether to train an employee to carry out the required accounting and analysis, or hire this expertise from outside the business. Even if outside help is utilized, the manager must have the skills and knowledge needed to read, interpret, and use this accounting information.

LABOR AND HUMAN RESOURCES

The farm manager of the twenty-first century will come to depend more and more on a team of employees or partners to carry out specific duties in the operation. Working with other people will become a more important factor in the success of the operation. Motivation, communication, evaluation, and training of personnel will become essential skills.

Farm businesses will have to offer wages, benefits, and working conditions that are competitive with nonfarm employment opportunities. They will likely have to follow more regulations with regard to worker safety in handling farm chemicals and equipment and see that employees are properly trained in the use of new technologies. Many of the most efficient farms and ranches in the twenty-first century will be those with a small number of operators or employees who have specialized responsibilities, and who have mastered the communication and decision-making techniques needed in such operations.

PRODUCING TO MEET CONSUMER DEMANDS

Agriculture has long been characterized by the production of "undifferentiated" products. Grain and livestock products from different farms are essentially treated alike by buyers as long as these products meet basic quality standards and grades. Because the trend is to offer more highly specialized and processed food products to the consumer, buyers are beginning to implement stricter product standards for producers.

For example, livestock processors want uniform animals with specific size and leanness characteristics to fit their processing equipment, packaging standards, and quality levels. Improved measuring devices and data processing will make it easier to pay differential prices based on product characteristics. As these processors invest in larger scale plants, they must operate at full capacity to remain competitive. Producers who can assure the packer of a continuous supply of high-quality, uniform animals will receive a premium price. Those who cannot may find themselves shut out of many markets or forced to accept a lower price.

In crop production, the protein and oil content of grain and forages will also become easier to measure, making differential pricing possible. Biotechnology research may allow plant characteristics to be altered with genetically engineered varieties produced for specific uses and areas.

So-called niche markets will also become more important. Organic produce, extra-lean meat, specialty fruits and vegetables, and custom grown products for restaurants and food services will be in greater demand. Foreign markets will also be more important as international trade barriers continue to fall. These markets may require products with different characteristics. Farm managers who seek out these markets and learn the production techniques necessary to meet their specifications will realize a higher return from their resources. Of course, the manager will have to evaluate the additional costs and increased risks associated with specialty markets and then compare them with the potentially higher returns.

ENVIRONMENTAL CONCERNS

As the availability of an adequate quantity of food becomes more and more taken for granted, concerns about food quality, food safety, and the present and future condition of our soil, water, and air will continue to receive high priority from the general public. Farmers and ranchers have always had a strong interest in maintaining the productivity of natural resources under their control. However, the off-farm and long-term effects that new production technologies have had on the environment have not always been well quantified nor understood.

As research and experience improve understanding of the interactions among and between various biological systems, education and regulation will be used to increase the margin of safety for preserving resources for future generations. Top agricultural managers of today recognize the need to keep abreast of the environmental implications of their production practices and are often leaders in developing innovative production systems. All farm managers in the twenty-first century must be aware of the effects their production practices have on the environment, both on and off the farm, and take the steps necessary to keep our agricultural resources productive and environmentally safe.

The value of agricultural assets, particularly farmland, will be affected by environmental conditions and regulations. Environmental audits are becoming routine when farms are sold or appraised to warn potential buyers of any costs which might have to be incurred to clean up environmental hazards. The crop production combinations and practices allowed by a farm's conservation plan also affect its value. Farm managers in the twenty-first century will have to evaluate every decision not only in terms of profitability but also in terms of how it affects the environment. The successful managers will be those who can generate a profit while conserving productive and environmentally safe resources on the farm and minimizing environmental problems off the farm.

NEW TECHNOLOGY

Agricultural technology has been evolving and advancing for many decades and will continue to change in the future. The field of biotechnology offers the possibilities of gains in production efficiency. This may include crop varieties that are engineered to fit growing conditions at particular locations, that are resistant to pesticide damage or certain insects and diseases, or which have a more highly valued chemical composition such as higher protein or oil content. Livestock performance may be improved by introducing new genetic characteristics or by using growth stimulants. New nonfood uses for agricultural products will open new markets but may also bring about changes in the desired characteristics or composition of products grown specifically for these uses.

One example of a technology just now coming into agricultural application is the use of satellites to pinpoint the exact location of equipment in a field. Combined with other technology, this may have widespread application in the twenty-first century. For example, by combining satellite reception with a yield monitor on harvesting equipment, the crop yield can be measured and recorded continuously for every point in the field. Variations in yield due to soil type, previous crops, different tillage methods, and fertilizer rates can be quickly identified and recommendations made to correct problems. This technology is now being used to automatically adjust the application rates of fertilizer and chemicals as the applicator moves across the field. With this equipment, fertilizer and chemicals are applied only at the rates and locations needed.

This technology and others yet to be developed will provide the farm manager of the twenty-first century with a continual challenge. Should this or any new technology be adopted? The cost of any new technology must be weighed against its benefits which may come in several forms. There may be increased yields, an improvement in product quality, less variation in yield, a reduced impact on the environment or, most likely, some combination of these benefits. Decisions about when and if to adopt a new technology can have a major effect on the profitability of a farm or ranch business.

SUMMARY

Farmers and ranchers in the twenty-first century will be making most of the same basic decisions as they have in the twentieth century, but they will be making them faster and with more accurate information. Farm businesses will continue to become larger, and their operators will have to acquire specialized skills in managing personnel, interpreting data, competing for resources with nonfarm businesses, and customizing products to meet the demands of new markets. All this must be done while balancing the need to earn a profit in the short run with the need to preserve agricultural resources and environmental quality into the future.

QUESTIONS FOR REVIEW AND FURTHER THOUGHT

1 What forces have caused farms and ranches to become larger? Is this trend likely to continue? Why or why not?
2 How will quick access to more information help farm managers in the twenty-first century make better decisions?
3 List two examples of specialty agricultural markets and the changes a conventional producer might have to make to fill them.
4 List other challenges not discussed in this chapter that you think farm and ranch managers may have to face in the twenty-first century.

REFERENCES

Barnard, Freddie L., Michael Boehlje, and J. H. Atkinson: *Agriculture 2000: A Strategic Perspective,* Purdue University, West Lafayette, IN, 1992.

Castle, Emery N., Manning H. Becker, and A. Gene Nelson: *Farm Business Management,* Macmillan Publishing Company, New York, 3d ed., 1987, chap. 18.

Choices, Third Quarter, 1992, contains a number of articles about agriculture in the 21st century and the forces bringing about change. American Agricultural Economics Association, Ames, IA.

Urban, Thomas N.: "Agricultural Industrialization: It's Inevitable," *Choices,* Fourth Quarter, 1991, American Agricultural Economics Association, Ames, IA.

APPENDIX

TABLE 1   AMORTIZATION FACTORS FOR EQUAL ANNUAL TOTAL PAYMENTS

Years	4%	5%	6%	7%	8%	9%	10%	11%	12%	13%	14%	15%	16%
							Interest Rate						
1	1.04000	1.05000	1.06000	1.07000	1.08000	1.09000	1.10000	1.11000	1.12000	1.13000	1.14000	1.15000	1.16000
2	0.53020	0.53780	0.54544	0.55309	0.56077	0.56847	0.57619	0.58393	0.59170	0.59948	0.60729	0.61512	0.62296
3	0.36035	0.36721	0.37411	0.38105	0.38803	0.39505	0.40211	0.40921	0.41635	0.42352	0.43073	0.43798	0.44526
4	0.27549	0.28201	0.28859	0.29523	0.30192	0.30867	0.31547	0.32233	0.32923	0.33619	0.34320	0.35027	0.35738
5	0.22463	0.23097	0.23740	0.24389	0.25046	0.25709	0.26380	0.27057	0.27741	0.28431	0.29128	0.29832	0.30541
6	0.19076	0.19702	0.20336	0.20980	0.21632	0.22292	0.22961	0.23638	0.24323	0.25015	0.25716	0.26424	0.27139
7	0.16661	0.17282	0.17914	0.18555	0.19207	0.19869	0.20541	0.21222	0.21912	0.22611	0.23319	0.24036	0.24761
8	0.14853	0.15472	0.16104	0.16747	0.17401	0.18067	0.18744	0.19432	0.20130	0.20839	0.21557	0.22285	0.23022
9	0.13449	0.14069	0.14702	0.15349	0.16008	0.16680	0.17364	0.18060	0.18768	0.19487	0.20217	0.20957	0.21708
10	0.12329	0.12950	0.13587	0.14238	0.14903	0.15582	0.16275	0.16980	0.17698	0.18429	0.19171	0.19925	0.20690
11	0.11415	0.12039	0.12679	0.13336	0.14008	0.14695	0.15396	0.16112	0.16842	0.17584	0.18339	0.19107	0.19886
12	0.10655	0.11283	0.11928	0.12590	0.13270	0.13965	0.14676	0.15403	0.16144	0.16899	0.17667	0.18448	0.19241
13	0.10014	0.10646	0.11296	0.11965	0.12652	0.13357	0.14078	0.14815	0.15568	0.16335	0.17116	0.17911	0.18718
14	0.09467	0.10102	0.10758	0.11434	0.12130	0.12843	0.13575	0.14323	0.15087	0.15867	0.16661	0.17469	0.18290
15	0.08994	0.09634	0.10296	0.10979	0.11683	0.12406	0.13147	0.13907	0.14682	0.15474	0.16281	0.17102	0.17936
16	0.08582	0.09227	0.09895	0.10586	0.11298	0.12030	0.12782	0.13552	0.14339	0.15143	0.15962	0.16795	0.17641
17	0.08220	0.08870	0.09544	0.10243	0.10963	0.11705	0.12466	0.13247	0.14046	0.14861	0.15692	0.16537	0.17395
18	0.07899	0.08555	0.09236	0.09941	0.10670	0.11421	0.12193	0.12984	0.13794	0.14620	0.15462	0.16319	0.17188
19	0.07614	0.08275	0.08962	0.09675	0.10413	0.11173	0.11955	0.12756	0.13576	0.14413	0.15266	0.16134	0.17014
20	0.07358	0.08024	0.08718	0.09439	0.10185	0.10955	0.11746	0.12558	0.13388	0.14235	0.15099	0.15976	0.16867
21	0.07128	0.07800	0.08500	0.09229	0.09983	0.10762	0.11562	0.12384	0.13224	0.14081	0.14954	0.15842	0.16742
22	0.06920	0.07597	0.08305	0.09041	0.09803	0.10590	0.11401	0.12231	0.13081	0.13948	0.14830	0.15727	0.16635
23	0.06731	0.07414	0.08128	0.08871	0.09642	0.10438	0.11257	0.12097	0.12956	0.13832	0.14723	0.15628	0.16545
24	0.06559	0.07247	0.07968	0.08719	0.09498	0.10302	0.11130	0.11979	0.12846	0.13731	0.14630	0.15543	0.16467
25	0.06401	0.07095	0.07823	0.08581	0.09368	0.10181	0.11017	0.11874	0.12750	0.13643	0.14550	0.15470	0.16401
30	0.05783	0.06505	0.07265	0.08059	0.08883	0.09734	0.10608	0.11502	0.12414	0.13341	0.14280	0.15230	0.16189
35	0.05358	0.06107	0.06897	0.07723	0.08580	0.09464	0.10369	0.11293	0.12232	0.13183	0.14144	0.15113	0.16089
40	0.05052	0.05828	0.06646	0.07501	0.08386	0.09296	0.10226	0.11172	0.12130	0.13099	0.14075	0.15056	0.16042

TABLE 2 COMPOUND INTEREST TABLE OR FUTURE VALUE OF A $1 INVESTMENT

Interest Rate

Years	4%	5%	6%	7%	8%	9%	10%	11%	12%	13%	14%	15%	16%
1	1.0400	1.0500	1.0600	1.0700	1.0800	1.0900	1.1000	1.1100	1.1200	1.1300	1.1400	1.1500	1.1600
2	1.0816	1.1025	1.1236	1.1449	1.1664	1.1881	1.2100	1.2321	1.2544	1.2769	1.2996	1.3225	1.3456
3	1.1249	1.1576	1.1910	1.2250	1.2597	1.2950	1.3310	1.3676	1.4049	1.4429	1.4815	1.5209	1.5609
4	1.1699	1.2155	1.2625	1.3108	1.3605	1.4116	1.4641	1.5181	1.5735	1.6305	1.6890	1.7490	1.8106
5	1.2167	1.2763	1.3382	1.4026	1.4693	1.5386	1.6105	1.6851	1.7623	1.8424	1.9254	2.0114	2.1003
6	1.2653	1.3401	1.4185	1.5007	1.5869	1.6771	1.7716	1.8704	1.9738	2.0820	2.1950	2.3131	2.4364
7	1.3159	1.4071	1.5036	1.6058	1.7138	1.8280	1.9487	2.0762	2.2107	2.3526	2.5023	2.6600	2.8262
8	1.3686	1.4775	1.5938	1.7182	1.8509	1.9926	2.1436	2.3045	2.4760	2.6584	2.8526	3.0590	3.2784
9	1.4233	1.5513	1.6895	1.8385	1.9990	2.1719	2.3579	2.5580	2.7731	3.0040	3.2519	3.5179	3.8030
10	1.4802	1.6289	1.7908	1.9672	2.1589	2.3674	2.5937	2.8394	3.1058	3.3946	3.7072	4.0456	4.4114
11	1.5395	1.7103	1.8983	2.1049	2.3316	2.5804	2.8531	3.1518	3.4785	3.8359	4.2262	4.6524	5.1173
12	1.6010	1.7959	2.0122	2.2522	2.5182	2.8127	3.1384	3.4985	3.8960	4.3345	4.8179	5.3503	5.9360
13	1.6651	1.8856	2.1329	2.4098	2.7196	3.0658	3.4523	3.8833	4.3635	4.8980	5.4924	6.1528	6.8858
14	1.7317	1.9799	2.2609	2.5785	2.9372	3.3417	3.7975	4.3104	4.8871	5.5348	6.2613	7.0757	7.9875
15	1.8009	2.0789	2.3966	2.7590	3.1722	3.6425	4.1772	4.7846	5.4736	6.2543	7.1379	8.1371	9.2655
16	1.8730	2.1829	2.5404	2.9522	3.4259	3.9703	4.5950	5.3109	6.1304	7.0673	8.1372	9.3576	10.7480
17	1.9479	2.2920	2.6928	3.1588	3.7000	4.3276	5.0545	5.8951	6.8660	7.9861	9.2765	10.7613	12.4677
18	2.0258	2.4066	2.8543	3.3799	3.9960	4.7171	5.5599	6.5436	7.6900	9.0243	10.5752	12.3755	14.4625
19	2.1068	2.5270	3.0256	3.6165	4.3157	5.1417	6.1159	7.2633	8.6128	10.1974	12.0557	14.2318	16.7765
20	2.1911	2.6533	3.2071	3.8697	4.6610	5.6044	6.7275	8.0623	9.6463	11.5231	13.7435	16.3665	19.4608
21	2.2788	2.7860	3.3996	4.1406	5.0338	6.1088	7.4002	8.9492	10.8038	13.0211	15.6676	18.8215	22.5745
22	2.3699	2.9253	3.6035	4.4304	5.4365	6.6586	8.1403	9.9336	12.1003	14.7138	17.8610	21.6447	26.1864
23	2.4647	3.0715	3.8197	4.7405	5.8715	7.2579	8.9543	11.0263	13.5523	16.6266	20.3616	24.8915	30.3762
24	2.5633	3.2251	4.0489	5.0724	6.3412	7.9111	9.8497	12.2392	15.1786	18.7881	23.2122	28.6252	35.2364
25	2.6658	3.3864	4.2919	5.4274	6.8485	8.6231	10.8347	13.5855	17.0001	21.2305	26.4619	32.9190	40.8742
30	3.2434	4.3219	5.7435	7.6123	10.0627	13.2677	17.4494	22.8923	29.9599	39.1159	50.9502	66.2118	85.8499
35	3.9461	5.5160	7.6861	10.6766	14.7853	20.4140	28.1024	38.5749	52.7996	72.0685	98.1002	133.1755	180.3141
40	4.8010	7.0400	10.2857	14.9745	21.7245	31.4094	45.2593	65.0009	93.0510	132.7816	188.8635	267.8635	378.7212

TABLE 3   AMOUNT OF AN ANNUITY OR FUTURE VALUE OF $1 INVESTED AT END OF EACH YEAR

Interest Rate

Years	4%	5%	6%	7%	8%	9%	10%	11%	12%	13%	14%	15%	16%
1	1.0000	1.0000	1.0000	1.0000	1.0000	1.0000	1.0000	1.0000	1.0000	1.0000	1.0000	1.0000	1.0000
2	2.0400	2.0500	2.0600	2.0700	2.0800	2.0900	2.1000	2.1100	2.1200	2.1300	2.1400	2.1500	2.1600
3	3.1216	3.1525	3.1836	3.2149	3.2464	3.2781	3.3100	3.3421	3.3744	3.4069	3.4396	3.4725	3.5056
4	4.2465	4.3101	4.3746	4.4399	4.5061	4.5731	4.6410	4.7097	4.7793	4.8498	4.9211	4.9934	5.0665
5	5.4163	5.5256	5.6371	5.7507	5.8666	5.9847	6.1051	6.2278	6.3528	6.4803	6.6101	6.7424	6.8771
6	6.6330	6.8019	6.9753	7.1533	7.3359	7.5233	7.7156	7.9129	8.1152	8.3227	8.5355	8.7537	8.9775
7	7.8983	8.1420	8.3938	8.6540	8.9228	9.2004	9.4872	9.7833	10.0890	10.4047	10.7305	11.0668	11.4139
8	9.2142	9.5491	9.8975	10.2598	10.6366	11.0285	11.4359	11.8594	12.2997	12.7573	13.2328	13.7268	14.2401
9	10.5828	11.0266	11.4913	11.9780	12.4876	13.0210	13.5795	14.1640	14.7757	15.4157	16.0853	16.7858	17.5185
10	12.0061	12.5779	13.1808	13.8164	14.4866	15.1929	15.9374	16.7220	17.5487	18.4197	19.3373	20.3037	21.3215
11	13.4864	14.2068	14.9716	15.7836	16.6455	17.5603	18.5312	19.5614	20.6546	21.8143	23.0445	24.3493	25.7329
12	15.0258	15.9171	16.8699	17.8885	18.9771	20.1407	21.3843	22.7132	24.1331	25.6502	27.2707	29.0017	30.8502
13	16.6268	17.7130	18.8821	20.1406	21.4953	22.9534	24.5227	26.2116	28.0291	29.9847	32.0887	34.3519	36.7862
14	18.2919	19.5986	21.0151	22.5505	24.2149	26.0192	27.9750	30.0949	32.3926	34.8827	37.5811	40.5047	43.6720
15	20.0236	21.5786	23.2760	25.1290	27.1521	29.3609	31.7725	34.4054	37.2797	40.4175	43.8424	47.5804	51.6595
16	21.8245	23.6575	25.6725	27.8881	30.3243	33.0034	35.9497	39.1899	42.7533	46.6717	50.9804	55.7175	60.9250
17	23.6975	25.8404	28.2129	30.8402	33.7502	36.9737	40.5447	44.5008	48.8837	53.7391	59.1176	65.0751	71.6730
18	25.6454	28.1324	30.9057	33.9990	37.4502	41.3013	45.5992	50.3959	55.7497	61.7251	68.3941	75.8364	84.1407
19	27.6712	30.5390	33.7600	37.3790	41.4463	46.0185	51.1591	56.9395	63.4397	70.7494	78.9692	88.2118	98.6032
20	29.7781	33.0660	36.7856	40.9955	45.7620	51.1601	57.2750	64.2028	72.0524	80.9468	91.0249	102.4436	115.3797
21	31.9692	35.7193	39.9927	44.8652	50.4229	56.7645	64.0025	72.2651	81.6987	92.4699	104.7684	118.8101	134.8405
22	34.2480	38.5052	43.3923	49.0057	55.4568	62.8733	71.4027	81.2143	92.5026	105.4910	120.4360	137.6316	157.4150
23	36.6179	41.4305	46.9958	53.4361	60.8933	69.5319	79.5430	91.1479	104.6029	120.2048	138.2970	159.2764	183.6014
24	39.0826	44.5020	50.8156	58.1767	66.7648	76.7898	88.4973	102.1742	118.1552	136.8315	158.6586	184.1678	213.9776
25	41.6459	47.7271	54.8645	63.2490	73.1059	84.7009	98.3471	114.4133	133.3339	155.6196	181.8708	212.7930	249.2140
30	56.0849	66.4388	79.0582	94.4608	113.2832	136.3075	164.4940	199.0209	241.3327	293.1992	356.7868	434.7451	530.3117
35	73.6522	90.3203	111.4348	138.2369	172.3168	215.7108	271.0244	341.5896	431.6635	546.6808	693.5727	881.1702	1120.7130
40	95.0255	120.7998	154.7620	199.6351	259.0565	337.8824	442.5926	581.8261	767.0914	1013.7042	1342.0251	1779.0903	2360.7572

TABLE 4 PRESENT VALUE OF $1 TO BE RECEIVED AT END OF A SPECIFIED TIME PERIOD

Interest Rate

Years	4%	5%	6%	7%	8%	9%	10%	11%	12%	13%	14%	15%	16%
1	0.96154	0.95238	0.94340	0.93458	0.92593	0.91743	0.90909	0.90090	0.89286	0.88496	0.87719	0.86957	0.86207
2	0.92456	0.90703	0.89000	0.87344	0.85734	0.84168	0.82645	0.81162	0.79719	0.78315	0.76947	0.75614	0.74316
3	0.88900	0.86384	0.83962	0.81630	0.79383	0.77218	0.75131	0.73119	0.71178	0.69305	0.67497	0.65752	0.64066
4	0.85480	0.82270	0.79209	0.76290	0.73503	0.70843	0.68301	0.65873	0.63552	0.61332	0.59208	0.57175	0.55229
5	0.82193	0.78353	0.74726	0.71299	0.68058	0.64993	0.62092	0.59345	0.56743	0.54276	0.51937	0.49718	0.47611
6	0.79031	0.74622	0.70496	0.66634	0.63017	0.59627	0.56447	0.53464	0.50663	0.48032	0.45559	0.43233	0.41044
7	0.75992	0.71068	0.66506	0.62275	0.58349	0.54703	0.51316	0.48166	0.45235	0.42506	0.39964	0.37594	0.35383
8	0.73069	0.67684	0.62741	0.58201	0.54027	0.50187	0.46651	0.43393	0.40388	0.37616	0.35056	0.32690	0.30503
9	0.70259	0.64461	0.59190	0.54393	0.50025	0.46043	0.42410	0.39092	0.36061	0.33288	0.30751	0.28426	0.26295
10	0.67556	0.61391	0.55839	0.50835	0.46319	0.42241	0.38554	0.35218	0.32197	0.29459	0.26974	0.24718	0.22668
11	0.64958	0.58468	0.52679	0.47509	0.42888	0.38753	0.35049	0.31728	0.28748	0.26070	0.23662	0.21494	0.19542
12	0.62460	0.55684	0.49697	0.44401	0.39711	0.35553	0.31863	0.28584	0.25668	0.23071	0.20756	0.18691	0.16846
13	0.60057	0.53032	0.46884	0.41496	0.36770	0.32618	0.28966	0.25751	0.22917	0.20416	0.18207	0.16253	0.14523
14	0.57748	0.50507	0.44230	0.38782	0.34046	0.29925	0.26333	0.23199	0.20462	0.18068	0.15971	0.14133	0.12520
15	0.55526	0.48102	0.41727	0.36245	0.31524	0.27454	0.23939	0.20900	0.18270	0.15989	0.14010	0.12289	0.10793
16	0.53391	0.45811	0.39365	0.33873	0.29189	0.25187	0.21763	0.18829	0.16312	0.14150	0.12289	0.10686	0.09304
17	0.51337	0.43630	0.37136	0.31657	0.27027	0.23107	0.19784	0.16963	0.14564	0.12522	0.10780	0.09293	0.08021
18	0.49363	0.41552	0.35034	0.29586	0.25025	0.21199	0.17986	0.15282	0.13004	0.11081	0.09456	0.08081	0.06914
19	0.47464	0.39573	0.33051	0.27651	0.23171	0.19449	0.16351	0.13768	0.11611	0.09806	0.08295	0.07027	0.05961
20	0.45639	0.37689	0.31180	0.25842	0.21455	0.17843	0.14864	0.12403	0.10367	0.08678	0.07276	0.06110	0.05139
21	0.43883	0.35894	0.29416	0.24151	0.19866	0.16370	0.13513	0.11174	0.09256	0.07680	0.06383	0.05313	0.04430
22	0.42196	0.34185	0.27751	0.22571	0.18394	0.15018	0.12285	0.10067	0.08264	0.06796	0.05599	0.04620	0.03819
23	0.40573	0.32557	0.26180	0.21095	0.17032	0.13778	0.11168	0.09069	0.07379	0.06014	0.04911	0.04017	0.03292
24	0.39012	0.31007	0.24698	0.19715	0.15770	0.12640	0.10153	0.08170	0.06588	0.05323	0.04308	0.03493	0.02838
25	0.37512	0.29530	0.23300	0.18425	0.14602	0.11597	0.09230	0.07361	0.05882	0.04710	0.03779	0.03038	0.02447
30	0.30832	0.23138	0.17411	0.13137	0.09938	0.07537	0.05731	0.04368	0.03338	0.02557	0.01963	0.01510	0.01165
35	0.25342	0.18129	0.13011	0.09366	0.06763	0.04899	0.03558	0.02592	0.01894	0.01388	0.01019	0.00751	0.00555
40	0.20829	0.14205	0.09722	0.06678	0.04603	0.03184	0.02209	0.01538	0.01075	0.00753	0.00529	0.00373	0.00264

TABLE 5   PRESENT VALUE OF AN ANNUITY OF $1 TO BE RECEIVED AT THE END OF EACH YEAR

Interest Rate

Years	4%	5%	6%	7%	8%	9%	10%	11%	12%	13%	14%	15%	16%
1	0.9615	0.9524	0.9434	0.9346	0.9259	0.9174	0.9091	0.9009	0.8929	0.8850	0.8772	0.8696	0.8621
2	1.8861	1.8594	1.8334	1.8080	1.7833	1.7591	1.7355	1.7125	1.6901	1.6681	1.6467	1.6257	1.6052
3	2.7751	2.7232	2.6730	2.6243	2.5771	2.5313	2.4869	2.4437	2.4018	2.3612	2.3216	2.2832	2.2459
4	3.6299	3.5460	3.4651	3.3872	3.3121	3.2397	3.1699	3.1024	3.0373	2.9745	2.9137	2.8550	2.7982
5	4.4518	4.3295	4.2124	4.1002	3.9927	3.8897	3.7908	3.6959	3.6048	3.5172	3.4331	3.3522	3.2743
6	5.2421	5.0757	4.9173	4.7665	4.6229	4.4859	4.3553	4.2305	4.1114	3.9975	3.8887	3.7845	3.6847
7	6.0021	5.7864	5.5824	5.3893	5.2064	5.0330	4.8684	4.7122	4.5638	4.4226	4.2883	4.1604	4.0386
8	6.7327	6.4632	6.2098	5.9713	5.7466	5.5348	5.3349	5.1461	4.9676	4.7988	4.6389	4.4873	4.3436
9	7.4353	7.1078	6.8017	6.5152	6.2469	5.9952	5.7590	5.5370	5.3282	5.1317	4.9464	4.7716	4.6065
10	8.1109	7.7217	7.3601	7.0236	6.7101	6.4177	6.1446	5.8892	5.6502	5.4262	5.2161	5.0188	4.8332
11	8.7605	8.3064	7.8869	7.4987	7.1390	6.8052	6.4951	6.2065	5.9377	5.6869	5.4527	5.2337	5.0286
12	9.3851	8.8633	8.3838	7.9427	7.5361	7.1607	6.8137	6.4924	6.1944	5.9176	5.6603	5.4206	5.1971
13	9.9856	9.3936	8.8527	8.3577	7.9038	7.4869	7.1034	6.7499	6.4235	6.1218	5.8424	5.5831	5.3423
14	10.5631	9.8986	9.2950	8.7455	8.2442	7.7862	7.3667	6.9819	6.6282	6.3025	6.0021	5.7245	5.4675
15	11.1184	10.3797	9.7122	9.1079	8.5595	8.0607	7.6061	7.1909	6.8109	6.4624	6.1422	5.8474	5.5755
16	11.6523	10.8378	10.1059	9.4466	8.8514	8.3126	7.8237	7.3792	6.9740	6.6039	6.2651	5.9542	5.6685
17	12.1657	11.2741	10.4773	9.7632	9.1216	8.5436	8.0216	7.5488	7.1196	6.7291	6.3729	6.0472	5.7487
18	12.6593	11.6896	10.8276	10.0591	9.3719	8.7556	8.2014	7.7016	7.2497	6.8399	6.4674	6.1280	5.8178
19	13.1339	12.0853	11.1581	10.3356	9.6036	8.9501	8.3649	7.8393	7.3658	6.9380	6.5504	6.1982	5.8775
20	13.5903	12.4622	11.4699	10.5940	9.8181	9.1285	8.5136	7.9633	7.4694	7.0248	6.6231	6.2593	5.9288
21	14.0292	12.8212	11.7641	10.8355	10.0168	9.2922	8.6487	8.0751	7.5620	7.1016	6.6870	6.3125	5.9731
22	14.4511	13.1630	12.0416	11.0612	10.2007	9.4424	8.7715	8.1757	7.6446	7.1695	6.7429	6.3587	6.0113
23	14.8568	13.4886	12.3034	11.2722	10.3711	9.5802	8.8832	8.2664	7.7184	7.2297	6.7921	6.3988	6.0442
24	15.2470	13.7986	12.5504	11.4693	10.5288	9.7066	8.9847	8.3481	7.7843	7.2829	6.8351	6.4338	6.0726
25	15.6221	14.0939	12.7834	11.6536	10.6748	9.8226	9.0770	8.4217	7.8431	7.3300	6.8729	6.4641	6.0971
30	17.2920	15.3725	13.7648	12.4090	11.2578	10.2737	9.4269	8.6938	8.0552	7.4957	7.0027	6.5660	6.1772
35	18.6646	16.3742	14.4982	12.9477	11.6546	10.5668	9.6442	8.8552	8.1755	7.5856	7.0700	6.6166	6.2153
40	19.7928	17.1591	15.0463	13.3317	11.9246	10.7574	9.7791	8.9511	8.2438	7.6344	7.1050	6.6418	6.2335

GLOSSARY

accounting A comprehensive system for recording and summarizing business transactions.

accounting period The period of time over which accounting transactions are summarized.

account payable An expense that has been incurred but not yet paid.

account receivable Income that has been earned but for which no payment has been received.

accrual accounting An accounting system that recognizes income when it is earned and expenses when they are incurred.

accrued expense An expense that has been incurred, sometimes accumulating over time, but has not been paid.

adjusted basis The current tax basis of an asset, equal to the original basis reduced by the amount of depreciation expense claimed and/or increased by the cost of any improvements made.

Agricultural Stabilization and Conservation Service (ASCS) An agency of the U.S. Department of Agriculture that administers farm production control, price stabilization, and conservation programs.

amortized loan A loan that is scheduled to be repaid in a series of periodic payments of both interest and principal.

annual percentage rate (APR) The true annual rate at which interest is charged on a loan.

annuity A series of equal periodic payments.

appraisal The process of estimating the market value of an asset.

appreciation An increase in the market value of an asset.

asset Physical or financial property which has value and is owned by a business or individual.

average fixed cost (AFC)   Total fixed cost divided by total output; average fixed cost per unit of output.

average physical product (APP)   The average amount of physical output produced for each unit of input used.

average total cost (ATC)   Total cost divided by total output; average cost per unit of output.

average variable cost (AVC)   Total variable cost divided by total output; average variable cost per unit of output.

balance sheet   A report summarizing the assets, liabilities, and net worth of the business at a point in time.

balloon loan   A loan amortization method in which a large portion of the principal is due with the final payment.

basis (marketing)   The difference between the local cash price and the futures contract price of the same commodity.

basis (tax)   The value of an asset for tax purposes.

bonus   A payment made to an employee in addition to the normal salary, based on superior performance or other criteria.

book value   The original cost of an asset minus the total depreciation expense taken to date.

break-even price   The selling price for which total income will just equal total expenses for a given level of production.

break-even yield   The yield level at which total income will just equal total expenses at a given selling price.

breeding livestock   Livestock owned for the primary purpose of producing offspring.

budget   An estimate of future income, expenses, or cash flows.

bushel lease   A leasing arrangement in which the rent is paid as a specified number of bushels of grain delivered to the landowner.

capital   A collection of physical and financial assets that have a market value.

capital asset   An asset that is expected to provide services through more than one production cycle, and can be used to produce other assets or services.

capital budgeting   A process for determining the profitability of a capital investment.

capital gain   The amount by which the sale value of an asset exceeds its cost or tax basis.

capitalization method   A procedure for estimating the value of an asset by dividing the expected annual net returns by an annual discount rate.

cash accounting   An accounting system that recognizes income when it is actually received as cash and expenses when they are actually paid.

cash expenses   Expenses that require the expenditure of cash.

cash flow   The flow of funds into and out of a business.

cash flow budget   A projection of the expected cash inflows and cash outflows for a business over a period of time.

cash rent   A rental arrangement in which the operator makes a cash payment to the owner for the use of certain property, pays all production costs, and keeps all the income generated.

coefficient of variation   A measure of the variability of the outcomes of a particular event; equal to the standard deviation divided by the mean.

collateral   Assets pledged as security for a loan.

Commodity Credit Corporation (CCC)   An entity controlled by the U.S. Department of

Agriculture which buys and sells surplus commodities and provides loans for certain government farm programs.

comparable sale   An actual land sale used in an appraisal to help estimate the market value of a similar piece of land.

comparative analysis   The comparison of the performance level of a farm business to the performance level of other similar farms in the same area or to other established standards.

competitive enterprises   Enterprises for which the output level of one can be increased only by decreasing the output level of the other.

complementary enterprises   Enterprises for which increasing the output level of one also increases the output level of the other.

compounding   The process of determining the future value of an investment or loan, in which interest is charged on the accumulated interest as well as the original capital.

control   The process of monitoring the progress of a farm business and taking corrective action when desired performance levels are not being met.

corporation   A form of business organization in which the owners have shares in a separate legal entity that itself can own assets and borrow money.

credit   The capacity or ability to borrow money.

crop share lease   A lease agreement in which crop production and certain input costs are divided between the operator and the owner.

cumulative distribution function   A graph of all the possible outcomes for a certain event, and the probability that each outcome or one with a lower value will occur.

current assets   Assets that are normally used up or sold within a year and which can be converted to cash quickly.

current liabilities   Liabilities that are normally paid within a year.

current ratio   The ratio of current assets to current liabilities; a measure of liquidity.

custom farming   An arrangement in which the landowner pays the operator a fixed amount to perform all the labor and machinery operations needed to produce and harvest a crop.

custom work   An arrangement in which an operator performs one or more machinery operations for someone else for a fixed charge.

cwt   An abbreviation for hundredweight, equal to 100 pounds. Many livestock products and some crops are priced by this unit.

debt   An obligation to pay, such as a loan or account payable.

debt/asset ratio   The ratio of total liabilities to total assets; a measure of solvency.

debt/equity ratio   The ratio of total liabilities to owner's equity; a measure of solvency.

debt service   The payment of debts according to a specified schedule.

decision tree   A diagram that traces out all the possible strategies and outcomes for a particular decision or sequence of related decisions.

deflation   A general decrease in the level of all prices.

depreciation   An annual, noncash expense to recognize the amount by which an asset loses value due to use, age, and obsolescence.

depreciation recapture   Taxable income that results from selling a depreciable asset for more than its adjusted tax basis.

diminishing returns   A decline in the rate at which total output increases as more and more inputs are used.

discounting   The process of reducing the value of a sum to be paid or received in the future by the amount of interest that would be accumulated on it to that point in time.

discount rate   The interest rate used to find the present value of an amount to be paid or received in the future.

diseconomies of size   A production relationship in which the average total cost per unit of output increases as more output is produced or is negatively correlated.

diversification   The production of two or more commodities for which production levels and/or prices are not closely correlated.

double-entry accounting   An accounting system in which changes in assets and liabilities as well as income and expenses are recorded.

down payment   The portion of the cost of purchasing a capital asset that is financed from owner's equity.

economic efficiency   The ratio of the value of output per physical unit of input or per unit cost of the input.

economies of size   A production relationship in which average total cost per unit of output decreases as output increases.

enterprise   Production of a single crop or type of livestock, such as wheat or dairy.

enterprise budget   A projection of all the costs and returns for a single enterprise.

equal marginal principle   The principle which states that a limited resource should be allocated among competing uses in such a way that the marginal value products from the last unit in each use are equal.

equity   The amount by which the value of total assets exceeds total liabilities; the amount of the owner's own capital invested in the business.

equity/asset ratio   The ratio of owner's equity to total assets; a measure of solvency.

expected value   The weighted average outcome from an uncertain event based on its possible outcomes and their respective probabilities.

extension service   An educational service for farmers and others provided jointly by the U.S. Department of Agriculture, state land grant universities, and county governments.

Farmers Home Administration (FmHA)   An agency of the U.S. Department of Agriculture that provides credit to beginning farmers and other operators who are unable to obtain it from conventional sources.

Farm Financial Standards Task Force (FFSTF)   A committee of agricultural financial experts that developed a set of guidelines for uniform financial reporting and analysis of farm businesses.

farm management   The process of making decisions about the allocation of scarce resources in agricultural production for the purpose of meeting certain management goals.

feasibility analysis   An analysis of the cash inflows generated by an investment compared to the cash outflows required.

feeder livestock   Young livestock that are purchased for the purpose of being fed until they reach slaughter weight.

field efficiency   The actual accomplishment rate for a field implement as a percent of the theoretical accomplishment rate if no time were lost due to overlapping, turning, and adjusting the machine.

financing   The acquisition of funds to meet the cash flow requirements of an investment or production activity.

fiscal year   An annual accounting period that does not correspond to the calendar year.

fixed assets   Assets that are expected to have a long or indefinite productive life.

fixed costs   Costs that will not change in the short run or with the level of production and exist even if no production takes place.

fringe benefits   Compensation provided to employees in addition to cash wages and salary.

future value (FV)   The value that a payment or set of payments will have at some time in the future, when interest is compounded.

gross income   The total income, both cash and noncash, received from an enterprise or business, before any expenses are paid.

gross margin   The difference between gross income and variable costs; also called income above variable costs.

gross revenue   The total of all the revenue received by a business over a period of time; same as gross income.

hedging   A strategy for reducing the risk of a decline in prices by selling a commodity futures contract in advance of when the actual commodity is sold.

implementation   The process of carrying out management decisions.

improvements   Repairs, renovations, or additions to capital assets which improve their productivity and/or extend their useful lives.

income   Economic gain resulting from the production of goods and services, including receipts from the sale of commodities, other cash payments, increases in inventories, and accounts receivable.

income statement   A report that summarizes the income and expenses of a business over a period of time.

inflation   A general increase in the level of all prices over time.

input   A resource used in the production of an output.

interest   The cash cost paid to a lender for the use of borrowed money, or the opportunity cost of investing equity capital in an alternative use.

internal rate of return (IRR)   The discount or interest rate at which the net present value of an investment is just equal to zero.

internal transaction   A noncash accounting transaction carried out between two enterprises within the same business.

inventory   A complete listing of the number, type, and value of assets owned at a point in time.

isoquant   A line on a graph connecting points that represent all the possible combinations of inputs that can produce the same output.

joint venture   Any of several forms of business operation in which more than one person is involved in ownership and management.

labor share lease   A leasing agreement in which the operator receives a share of the production in exchange for contributing only labor.

land contract   An agreement by which a land buyer makes principal and interest payments to the seller on a regular schedule.

law of diminishing returns   A relationship observed in many physical and biological production processes in which the marginal physical product declines as more and more units of a variable input are used in combination with one or more fixed inputs.

lease   An agreement that allows a person to use and/or possess someone else's property in exchange for a rental payment.

lessee   An operator who leases property from the owner; same as tenant.

lessor   An owner who leases property to a lessee.

leverage   The practice of using credit to increase the total capital managed beyond the value of the owner's equity.

liabilities   Financial obligations which must be paid at some future time.

limited partnership   A form of business in which more than one person has ownership, but some (the limited partners) do not participate in management and have liability limited to the amount of their investment.

linear programming   A mathematical technique used to find a set of economic activities that maximizes or minimizes a certain objective, given a set of limited resources and/or other constraints.

line of credit   An arrangement by which a lender transfers funds to the borrower as they are needed, up to a maximum amount.

liquidate   To convert the assets of a business into cash.

liquidity   The ability of a business to meet its cash financial obligations as they come due.

livestock share lease   A lease agreement in which both the owner and operator contribute capital and share the production of crops and livestock.

long-term liabilities   Liabilities that are scheduled to be repaid over a period of 10 years or longer.

lumpy input   A resource that can be obtained only in certain indivisible sizes, such as a combine or a full-time employee.

marginal cost (MC)   The additional cost incurred from producing an additional unit of output.

marginal input cost (MIC)   The additional cost incurred by using an additional unit of input.

marginal physical product (MPP)   The additional physical product resulting from the use of an additional unit of input.

marginal revenue (MR)   The additional income received from selling one additional unit of output.

marginal tax rate   The additional tax that results from an additional dollar of taxable income.

marginal value product (MVP)   The additional income received from using an additional unit of input.

marketable securities   Stocks, bonds, and other financial instruments that can be bought and sold easily.

market livestock   Animals that are fed for eventual slaughter.

market value   The value for which an asset would be sold in an open transaction.

Modified Accelerated Cost Recovery System (MACRS)   A system for calculating tax depreciation as specified by federal income tax regulations.

mortgage   A legal agreement by which a lender receives the right to acquire a borrower's property to satisfy a debt if the repayment schedule is not met.

net farm income   The difference between total revenue and total expenses, including gain or loss on the sale of all capital assets; also the return to owner equity, unpaid labor, and management.

net farm income from operations   The difference between total revenue and total expenses, *not* including gain or loss on the sale of certain capital assets.

net operating loss (NOL)   A negative net farm profit for income tax purposes, which can be used to offset past and/or future taxable income.

net present value (NPV)   The present value of the net cash flows that will result from an investment, minus the amount of the original investment.

net worth   The difference between the value of the assets owned by a business and the value of its liabilities. Also called equity.

noncash expense   An expense that does not involve the expenditure of cash, such as depreciation.

noncurrent asset An asset which will normally be owned or used up over a period longer than a year.

noncurrent liability A liability which will normally be paid over a period longer than a year.

non-real estate All assets other than land and items attached to land such as buildings and fences.

operating costs Costs for the purchase of inputs and services that are used up relatively quickly, usually in one production cycle.

operating profit margin ratio The value represented by net farm income from operations plus interest expense minus opportunity cost of operator labor and management expressed as a percentage of gross revenue.

opportunity cost The income that could be received by employing a resource in its most profitable alternative use.

option A marketing transaction in which a buyer pays a seller a premium to acquire the right to sell or buy a futures contract at a specified price.

organizational chart A diagram that shows the workers involved in a business and the lines of authority and communication among them.

output The result or yield from a production process, such as raising crops and livestock.

overhead costs Costs that are not directly related to the type and quantity of products produced; a type of fixed cost.

owner's equity The difference between the total value of the assets of a business and the total value of its liabilities; also called net worth or equity.

ownership costs Costs that result simply from owning assets, regardless of how much they are used.

partial budget An estimate of the changes in income and expenses that would result from carrying out a proposed change in the current farm plan.

partnership A form of business organization in which more than one operator owns the resources and/or provides management.

payback period The length of time it takes for the accumulated net returns earned from an investment to equal the original investment.

payoff matrix A contingency table that illustrates the possible outcomes for a particular occurrence and their respective probabilities.

person-year equivalent A total of 12 months of labor contributed by one or more persons.

physical efficiency The ratio of output received per unit of input used all in physical units.

prepaid expense A payment made for an input or service prior to the accounting period in which it will be used.

present value (PV) The current value of a set of payments to be received or paid out over a period of time.

price ratio The ratio of the price of the input being added to the price of the input being replaced, or the ratio of the price of the output being gained to the price of the output being lost.

principal The amount borrowed or that part of the original loan which has not yet been repaid.

probability distribution A set of possible outcomes to a particular event and the probability of each one occurring.

production function A physical or biological relationship showing how much output results from using certain quantities of inputs.

production possibility curve A line on a graph connecting points representing all the possible combinations of outputs which can be produced from a fixed set of resources.

profit (economic) The value that remains after all costs, including opportunity costs, have been subtracted from gross income.

profitability The degree to which the value of the income derived from a set of resources exceeds their cost.

promissory note A legal agreement that obligates a borrower to repay a loan.

real estate Land or assets permanently attached to land.

retained farm earnings Net income generated by a farm business that is used to increase owner's equity rather than withdrawn to pay for living expenses, taxes, or dividends.

return on assets (ROA) The value represented by net farm income from operations plus interest expense minus the opportunity cost of operator labor and management. It is usually expressed as a percentage of the average value of total assets.

return on equity (ROE) The net return generated by the business before gains or losses on capital assets are realized, but after the value of unpaid labor and management is subtracted. Usually expressed as a percent of the average value of owner's equity.

return to management The net return generated by a business after all expenses have been paid and the opportunity costs for owner's equity and unpaid labor have been subtracted.

risk A situation in which more than one possible outcome exists, some of which may be unfavorable.

rule of 72 A relationship used to estimate the time it will take for an investment to double in value; found by dividing 72 by the percent rate of return earned on the investment.

salvage value The market value of a depreciable asset at the time it is sold or removed from service.

secured loan A loan for which the borrower agrees to let the lender take possession of and sell certain assets if the repayment terms are not met.

self-liquidating loan A loan which will be repaid from the sale of the assets originally purchased with the loan funds.

sensitivity analysis A procedure for assessing the riskiness of a decision by using several possible price and/or production outcomes to budget the results and comparing them.

shadow price Values obtained from a linear programming solution that show the amount by which total gross margin would be increased if one more unit of a limiting input were available, or the amount by which total gross margin would be reduced if one unit of an enterprise not in the solution were carried out.

short-term loan A loan scheduled to be repaid in less than a year.

signature loan A loan for which no collateral is pledged.

simplified programming A mathematical procedure used in whole farm planning to find the set of enterprises that maximizes total gross margin given a set of limited resources.

single-entry accounting An accounting system in which income and expenses are recorded but changes in assets and liabilities are not.

Soil Conservation Service (SCS) An agency of the U.S. Department of Agriculture that provides technical and financial assistance for carrying out soil and water conservation practices.

sole proprietorship A form of business organization in which one operator or family owns the resources and provides the management.

solvency The degree to which the liabilities of a business are backed up by assets; the relationship between debt and equity capital.

standard deviation   A measure of the variability of possible outcomes for a particular event; equal to the square root of the variance.

statement of cash flows   A summary of the actual cash inflows and cash outflows experienced by a business during some past time period.

subjective probability   Probabilities based only on individual judgment and past experiences.

substitution ratio   The ratio of the amount of one input replaced to the amount of another input added, or the amount of one output lost to the amount of another output gained.

sunk cost   A cost that can no longer be reversed, changed, or avoided; a fixed cost.

supplementary enterprises   Enterprises for which the level of production of one can be increased without affecting the level of production of the other.

sustainable agriculture   Agricultural production practices that maximize the long-run social and economic benefits from the use of land and other agricultural resources.

systems analysis   An evaluation of individual enterprises and technologies that takes into account their interactions with other enterprises and technologies.

tax-free exchange   A trade of one piece of farm property for another similar piece of property, such that any taxable gain is reduced or eliminated.

technical coefficient   The rate at which units of input are transformed into output.

technology   A particular system of inputs and production practices.

tenant   A farm operator who rents land, buildings, or other assets from their owner; same as lessee.

tenure   The manner by which an operator gains control and use of real estate assets, such as renting or owning them.

tillable acres   Land that is or could be cultivated.

total cost (TC)   The sum of total fixed cost and total variable cost.

total fixed cost (TFC)   The sum of all fixed costs.

total physical product (TPP)   The quantity of output produced by a given quantity of inputs.

total revenue (TR)   The income received from the total physical product; same as total value product.

total value product (TVP)   Total physical product multiplied by the selling price of the product.

total variable cost (TVC)   The sum of all variable costs.

trend analysis   Comparison of the performance level of a farm business to the past performance of the same business.

unsecured loan   A loan for which the borrower does not give the lender the right to possess certain assets if the repayment terms are not met; there is no collateral.

value of farm production   The market value of all crops and livestock and other income generated by a farm business as measured by accrual accounting, after subtracting the value of purchased livestock and feed.

variable cash lease   A leasing arrangement in which a cash payment is made in return for the use of the owner's property, but the amount of the payment depends on the actual production and/or price received by the tenant.

variable costs   Costs that will occur only if production actually takes place, and which tend to vary with the level of production.

variable interest rate   An interest rate that can change during the repayment period of a loan.

variance   A measure of the variability of the possible outcomes of a particular event.

whole farm budget   A projection of the total production, income, and expenses of a farm business for a given whole farm plan.

whole farm plan A summary of the intended kinds and volume of enterprises to be carried out by a farm business.

Workers' Compensation An insurance plan required by law in most states which protects employees from job-related accidents or illnesses, and sets maximum compensation limits for such occurrences.

working capital The difference in value between current assets and current liabilities; a measure of liquidity.

INDEX